Emerging Trends in Leather Science and Technology

Sreeram Kalarical Janardhanan · Luis A. Zugno
Editors

Emerging Trends in Leather Science and Technology

Springer

Editors
Sreeram Kalarical Janardhanan
CSIR-Central Leather Research Institute (CLRI)
Chennai, Tamil Nadu, India

Luis A. Zugno
International Union of Leather Technologists and Chemists Societies (IULTCS)
Basel, Switzerland

ISBN 978-981-99-9753-4 ISBN 978-981-99-9754-1 (eBook)
https://doi.org/10.1007/978-981-99-9754-1

This Springer imprint is published by the registered company Springer Nature Singapore Pte Ltd.
The registered company address is: 152 Beach Road, #21-01/04 Gateway East, Singapore 189721, Singapore

Paper in this product is recyclable.

Foreword by Dr. Michael Redwood

A long time ago we used to learn that the leather industry was somehow "special". It battles with "inelasticity" because its raw material supply cannot grow to meet demand. The raw material is complex, impacted not only by species but by climate, husbandry, age and sex. The processing is difficult with different layers and sections often requiring separate treatment and the science is not always fully understood.

As a natural material surface defects can sometimes be difficult to sell and making every part of every piece perfect for its intended purpose is not easy. There is a great deal of skilled craft associated with making a piece of leather that is both beautiful and high performance.

Yet while leather itself is certainly special the factories that make and use it are buffeted by the winds of geopolitics and changing trends as much or more than any other material. Perhaps what has been really special has been leather's ability to endlessly adapt to serve a changing world over many millennia.

In the beginning, hides and skins were all that was available for covering and containing the multitude of items needed for the daily life of early humans. They were excellent in the raw but with a little treatment, they could be protected from the most aggressive bacteria and made more suited for various purposes. This is tanning. And the thrill for the tanner is the magic and mystery of managing the chemistry and the physical structure, along with mixing art and science.

The arrival of textiles, pottery, glass, paper and metals took end uses away from leather but increasing populations and the growth of society put new, more sophisticated demands on leather. Improved husbandry also means that supply will never catch up with demand.

The discovery and exploitation of plastics changed this balance and now no end uses exist where leather cannot be replaced, but usually elegance and functionality are sacrificed. Now we are learning the environmental damage caused by plastics and we better understand the role leather can play to fight climate change and biodiversity loss. Leather needs the world, and the world needs leather.

Hides and skins are present in nearly every part of the globe and have been traded internationally since the earliest days. As we face the challenges of climate change, poverty and diminishing natural resources using hides and skins well becomes essential. We must add value, make good use of the many jobs leather creates and teach the world the value of producing things that last when looked after and repaired.

I have had a long and happy career working with leather and am very honoured to have worked with many and met most of those involved in the production of this Monograph. It is both timely and prescient.

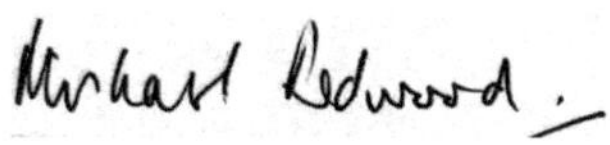

October 2023

Dr. Michael Redwood
Spokesman for Leather Naturally
ILM Opinion Writer
Leather Conservation Trustee
London, UK

Preface

Leather is a unique commodity that links the product of rural farmers to the fashion world. Leather and its products are among the most widely traded products worldwide, as they are produced from a renewable and readily available resource, a byproduct from the meat industry. A key factor for the development of the leather sector is undoubtedly the strong raw material base. But when talking about sustainability in the leather industry, it does not only mean the sustainable manufacture of leather goods alone. An ideal leather industry network comprises of the following: Livestock (cattle, buffalo, sheep, pig and goat), slaughterhouses, raw hides and skin collection, tanning industry, leather goods manufacturing industries, export market and other allied industries. The other small industries that can co-exist within this network would include glue and manure manufacturing industries, processing of meat into various by-products, etc.

Production and supply have gradually moved from industrialized to developing countries and emerging economies, which are fast becoming the most important suppliers of value-added leather products. It is fascinating to note that more than half of the world's supply of leather raw materials comes from developing countries. As such, leather processing is an important economic activity in several developing countries, and it coincides with the fact that the availability of this raw material is high in developing countries. It is estimated that the international trade of this important material exceeds US$ 80 billion annually and is expected to continue growing alongside the increase in population and urbanization of developing countries. From the above, it is evident that the leather sector occupies a very important place in the economic development of a country on account of its substantial export earnings, the potential for the creation of employment opportunities and favourable conditions for its sustained growth.

However, one of the major challenges of leather processing operations is the potential environmental impact that could have devastating socio-economic consequences. Pollution from the tanneries has a negative long term impact on the growth potential of a country, irrespective of immediate economic benefits. Polluted water, air and soil affect peoples' health and damage the ecological processes that sustain the production of food.

Demand for leather goods is growing, but so is the criticism of its severe environmental impacts, which is driving a keen interest in sustainable alternatives. On this account, the three important aspects of the leather producing industry are environmental sustainability, ethical and social sustainability and economic sustainability. Because of the heterogeneity of this sector and also as sustainability has become unavoidable for any industry nowadays, it becomes essential to analyze several issues and factors that can facilitate the attainment of sustainability in the leather industry.

In addition, the leather goods industry has a history with very distinct shifts in end uses and materials from the use of leather products for specific functional purposes to the current market of luxury goods. Raw materials used in the leather goods industry are as diverse as the products. Leather remains important, but the use of materials such as nylon, polyester, polyurethane fabric and even polypropylene is growing. The leather clothing market is highly volatile because the demand for such products depends largely on consumers' disposable income and, on fashion trends. Historically, leather garments have been less of a luxury and more of a necessity, as other materials were not available for protection against cold. In many regions of the world, leather garments became the primary cold-weather outerwear. However, rising wealth and the emergence of excellent synthetic fabrics for waterproof and insulating garments at affordable cost have steadily weakened the position of leather in the traditional outerwear market in most countries.

Significant changes have also occurred in the industries that supply machinery, components, software and chemicals to the footwear and other leather product industries. The demand for leather footwear has increased over the years. The growing demand for leather footwear, as well as the fact that this demand competes for raw materials with other products, appears to be a significant business opportunity for developing countries including the African continent. The footwear industry is a valuable source of employment for developing countries. It is estimated that 10 million people are employed worldwide in the direct production of footwear and there are significant additional numbers employed in the support industries. Despite the value of the industry in improving the living standards of people in the developing world, there is little coordinated information on the industry as a whole available, particularly in developing countries.

The range of synthetic materials used in the leather goods industry is very wide. Increasingly, materials are being used in combination with leather to achieve a certain look or price range. Synthetic materials can be found in all types of leather goods, in luxury items as well as in economical casual articles. As a rule, the quality of non-leather materials follows the trends of the respective market segments. Today, significant progress has been made to make it possible to manufacture accessories at more competitive prices. In addition, one of the more challenging issues in the leather sector is the design of leather products and accessories. Today digital techniques and artificial intelligence are playing a major role in designing of a wide array of leather products. To sum it up, *"Emerging Trends in Leather Science and Technology"* is a very challenging subject and a comprehensive publication addressing all such major issues is not available in the market for reference by the scientists, technologists, policymakers and other stakeholders from both Global South as well as Global North.

This book has been developed in the above backdrop; thus, its major goal is to channel the concurrent knowledge from the leading scientists and technologists working in the field to the scientific public in developing countries. The contents of the book will also be a very valuable source of information for academics, researchers, industrialists and business communities in developed countries. The monograph will also be a very useful handbook for early career researchers and engineers involved with leather related research and consultancy, irrespective of their location.

The concept of developing such a book emerged during the discussion between Dr. Amitava Bandopadhyay, Director General, NAM S&T Centre, New Delhi and Dr. S. K. Janardhanan, Director, CSIR-Central Leather Research Institute (CLRI), Chennai in early 2021. Since 2018, the NAM S&T Centre has been collaborating with CSIR-CLRI and has so far organized two International Workshops on "Sustainability of Leather Sector" and "Trends in Materials, Design, Innovation and Intelligent Manufacturing of Footwear and Leather Products" for the benefits of scientists, technologists and policymakers from the developing countries.

This book contains 17 chapters that cover various challenges and opportunities of leather manufacturing, product making, environmental management, country status and policies, which the developing world needs to understand, manage and improve in regard to their processing, manufacturing infrastructure and export requirements to achieve sustainability. The chapters in the Monograph have been contributed by scientists and experts from 12 countries, namely Australia, Croatia, Ethiopia, France, Germany, India, Kenya, Switzerland, Türkiye, UK, United States and Vietnam to share their knowledge and expertise in various facets of leather and leather product making and related subjects.

We would like to express our heartfelt gratitude to all the authors who have kindly accepted our invitation to write the chapters, revise the contents based on reviewer's comments and submit the final version well in time in spite of their busy schedule.

We would also like to thank Dr. P. Thanikaivelan, Chief Scientist, CSIR-CLRI for his significant contributions in taking this publication project forward. We are also thankful to Dr. T. Ramasami, Former Director, CSIR-CLRI and former Secretary, Department of Science and Technology, Government of India for his guidance and support during this publication project. Thanks are also due to Dr. Amitava Bandopadhyay and his team at the NAM S&T Centre for all the support rendered towards publication of this Monograph.

We are grateful to Dr. Michael Redwood, Spokesman for Leather Naturally, ILM Opinion Writer and Leather Conservation Trustee for writing the "Foreword" of this book in spite of his extremely busy schedule. We also express our gratitude to Dr. Loyola D'Silva, Executive Editor, Springer Nature, Singapore and his team for making this endeavour a success.

Chennai, India Sreeram Kalarical Janardhanan
Basel, Switzerland Luis A. Zugno

Introduction

Leather and its products are one of the most traded goods globally due to their greater durability, good hydrothermal stability, good mechanical properties and resistance to chemical, biological and environmental degradation. The leather industry is a multi-billion dollar global business which has emerged as an important economic activity in several developing countries. The global leather goods market is projected to grow from USD 468.5 billion in 2023 to USD 738.6 billion by 2030, exhibiting a Compound Annual Growth Rate (CAGR) of 6.7% during the forecast period. The process of leather manufacturing involves several stages, including the procurement of raw hides and skins, tanning and processing of the hides and production of finished leather products.

The leather manufacturing industry has faced criticism in the recent years due to concerns over animal welfare, severe environmental impacts and poor working conditions in some countries. Efforts are being made to resolve these issues through improved sustainability. The closedown of manufacturing units, supply chain disruption and economic slowdown during COVID-19 pandemic in 2020, had also severely impacted the global economy of leather industry. However, as the infection rate and spread was reduced, gradually everything returned to track and improved the sustainable consumer spending.

In spite of the implementation of several advanced processing techniques and treatment systems, leather industry is still facing serious challenges from the public and the government authorities. Hence, there is an urgent need to revamp the existing leather processing methods for the sustainability of the leather industry in the future.

Some of the environmental concerns associated with leather production are sourcing of raw materials, use of hazardous chemicals, improper waste management and high water usage as well as air and water pollution. It is crucial for the industry to prioritize these environmental concerns and take necessary actions to address them in order to ensure a more viable future for leather production. Addressing these challenges requires collaboration between industry stakeholders, including manufacturers, suppliers, consumers, R&D institutions and regulatory authorities, to implement measures that promote sustainability, ethical practices and responsible operations throughout the leather supply chain.

In order to address the above mentioned issues, the *Centre for Science and Technology of the Non-aligned and Other Developing Countries (NAM S&T Centre), New Delhi* has brought out this Monograph for dissemination of relevant knowledge and information to the scientists, researchers and managers from the leather sector and other stakeholders interested in the leather industry. The book through its seventeen chapters underscores the various challenges and opportunities of leather manufacturing, product making, environmental management and policies regarding leather processing, which the developing world needs to understand, manage and improve in regard to its processing, manufacturing infrastructure and export requirements to achieve sustainability. The book intends to provide information on smart as well as sustainable leather manufacturing practices to the professionals from the developing countries.

The book brings together scientific communities from Australia, Croatia, Ethiopia, France, Germany, India, Kenya, Switzerland, Türkiye, UK, United States and Vietnam to share their knowledge and expertise to provide a detailed insight into the leather manufacturing and processing, current trends in leather science and technology and policies related to the improvement of leather trade.

I am immensely grateful to Dr. Michael Redwood, Spokesman for Leather Naturally, ILM Opinion Writer and Leather Conservation Trustee, UK for kindly agreeing to write the *Foreword* of this book in spite of his very busy schedule.

I am thankful to Dr. Loyola D'Silva, Executive Editor, Springer Nature, Singapore for his kind support and guidance towards bringing out this Monograph and Ms. Niraja Deshmukh, Production Editor, Springer Nature, India for managing all the technical and administrative tasks for the publication process.

I would like to express my sincere gratitude to the Editors of this book, Dr. S. K. Janardhanan, Director, CSIR-Central Leather Research Institute (CLRI), Chennai, India and Dr. Luis A. Zugno, Secretary and Former President, International Union of Leather Technologists and Chemists Societies (IULTCS), Switzerland, for their initiatives and efforts and sharing their valuable time in reviewing the papers for this book and taking charge of the project. I am especially thankful to Dr. P. Thanikaivelan, Chief Scientist, CSIR-Central Leather Research Institute, Chennai, India for coordinating the entire publication project. The publication of this book would not have been possible without his support throughout the process.

I also acknowledge the valuable support of the entire team at the NAM S&T Centre and am especially thankful to Mr. Madhusudan Bandyopadhyay, Senior Adviser for his support and guidance; and Mr. Pankaj Buttan, Data Processing Manager; Ms. Nidhi Utreja, Former Programme Officer and Ms. Abhirami Ramdas, Research Associate for their contributions in taking this publication project forward and bringing it to a successful conclusion.

I also record my appreciation for the assistance and support rendered by my colleague Mr. Rahul Kumra, Assistant Administrative Officer towards bringing out this Monograph.

I believe that this Monograph would serve as a valuable resource material for scientists and researchers from R&D institutions, professionals from the leather sector, owners of leather industries, technical institutions, government officials, policymakers and students who are actively engaged in the area of leather science and technology.

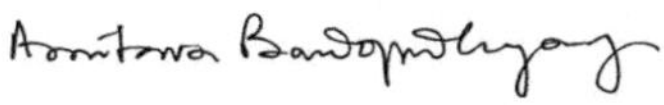

Amitava Bandopadhyay, Ph.D.
Director General
NAM S&T Centre
New Delhi, India

Contents

Policies for Promoting Leather Trade

Country Status and Case Studies

Epilogue

Science and Technology of Leather Manufacture

Hides and Skins: Unravelling Nature's Marvel

John A. M. Ramshaw and Balaraman Madhan

1 Introduction

Undoubtedly, the evolution of skin across the animal kingdom is one of nature's marvels. It comes in many formats and is often referred to by other names when it is adapted for human use. These include hide, where the skin is from a large animal and tradable, such as a bovine, or fur for some species when dense hair is retained on the product. In humans, skin is often quoted as being the largest organ of the body, but this is not an opinion held by all (Sontheimer, 2014). Nevertheless, whatever the discussion, skin is a large organ. For a 70 kg person, the skin comprises around 3.9 kg (5.5%) of the body weight, with an area of around 1.7 m^2 (Goldsmith, 1990). And the key component of skin is collagen, especially type I collagen (James et al., 2006; Ramshaw & Glattauer, 2020). Recent advances in genomics have identified 28 genetically distinct collagen types in humans (Ricard-Blum, 2011). Of these, at least 18 are found in human skin (Table 1) (Nystrom, 2016), perhaps more when the roles and locations of the more recently discovered collagens become better defined. The most abundant collagen types that are present (i.e.: types I and III) are involved in the overall structure and organisation of the skin, whereas minor collagens are often associated with the dermal-epidermal junction (e.g.: types IV and VII) (Abreu-Velez & Howard, 2012) or are possibly involved during initial tissue formation (e.g.: types V and VI) (White et al., 2014), subsequently decreasing in their relative amounts or becoming masked by other components (White et al., 2014).

Skin has evolved to fulfil a range of functions, principally as a flexible outer tissue enclosing the body that can provide protection through acting as a barrier against

J. A. M. Ramshaw (✉)
School of Engineering, Swinburne University of Technology, Hawthorn, VIC 3122, Australia
e-mail: johnamramshaw@gmail.com; jramshaw@swin.edu.au

B. Madhan
Centre for Academic and Research Excellence, CSIR-Central Leather Research Institute, Sardar Patel Road, Adyar, Chennai 600020, India

S. K. Janardhanan and L. A. Zugno (eds.), *Emerging Trends in Leather Science and Technology*, https://doi.org/10.1007/978-981-99-9754-1_1

Table 1 Collagens involved in the structure of skin

Type	Gene(s)*	Roles	References
I	COL1A1, COL1A2	The most abundant collagen in skin, forming the major fibrillar and fibre bundle structures of the skin	
III	COL3A1	An abundant fibril forming collagen, particularly in young skin and the papillary dermis	Byers et al. (1974)
IV	COL4A1, COL4A2,	The principal collagenous component of basement membranes, found in the dermal-epidermal junction	Yaoita et al. (1978)
V	COL5A1, COL5A2	A low abundance fibril forming collagen found in most tissues, including skin, involved in fibril initiation	Birk et al. (1990)
VI	COL6A1, COL6A2	Also found in most tissues, as a distinct beaded network and possibly involved in the formation of new tissue	Werkmeister et al. (1993)
VII	COL7A1	Present as anchoring fibrils at the dermal-epidermal junction	Rattenholl et al. (2002)
VIII	COL8A1	Part of the collagen formed in association with most blood vessels, including in the skin	Plenz et al. (2003)
XII	COL12A1	A fibril associated collagen that is present with type I containing fibrils, mainly in the papillary dermis	Kronick et al. (1991)
XIII	COL13A1	A transmembrane, epidermal collagen in basal keratinocytes that interacts with basement membranes	Peltonen et al. (1999)
XIV	COL14A1	A fibril associated collagen that is present with type I containing fibrils, mainly in the papillary dermis	Kronick and Landola (1997)
XV	COL15A1	A chondroitin sulfate-proteoglycan, with multiple non-collagenous interruptions, that is found associated with the basement membrane and papillary dermis	Fukushige et al. (2005)
XVI	COL16A1	A fibril associated collagen that is present with type I containing fibrils, mainly in the papillary dermis	Akagi et al. (1999)
XVII	COL17A1	A transmembrane, epidermal collagen in basal keratinocytes that interacts with basement membranes	Zimina et al. (2005)
XVIII	COL18A1	A heparan sulfate-proteoglycan found associated with the basement membrane	Seppinen and Pihlajaniemi (2011)

(continued)

Table 1 (continued)

Type	Gene(s)*	Roles	References
XIX	COL19A1	A fibril associated collagen expressed at the dermal-epidermal junction during development but not in adults	Myers et al. (1997)
XX	COL20A1	A fibril associated collagen expressed weakly in the dermis during development but not in adult skin	Koch et al. (2001)
XXII	COL22A1	A collagen found at tissue junctions, including associated with hair follicles	Koch et al. (2004)
XXIII	COL23A1	A transmembrane, epidermal collagen involved in keratinocyte cell-cell junctions	Koch et al. (2006)
XXVII	COL27A1	A fibril forming collagen that is only expressed in developing pre-natal dermis	Plumb et al. (2007)
XXVIII	COL28A1	Expressed in skin, possibly just in the papillary dermis	Veit et al. (2006)

*Additional genes for some of the collagen types have been described, but these are not found in skin or are only minor compared to the genes noted

mechanical forces and deleterious entities including chemicals and microorganisms, and sensation through a nerve system that detects changes in the environment. And in more complex forms, it is involved in physiological regulation, for example of body temperature via the presence of hair and sweat glands, and in colour.

Skin is seen as a specific tissue of vertebrates, although its form varies between fish, amphibians, reptiles, birds and mammals (Alibardi, 2003). Among the invertebrates, a visible outer coating, an exoskeleton, comprising the polysaccharide chitin or a mineralised shell, has evolved in some lineages. But even in the most primitive extant species there is still tissue that defines and maintains shape and form. The transition from unicellular to pluricellular (colony forming) and thence to multicellular organisms is not understood and a range of theories have been proposed (Niklas, 2014; Tong et al., 2022) to take us through or account for the missing evidence. And it is probable that many different routes led to our different present-day kingdoms, animals, fungi and plants (Parfrey & Lahr, 2013).

The development of multicellularity, especially complex multicellularity, provided advantages or useful trade-offs and diversity for the competition of life, albeit that a significant abundance of lifeforms is still unicellular (Bar-On et al., 2018). For example, multicellularity allows diffusions limits on size to be overcome, it provides for longer lifespans as a single cell can die, (and potentially be replaced), without the entire organism dying, and it provides for complexity as cells with different, distinct functionalities can be formed within a single organism (Michaud & Roze, 2001).

With the evolution of multicellular organisms, systems beyond cell-to-cell adhesion needed to emerge to stabilise the new structures. Collagen has been considered as the system that emerged for the animal lineage. Genomic data indicated that (Gly-Xaa-Yaa)$_n$ sequences, which are characteristic of all collagen types, were present in certain bacteria (Rasmussen et al., 2003), and subsequent studies showed that many form the three chain, triple-helical tertiary structure (Yu et al., 2014) that is also characteristic of all collagens (Brodsky & Ramshaw, 1997). The collagen family is well represented in the genomes of the most primitive of animal species, including *Trichoplax adhaerens,* which is considered as probably the simplest known extant animal. This species is a widely distributed marine invertebrate that possesses only four types of somatic cells and has one of the smallest known animal genomes (Schierwater, 2005). Although it may lack an extracellular matrix (ECM) as seen in higher organisms, its genome, nevertheless, contains some putative collagen genes (Srivastava et al., 2008), while collagen type IV was identified by proteomics (Ringrose et al., 2013). Just a little further up the animal phylogeny, considered as a monophyletic lineage (Müller, 1995), are species with well-defined collagenous ECM structures. These include Porifera (sponges) where putative genes for as many as 13 of the 28 collagen types are represented (*Amphimedon*), of which 8 are found in skin (Srivastava et al., 2010). Similarly, Cnidarians, such as jellyfish and corals, have genomes where putative genes for at least 16 of the 28 collagen types are represented (*Dendronephthya*), of which 9 are found in skin (Jeon et al., 2019). Thus, even at the earliest stages of animal evolution, the genetic potential of the collagen genes that are essential to the later evolution in vertebrates of skin were present. And the evolution of skin has enabled many current industries, notably gelatin and leather production, to emerge during hominid evolution.

Leather is among the oldest materials manufactured by mankind. Other early manufactures, such as bone, antler and stone tools, have since been superseded, yet leather remains a crucial material, retaining applications, for example, in footwear and clothing, and now in furnishings and industrial products. The overall leather products market continues to grow (Anon, 2021a, 2021b, 2022a), and in 2021 is valued around 419 billion US$ and it is estimated to grow at CAGR of above 6% from 2022 to 2030 (Straits Research, 2023).

It is generally accepted that hominids emerged out of Africa. During this evolutionary process, these early people lost body hair and started to migrate to different, more variable and extreme climatic regions. Both these contributed to the need for some form of clothing and protection (Gilligan, 2010). Early hominids lacked the powerful canine teeth that other animals, especially predators, possessed for puncturing and shredding animal hides/skins and the poor ability to digest hair keratin. So, as meat became an increasing feature of diet, so too did the availability of various hides and skins removed from the kill to fill the need for clothing and other lifestyle requirements. Tools, such as stone cutters and scrapers, were available to remove and clean these tissues, not just for butchering.

Early evidence of leather production is limited. Early users of hides and skin would have recognised that the natural material deteriorated, and with the development of cognitive abilities (Epstein, 2002), also noted that accidental treatments, such as by smoke, or immersion in water with plant material, such as bark or leaves, led to an extended lifetime compared to untreated material. The date of this transition from a raw hide to leather-like material is uncertain, as is the transition to a defined tanning process. Early hominids would also have seen that treating dried proto-leather with animal fats could restore the flexibility of the material. Possibly the earliest evidence of processing comes from stonc tools found at Hoxne, UK, dated to around 400,000 years ago. When examined for microscopic wear patterns, some indicated their use on hides and skins (Keeley, 1977). Also, later, bone tools from around 120,000–90,000 years ago were found in Morocco and identified as having been used in hide and skin processing (Hallett et al., 2021).

The earliest leather artefacts, however, are relatively recent, dating from the Copper (Chalcolithic) era, with further evidence accumulating during subsequent eras and within specific cultures. The earliest artefact is a shoe, found in a pit in Armenia, dated to between 3627 and 3377 BCE (Pinhasi et al., 2010). Particularly interesting is the wide range of artefacts, from around 2 centuries later, that were found with Ötzi, 'The Iceman', who was found in 1991 in the Tyrolean Alps, adjacent to the present Austria/Italy border (O'Sullivan et al., 2016). The leather was identified as coming from several different species, where a clear selection had been made of optimal materials that could provide for distinct functions (O'Sullivan et al., 2016).

Subsequently, considerable progress has been made with developing and understanding the technology of leather production, as detailed later in this volume. Improvements, documentation, and use of leather making technology appeared in all ancient civilisations. For example, in India around 3000 BCE in the holy scripture of The Rig-Vega, the use of leather in bottles and other goods is first documented, while later, around 2000 BCE leather goods are mentioned in the Sankhya and Likhite law books (Anon, 2022b). In Egypt, artwork from about 5000 BCE indicates use of leather goods, while from around 2300 BCE, artefacts have indicated the wide range of leather goods that were used, including sandals, chariot harnesses and couplings, as well as jewellery. It is also thought that consistent methods for vegetable tanning were developed in Egypt later during the civilisation period (Anon, 2022b). Also, in China, from 1600 BCE or earlier, leather goods included shoes and tunics as armour (Anon, 2022c). The use of leather in Ancient Greece and then Ancient Rome, from about 800 BCE, also included footwear and bags as well as a steadily increasing use in military equipment, such as shields, armour, straps and equestrian fittings, incorporating refinements to the vegetable tanning approach and high-volume production methods. The Romans also developed an alum tawing to produce softer leather that could be dyed, with all forms of leather becoming plentiful, useful materials (Anon, 2022c).

Some significant changes have also happened in more recent times, such as the development of chrome tanning in the mid-nineteenth century. The approach was initially described by Friedrich Knapp (Germany) in 1847, and later, in 1853 by Hylten Cavalin (Sweden) and Rene de Kercado (France) (Nriago & Nieboer, 1988). Subsequently, in 1858, Knapp elaborated on his studies and on batch tanning. Its commercial introduction, however, was much later when Augustus Shultz (USA) patented, in 1884, a two-bath process (Shultz, 1884), followed by Martin Dennis (USA) in 1893 who patented a commercially relevant one-bath process (Dennis,

1893). Subsequently, notable improvements were introduced by Henry Proctor (UK), some of which are retained in current use (Nriago & Nieboer, 1988).

And so, to the present day, where global developments are being focussed on green chemistry and minimising environmental impacts. Fortunately, technical innovation persists and will continue to bring significant benefits to the leather industry.

1.1 Structure and Functions of Hides and Skins

The skins of different mammals are similar, sharing many characteristics. The skin has three main functions, these being protection, temperature regulation and sensation. Beyond this, the skin also is involved in many key physiological functions (Kolarsick et al, 2011). However, many variations also exist, for example within an individual across different parts of the body as well as further changes during aging. For example, in humans skin thickness varies from less than 0.1 mm in the eyelids up to 1.5 mm or more in the palms and in the soles of the feet (Kolarsick et al, 2011), while in other mammals, such as the rhinoceros or hippopotamus, the skin can be up to 50 mm thick. Thus, as well as variations in individuals and between individuals in a species, there are further, and often distinct differences between species. Nevertheless, the key functions including many physiological roles remain constant as do the major structural elements, the epidermis, the dermis, their intervening basement membrane and the subcutaneous zone (Fig. 1). Equally, the various appendages and many distinct specialised cell types are also common themes.

The skin provides protection to the body in its role as the interface with the external environment. This protection is by providing the first defence against external challenges. These can be biological, such as by pathogens, harmful chemicals, or physical through thermal or mechanical injury. In this case the skin repairs through formation of scar tissue, which has composition and characteristics distinct from the skin prior to damage, including for example a greater content of type III collagen (Barnes et al., 1976). Mammalian skin also provides a barrier preventing excessive loss of water, and solutes, while also providing some protection against the adverse effects of UV irradiation (Madison, 2003).

The skin in mammals also plays a key role in body temperature regulation to maintain a constant core temperature. It contains sweat glands allowing evaporation and blood vessels that can dilate to provide some heat loss, while constriction of blood vessels conserves body heat (Fig. 1). Also, all mammals, even including marine mammals such as whales that appear hairless, have hair, which when particularly dense is called fur or wool, providing some (but not humans) with effective insulation when the temperature is cold. Classically, fat is believed to play a role for insulation against cold, although it has been shown that lipolysis is not essential for cold-induced thermogenesis (Shin et al., 2017). Within the skin are arrector pili muscles that link to the hair follicles and can influence the angle of the hair fibres with respect to the skin surface, providing a method for control of the effectiveness of insulation properties

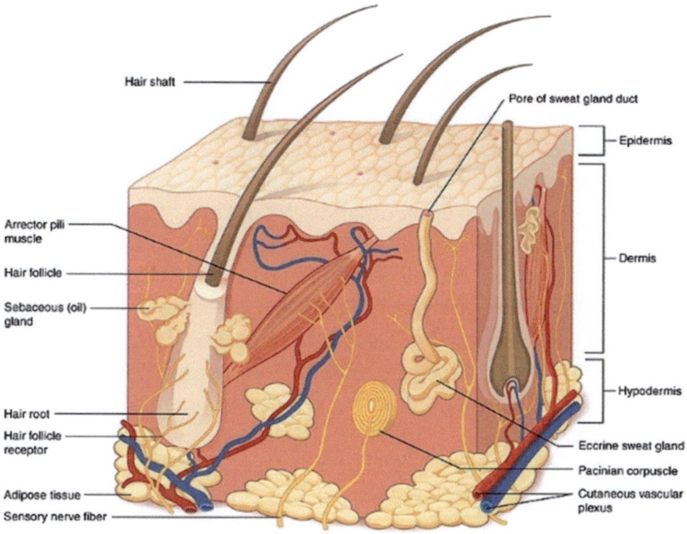

Fig. 1 A schematic representation of mammalian skin (From Betts et al., 2013) (Reproduced under the CREATIVE COMMONS attribution 3.0 unported licence)

(Fig. 1). In some animals, where the hair has evolved into tough, sharp spines, or quills, the arrector pili muscles control these in a host-defence mechanism.

A further primary function for skin is sensation, whereby the state and changes in the external environment can be detected. The skin contains many nerves and receptors that can perceive heat or cold, touch, vibration, pressure and pain from tissue injury or disease. Beyond this, mammalian skin acts as a water-resistant barrier, in part through accumulation of oils, so that internal nutrients and associated compounds are not lost. Other functions of the skin, found in most mammals, include acting as a storage centre for lipids and water. Various gases, including oxygen, may diffuse in small amounts into the mammalian epidermis, and while small compared to overall respiration can be important for the outermost cells.

Depending on the species, some or all of range of secondary functions provided by skin may be present. These include acting to detect infections, via Langerhans cells as part of the immune system (Madison, 2003). The most important biochemical process is the formation of vitamin D by sunlight, essential for the normal absorption of calcium and phosphorous for healthy bones, although for some species, notably dogs and cats, this does not occur (Zafalon et al., 2020). Also, for some species, the hair or fur can be adapted to allow camouflage. Necessarily, as the outermost organ, skin condition is important for social interactions and mate selection as it provides a visible indication of the health and fitness of an individual.

Mammalian skin consists of three principal zones, the epidermis and the dermis, separated by a thin basal lamina or basement membrane, with a hypodermis below, often termed a subcutaneous fatty layer (Fig. 1), with each layer having specific cells and different collagen types present (Table 1). Each of these three zones consists of further subzones, with different structural and functional characteristics and with different cell compositions (Table 2).

The epidermis is the outmost layer of the skin, plays a significant role towards the principal functions of skin, providing the physical and biological barrier, while preventing water loss (Lim, 2021). It is an elastic, stratified layer, with either 4 or 5 sub-layers, depending on the body location. The uppermost of these is the tough, horny layer, the stratum corneum, which consists of layers of flattened dead cells, corneocytes, that are continually abraded or shed. Associated with these cells are various lipids and other cellular components that are converted into an organised barrier which together with the corneocytes provides a barrier against water loss and unwanted chemical or biological penetration (Baroni et al., 2012). The corneocytes are continually being regenerated from keratinocytes in the underlying

Table 2 Structure, composition and function of skin

Skin layer	Sub-layer/ component	Function
Epidermis	Stratum corneum Stratum lucidum Stratum granulosum Stratum spinosum Stratum basale Keratinocytes Langerhans cells Melanocytes Merkel cells	A barrier that prevents water loss or entry of water, pathogens Present in areas where the skin is thickest, i.e., soles and palms. Consist of 3–4 layers of dead keratinocytes, which adds elasticity and serves as an additional barrier Acts as transition layer where the keratinocytes skin cells develop and die. These layers protect the underneath skin against physical damage, infection and water loss Consist of keratinocytes and desmosomes, which supports for the strength and flexibility of skin By continuous mitosis produces new cells to replace dying cells Strengthen epidermis, protect against abrasion and prevent water loss These cells recognise pathogens or foreign materials and determine the appropriate adaptive immune response system Produce melanin to protect skin from UV exposure Specialised neural pressure receptor cell, responding to sensation
Dermis	Collagen Papillary layer Reticular layer Fibroblast cells Sebaceous glands Sweat glands Mast cells Lymphocytes Macrophages Hair follicle	Extracellular matrix protein, the major component of skin that provides physical integrity around which other appendages exist Keeps the body temperature constant through the action of the sweat and fat glands and nourishes the epidermis Strengthens the skin and provides elasticity Synthesis of collagen, elastin, fibronectin and collagenase Produces sebum, a fatty substance that lubricates skin To regulate the body temperature in a hot environment Control and respond to certain bacteria and parasites Immunosurveillance or recognizing foreign pathogens Eliminate foreign substances by engulfing the foreign materials and initiating an immune response Supports growth of hair, that provides warmth
Hypodermis	Subcutaneous fat	Insulation, calorie reservation, cushioning and shock absorption

epidermal tissue and, as they move up lose their nuclei and proliferative ability to become dead flatted cells. The keratinocytes, about 95% of the cells in the epidermis, produce the keratin protein that eventually forms the tough, horny outmost surface. These cells are derived from cell divisions at the base of the epithelium, and move towards the outer surface, undergoing sequential stages of differentiation (Darmon & Blumenberg, 2012), eventually becoming flat and dying as they progress to form the stratum corneum. Lower numbers of other cells are present including melanocytes, Langerhans cells and Merkel cells (White & Yager, 1995). Of these, in mammals, melanocytes produce melanin, a pigment that gives protection against UV damage, while Langerhans cells are part of the immune system as antigen-presenting cells (Dumay et al., 2001), and Merkel cells which are associated with nerves and have a role in sensation. Nutrition and waste removal for all these cells must occur through the underlying dermis or via the environment as the epidermis is entirely avascular. The collagen content of the epidermis is minimal and restricted to membrane-associated collagens, such as XIII, XVII and XXIII that are present at cell junctions, but these are not components of leather as the entire epidermal layer is lost during the processing of hide/skin into leather.

Beneath the epidermis is a thin, yet layered structure, the basement membrane (the dermo-epidermal junction) which links and attaches the dermis to the epidermis. This layer is continuous and has undulations which increase the surface area for exchange of nutrient and waste products between the dermis and the avascular epidermis (Arumugasaamy et al., 2019). Transmission electron microscopy (TEM) suggests the basement membrane consists of two zones, the innermost (dermal) lamina densa which stains well and the outermost (epidermal) lamina lucida which stains lightly (Fig. 2), although the images obtained can be affected by the preparation method (Chan & Inoue, 1994). The main molecular components of the basement membrane are collagen type IV, laminin, nidogen and a heparan sulfate proteoglycan, perlecan (Randles et al., 2017). Type VII collagen is closely associated with this layer and forms a linking network beneath that is critical in attaching the basement membrane to the dermal layer. The structure and distribution of these key components is not completely understood, despite their important medical and commercial roles. Thus, in the leather industry, liming is used and this removes the epidermal layer. Further bating/protease treatment during leather manufacture facilitates the removal of inter-fibrillary materials. The remaining components of the basement membrane and adjacent proteins then comprise the 'grain surface' of leather, which contributes significantly to its commercial value through the required aesthetic appeal of the finished product. The use of bating and proteases can lead to degradation of this grain surface if not adequately controlled. The use of immunohistology can give an indication of the biochemistry of the grain (Stephens et al., 1993). With liming alone, followed by deliming and pickling, collagen types IV and VII, and laminin were retained (Fig. 2), but if bating, or alkaline protease in particular, was employed, these components were often degraded and partly lost, consistent with observed grain damage in these samples (Stephens et al., 1993). The other minor collagens that are found proximal to the basement membrane are both proteoglycan collagens, types XV and XVIII (Table 1).

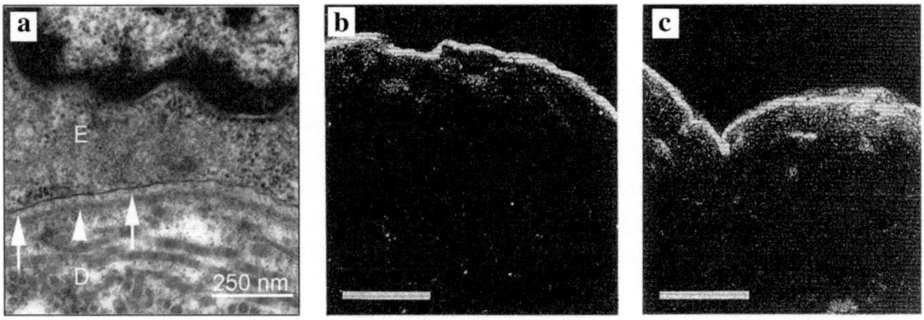

Fig. 2 The basement membrane that forms the dermal-epidermal junction. **a** TEM showing the epidermis, E, the lamina lucida (arrows) and lamina densa (arrowhead) of the basement membrane, and the underlying dermis, D. Bar = 250 nm. (Reprinted from Fig. 4 of Saikia et al., 2018, by permission from Springer Nature). Immunofluorescence, using specific monoclonal antibodies (white is positive) on samples after liming, deliming and pickling, showing **b** type VII collagen, with no bating, **c** laminin, with no bating. Bars = 100 μm (Reproduced from Stephens et al., 1993, by permission of the Society of Leather Technologists and Chemists)

The dermis is the most substantial, inner part of the skin, and supports and provides nutrition to the epidermal layer. It also provides protection to deeper structures in the body through its strength and shock absorbing properties and provides for the initial responses in wound healing (Montagna, 2012). The dermis comprises two zones, the uppermost, thinner papillary dermis, and the deeper reticular dermis which can provide the main bulk of the skin depending on the animal age and species.

The papillary dermis consists of looser connective tissue, containing some elastin and capillaries, and a collagen matrix that is richer in type III collagen compared to elsewhere in the skin (Ramshaw, 1986). There are several collagens associated with the dermal–epidermal junction, but all in small amounts. Nevertheless, despite the low amounts they can have critical roles. Type VII collagen forms the critical anchoring fibril structures that maintain the integrity of the dermal-basement membrane junction, and its absence or mutation leads to blistering diseases (Tamai et al., 2009). Several other fibril-associated collagens are found with the fibrils predominantly in the papillary dermis, including collagen types XII, XIV and XVI, where they may play a role in controlling fibril size (Table 1).

The lower, reticular dermis consists of thicker bundles of collagen, predominantly type I collagen. The type III collagen, which forms mixed fibrils with type I collagen (White et al., 2002), is probably retained but lowers in proportion as type I collagen accumulates. This tissue also contains type V collagen, which plays a key role in the initial fibril and fibre bundle formation (Wenstrup et al., 2004). It is not always evident as it forms the core of the fibrils and becomes masked as tissue accumulates additional type I and type III collagen. The tissue also contains a network of type VI collagen throughout, which may reflect a role in the initial formation of the tissue (White et al., 2014). The underlying reticular dermis contains a variety of cell types, mainly fibroblasts. It also contains elastic fibres, blood vessels that provide for nutrition and waste control, and it supports various appendages.

The appendages are important, integral parts of skin structure (Fig. 1), and comprise the sweat glands with pore exits and the hair follicles, both involved with temperature regulation, and the sebaceous glands that produce oils that are part of the surface film. The sweat and sebaceous glands are both unique to mammals. Fine type III collagen fibrils seem to be in greater concentration near the various appendages, while the minor collagen, type XXII, is associated with hair follicles (Table 1).

Beneath the dermis is the hypodermis or subcutaneous layer (Fig. 1) that consists of connective tissue and fat, which acts as an insulator, along with blood vessels and nerves, and which varies in importance between species, some of which use the skin as a main fat storage depot.

The skins of animals that appeared earlier in evolution than mammals can be used in many cases to make leather, and these skins have distinct characteristics. For example, the marsupials, mainly kangaroo or wallaby, have a skin structure similar to mammals, and produce a leather that is stronger for its thickness than any other leather (Looney et al., 2002). This arises from fine collagen fibres that are arranged more uniformly, parallel to the grain, with a uniform elastin content and a lack of sweat glands and arrector pili muscles. Kangaroo skins frequently have scars, as expected for a wild animal, that prevent its use in large areas in fashion applications. The scars are rich in type III collagen, and generally lead to significant blemishes during processing as the type III collagen in kangaroo is particularly susceptible to loss by proteolytic digestion (Stephens et al., 1988, 1991). In addition, while the main collagenous fibrils in mammals generally have no preferred orientation, in more primitive species the collagen fibrils are often present in layers where the fibres in successive layers are arranged orthogonally to each other (Hawkes, 1974). In other species, significant differences exist with respect to the appendages that are present. For example, there is not hair but rather scales and in birds there are feathers, made of hard keratins. Melanin no longer provides colour, which instead can arise from chromatophores in the dermis, which allow for colour adjustment (Bagnara et al., 1968). Amphibian skin is relatively porous and gases in respiration as well as various chemicals, can be readily absorbed. On the other hand, they frequently have additional glands for host defence, including poison glands in some species (Saporito et al., 2012) and adhesive producing glands in others (Graham et al., 2005). In many species, these variations lead to very distinctive patterning of the surface of a resulting leather, for example with ostriches, with crocodilians, and some fish such as stingrays, which makes these leathers attractive for speciality applications, especially for fashion accessories. These leathers, when properly produced and labelled are often not an issue for those observing some religions where certain mammalian skins can be an issue.

1.2 The Structure and Biosynthesis of Collagen

1.2.1 Structural Considerations

The major protein of skin is collagen. Until 1969, there was thought to be a single collagen protein, now called type I, until a further type, type II was identified from cartilage (Miller & Matukas, 1969). Subsequently, a total of 28 distinct types have been described (Ricard-Blum, 2011). Early studies, necessarily on type I collagen, indicated an amino acid composition for collagen that was rich in Gly, about one-third, as well as Pro and 4-hydroxyproline (Hyp). In 1932, this led Astbury to suggest that a Gly-Pro Hyp repeating unit was a structural element in collagen (Astbury, 1933). It was only some 4 decades later that amino acid sequencing technology had been developed to a stage where data for collagen could be obtained. Combinations of different approaches, including acid and enzymatic hydrolyses, and CNBr cleavage, along with manual and automated Edman sequencing (Edman, 1970) enabled a complete sequence for the α1[I] chain, just over 1000 amino acids long, to be obtained in the mid-1970s, shortly followed by sequences for other fibril forming collagen chains (Fietzek & Kühn, 1975, 1976). These sequences all showed a highly repeating sequence, $(Gly\text{-}Xaa\text{-}Yaa)_n$, as a characteristic feature, with Pro frequently in the Xaa position while Hyp was generally found in the Yaa position. These sequence data also showed the distribution of the other amino acids, indicating that of the 400 potential Gly-Xaa-Yaa triplets, only a few are frequent, while many, including those with Cys and Trp residues are rare or never found; there were also some strong preferences for certain amino acids in either the Xaa or Yaa positions (Ramshaw et al., 1998). For example, due to steric constraints, amino acids such as Phe, Tyr, or Leu are rarely found in the Yaa position (Ramshaw et al., 1998). While the original amino acid sequencing was slow and tedious, the emergence of DNA sequencing technology and genome analyses has led to a massive increase of the sequence data that are available for study, for all collagen types from numerous species across the full spectrum of the animal kingdom. These data are now readily accessible through, for example, the US National Library of Medicine Databases.

Initial progress in determining the tertiary structure for collagen also went slowly. An initial fibre diffraction was reported in 1921 (Herzog & Jancke, 1921). In 1933, Astbury noted the distinct diffraction pattern obtained from rat tail tendon, (Astbury, 1933). However, it was only at a conference in 1953 that appropriate diffraction pattern photographs emerged, from stretched rat tendon (Cowan et al, 1953). These diffraction studies had clearly indicated that a helical element was present, and a combination of biochemical and biophysical considerations led to the proposal, by Ramachandran, that the tertiary structure for collagen was a triple-helix of three separate chains (Ramachandran & Kartha, 1954) (Fig. 3a). The three-chain model was later supported by additional chemical evidence (Boedtker & Doty, 1956). This broad concept of the triple-helix proved correct, although various refinements were needed before a structure (termed RCII) that accommodated all the experimental

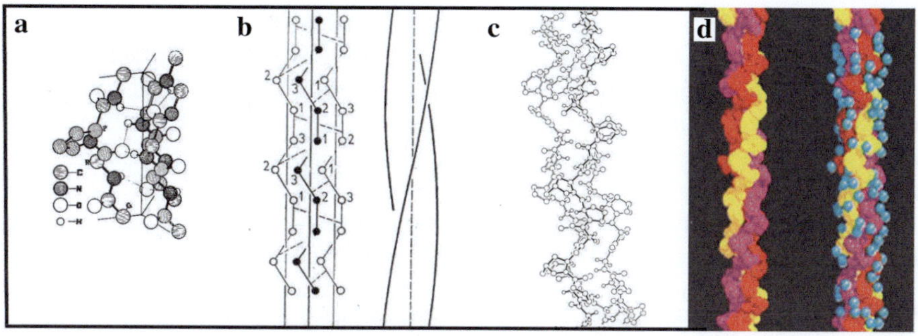

Fig. 3 Key advances in the understanding of the tertiary structure of collagen. **a** The initial proposal of a triple-helical structure (Reproduced from Fig. 1 of Ramachandran & Kartha, 1954, by permission from Springer Nature). **b** A refined structure for collagen consistent with observed properties, (Reproduced from Fig. 2d, e of Rich & Crick, 1961, with permission from Elsevier). **c** A refined molecular model for the collagen-triple helix from X-ray fibre diffraction. (Reproduced from Fig. 3 of Fraser et al., 1979, with permission from Elsevier). **d** A molecular model from an X-ray crystal structure of a collagen-like peptide with well-defined inner hydration. (Reproduced from Fig. 6a, b of Bella et al., 1995, with permission from Elsevier)

evidence, particularly stereo-chemical and hydrogen bonding issues, was achieved (Rich & Crick, 1961) (Fig. 3b).

The structure that emerged, the triple-helix, comprised a three-stranded model in which each chain was in a left-handed helix, and these chains then wound in a right-handed, three-stranded rope-like super-helix, supercoiled about a common axis, with parallel chains staggered by one residue with respect to each other. All three chains in this structure have *trans* peptide bonds and the chain conformation resembles that of the poly-L-proline II structure. This structure allows all the Pro residues, in both the Xaa and Yaa positions, to be included without any distortion. A further feature of this structure was that it placed Gly residues in the centre of the structure, as only Gly, the sole amino acid that lacks a C_β atom, was small enough to fit. This suggested a repetitive amino acid sequence pattern $(\text{Gly-Xaa-Yaa})_n$ that was recognised in early sequence studies and which has become recognised as a characteristic of all collagen molecules (Brodsky & Ramshaw, 1997). Thus, while all Gly residues are buried in the core of the triple-helix, all residues in the Xaa and Yaa positions have considerable exposure to solvent, with the Xaa position being more exposed than the Yaa position. These positions have a high percentage of imino acids and charged residues, although any amino acid could fit these locations. This structure is present in every native collagen, whatever the source, whatever the type and not just those in mammalian tissues.

This model for collagen served well in understanding the structural, biochemical and functional properties of collagen, with the spatial and molecular parameters being defined by Fraser and colleagues in 1979 (Fraser et al., 1979) (Fig. 3c). From quantitative X-ray diffraction data collected from stretched kangaroo tail tendon, the best solution closely resembled the RCII structure. These data gave a value for the unit twist of the native molecular helix of $107.1° \pm 0.6°$. This value was close to the value expected (108°) for a 10/3 helix (a helix with 10 units in 3 turns) (Fraser et al., 1979), which is now used as the way the native triple-helix is described, although there is no fundamental reason why a naturally occurring helix should have a simple numerical format. (Current crystallographic notation describes a 10/3 helix, with 3.3 units/turn, as a 10_7 helix, while a 7/2 helix, with 3.5 units/turn is described as a 7_5 helix (Bella, 2016)).

Subsequently, a crystal structure for a 30-residue synthetic peptide trimer was determined (Bella et al., 1994) (Fig. 3d). While this followed the expected outline for a triple helical structure, a conundrum emerged in that the helical structure was a 7/2 helix rather than the expected 10/3 helix (Bella et al., 1994). This 7/2 helix was also seen in subsequent peptide-based structures (Bella, 2016). Although the structural difference between 10/3 and 7/2 helices is only small, the difference would lead to major shifts in amino acid positions over the distances of a collagen chain of around 1000 amino acids. However, in all cases, the peptides used for these structures had an unnaturally high imino acid content, much greater than found in any natural sequence of a similar length. But when a peptide structure where a 12-residue naturally occurring segment was sandwiched between $(Gly\text{-}Pro\text{-}Hyp)_3$ terminal segments was described, the central, natural sequence had a 10/3 conformation, while the imino acid rich terminals had 7/2 structures (Kramer et al., 2001). Thus, it is likely that the overall collagen structure is predominantly close to a 10/3 helix, as indicated by fibre diffraction (Fraser et al., 1979). However, it is likely that some regions that are rich in imino acid content, such as the C-terminal section of the molecule, will tend to transition into a tighter 7/2 helical segment, providing subtle variations along the length of the collagen triple-helical molecule.

The high-resolution crystal structures of the peptide-based triple-helix models for collagen also allowed other features to be examined that could not be established on whole molecules, for example by fibre diffraction. One feature was the hydrogen bonding network, where interchain intermolecular amide (Gly) NH…O=C (Pro) bonds form the innermost and probably most stable H-bonding set, while the presence of regular (Gly) CH…O-C bonds was shown and, in peptides where Pro did not occupy the Xaa position, an additional interchain hydrogen bond, NH (Xaa)…CO (Gly), was observed, formed through one mediating water molecule (Brodsky, 1999) (Fig. 4). While only limited information was obtained on potential side chain interactions that were independent of crystal packing limitations, direct contacts to backbone carbonyl groups the guanidino of Arg side chains in the Yaa-position were seen, potentially accounting for the stability provided by this sequence motif (Yang et al., 1997).

The other important feature to be examined was the hydration network, where ordered water was known to be present that was distinct from bulk water as structural feature that is important in collagen stability (Bella, 2016) (Fig. 3d). The first collagen-like peptide structure that was solved (Bella et al., 1994), had an extensive, complex yet regular hydration network that exists in part as an intrinsic feature of the triple helix (Bella et al., 1995). Thus some 168 water molecules were identified associated with the trimeric peptide structure of 90 residues total, or 10 triplets in length. It was shown that every Hyp OH group along with the C=O of the Gly

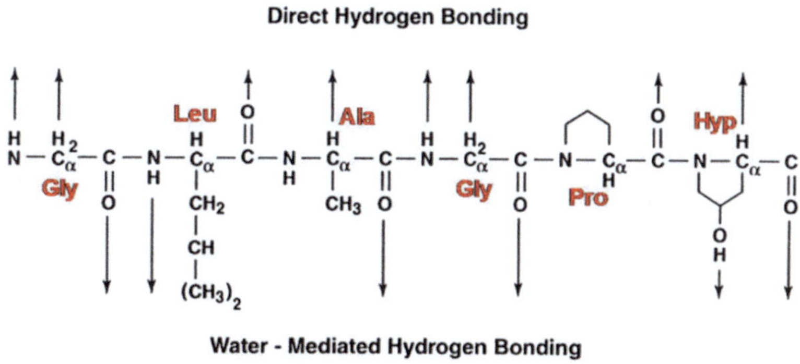

Fig. 4 A diagram to show the observed hydrogen bonding in a collagen triple-helix (Based on Brodsky, 1999) Direct hydrogen bonding is shown with an upward arrow (↑), while water-mediated hydrogen bonding is shown with a downward arrow (↓) (From Ramshaw & Glattauer, 2020 © IOP Publishing. Reproduced with permission. All rights reserved. https://doi.org/10.1088/978-0-7503-2096-2)

and Yaa (Hyp) residue positions were all H-bonded to water. Hence, every possible H-bonding donor and acceptor is involved (Bella et al., 1995). The stabilisation that Hyp provides to the triple-helix is through its pivotal position in the ordered water network. In parts of a sequence where there are no imino acids, the water-mediated hydrogen bonds contribute to local stability, maintaining the triple-helical conformation. (Bella, 2016) (Fig. 3d).

1.2.2 Biosynthesis

The main data from the biosynthesis of collagens have been obtained from studies on type I collagen and the other fibril forming, interstitial collagens. However, it is likely that most of the intracellular steps are common to all collagen types, while the extracellular steps, involved with fibril formation for some collagens and tissue formation more generally may differ for specific collagen types.

Fibril-forming collagens, for example, types I and III collagens, are synthesised with globular N-propeptide and C-propeptide domains flanking the triple-helix domain, which are removed during mature ECM formation (Fig. 5). Throughout the biosynthetic process, newly formed collagen chains interact with various other molecules that assist in folding and are involved in secondary modifications. Thus, proline hydroxylation and lysine hydroxylation only occur during this intermediate unfolded stage (Pihlajaniemi et al., 1991). Formation of Hyp is important for both the stability and the interactions of the final triple helix. Hyp residues make a critical contribution to the stability of the triple-helix through stereo-electronic effects and/or hydration, and also appear essential for collagen self-association and for some receptor interactions (Bella, 2016; Ramshaw & Glattauer, 2020). The specific enzyme involved, prolyl-4-hydroxylase (P4H), is a 2-oxoglutarate-dependent dioxygenase, that is located in the endoplasmic reticulum and consists of an $\alpha 2\beta 2$ tetramer containing 2 distinct subunits. P4H is an iron-containing enzyme, and ascorbic acid (vitamin C) is essential for proper production of collagen, with a deficiency leading to scurvy (Kipp et al., 1996). The specificity is to Pro in a Xaa-Pro-Gly motif, limiting its activity to Pro residues in the Yaa position in a sequence.

The fibril-forming collagen molecules after propeptide cleavage are left with small non-helical telopeptides at each end of the large central triple-helix, and are able to self-associate, and progress to form fibrils (Fig. 5). Specific cross-linking then occurs via the action of lysyl oxidase on the telopeptides and imparts stability and strength to tissue. The initial formation of crosslinks makes the collagen insoluble, although some crosslink types can be reversed in acid conditions. The lysyl oxidase acts on specific lysine resides, to convert the amino acid to allysine which has a reactive aldehyde group. This enzyme is a Cu^{++} and lysyl tyrosyl quinone dependent amine oxidase. The resulting aldehydes are highly reactive and lead to the formation of the native collagen cross-links, apparently without any subsequent enzyme involvement, that involves intramolecular telopeptide to telopeptide bonds and intermolecular telopeptide to triple helix bonding (Ramshaw & Glattauer, 2020).

Each of the 28 different types of collagen has a distinct function and distribution, and is characterised by $(\text{Gly-Xaa-Yaa})_n$ segments, sometimes with non-repetitive intervening sequences, enabling a three chain, triple helical structure to be formed.

In some cases, the molecules are homotrimers with three identical chains, such as type III collagen, $\alpha 1[\text{III}]_3$, while in other cases the molecules are heterotrimers, such as the abundant type I collagen, $\alpha 1[\text{I}]_2\alpha 2[\text{I}]$, typically two chains of one sequence and one of another. In a few cases, a heterotrimer with three different chains may be found.

These different collagen types can be categorised as belonging to a small subset of subgroups, depending on shared functions or packing systems (Fig. 6). In skin (Table 1), the predominant collagens come from the fibril-forming group, including types I, III and V collagens. Along with these are some of the fibril-associated collagens, which have interrupted triple helices, such as collagen types IX, XII, XIV, XVI and XIX. Also important in skin are the collagens that are associated with the dermal-epidermal junction, including type IV collagen that forms a key network within the basement membrane, and type VII collagen which forms anchoring fibrils that link the basement membrane to the underlying dermis (Fig. 6). Type IV collagen also has an interrupted triple helix, with numerous interruptions with a variety of forms and lengths.

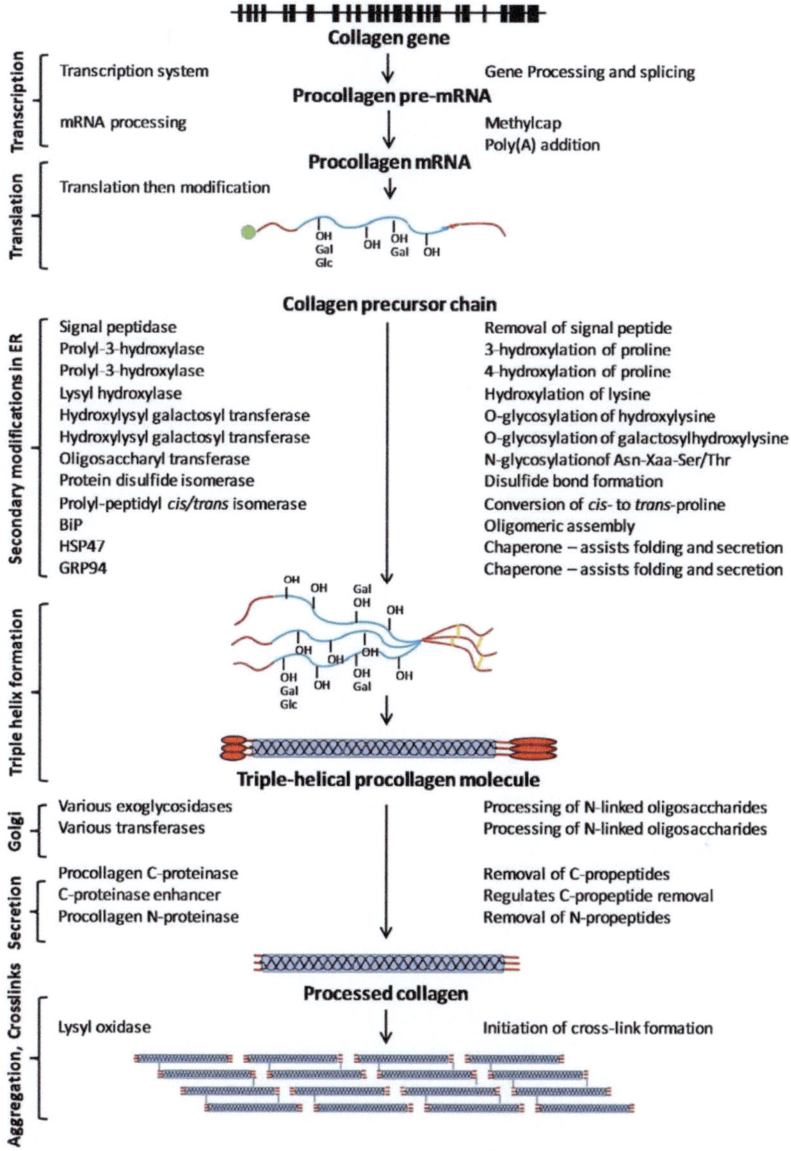

Fig. 5 An overview of some of the key steps and enzyme and chaperone involvement in the biosynthesis pathway for collagen (From Ramshaw & Glattauer, 2020 © IOP Publishing. Reproduced with permission. All rights reserved. https://doi.org/10.1088/978-0-7503-2096-2)

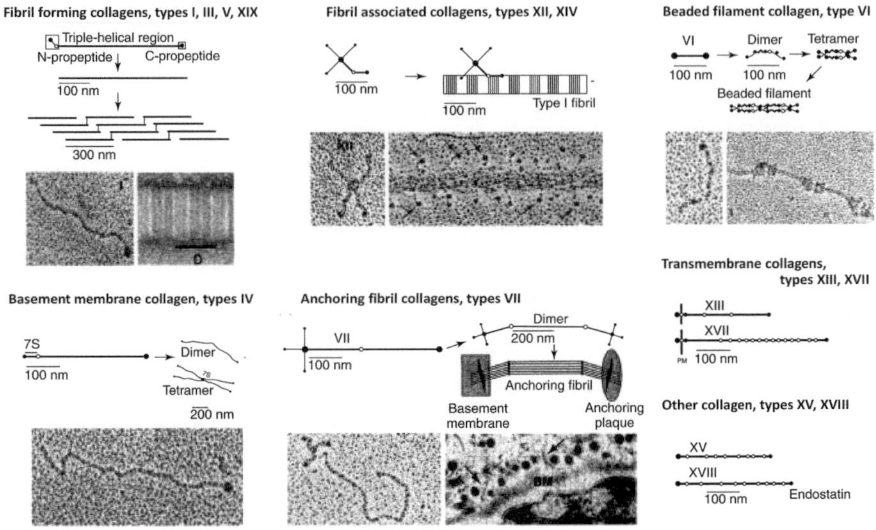

Fig. 6 Schematic illustrations of the structures and sizes of the principle collagens in skin, along with rotary shadowing images of molecules and transmission electron microscopy images of molecular assemblies [Closed circles indicate N- and C-terminal non-collagenous domains while open circles indicate non-collagenous domains that interrupt the triple-helical domains. GAG is glycosaminoglycan and PM is plasma membrane] (Adapted from Ramshaw & Glattauer, 2020.© IOP Publishing. Reproduced with permission. All rights reserved. https://doi.org/10.1088/978-0-7503-2096-2)

1.3 Assembly of the Collagen Triple Helix into Fibres and Tissue

The principal collagens of skin, the interstitial or fibril forming collagens, especially types I and III, have been the best studied for ascertaining how individual molecules assemble into higher order structures (Fig. 7). For type I collagen, the emergence of TEM in the late 1930s eventually showed that collagen fibril had an asymmetric banded pattern with repeating striations at 600–700 Å (Hall et al., 1942), now called the D-period. This was noted as similar to the periodicity deduced from X-ray studies. Subsequent X-ray data from tendons gave the generally accepted 670 Å D-period, while skin gave a shorter period of around 652–655 Å (Brodsky et al., 1980), suggesting that slight changes in the packing mode may be present.

These, and later studies, indicated that the collagen fibril in tendon consisted of a parallel array of staggered molecules where there are 4.4 D-periods per molecular length with a gap following, prior to the start of the next axially aligned collagen molecule (Hodge, 1967) (Fig. 7c). The differential stain uptake between the overlap (less) and gap (more) zones that gives the banded pattern that is seen by TEM. The consensus for the 3-dimensional arrangement of molecules in a tendon fibril that emerged is that of a quasi-hexagonal lattice that was consistent with all of the

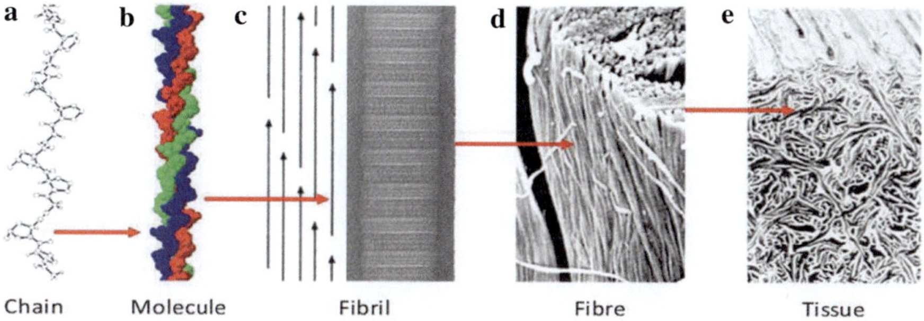

Fig. 7 Structural steps in the establishment of tissues from collagen chains (Adapted from Ramshaw & Glattauer, 2020.© IOP Publishing. Reproduced with permission. All rights reserved. https://doi.org/10.1088/978-0-7503-2096-2)

observed X-ray reflections, with slight fluctuations from linearity along the length of each molecule (Fraser et al., 1983). This model can be extended to provide concepts on how the larger scale packing of a collagen fibril could occur. However, while these data may fit for a tendon, in skin and other tissues that contain type III collagen, a shorter D-period is observed, suggesting variations in the packing (Brodsky et al., 1980).

In vivo evidence for fibril formation has been obtained using TEM, which has shown new fibrils being formed in developing tissues and in a wound healing model (White et al., 2002, 2014). These data indicated that most new fibrils were unusual with both ends as asymmetric N-terminal sequences either side of a short reversal zone (Fig. 8), which was displaced to one end with a longer, sharper end and a shorter, blunter end either side of the reversal zone. Immuno-TEM studies showed that fibrils were heterotypic, with type I and type III collagens both present (White et al., 2002) (Fig. 8c), while other, mainly in vitro studies showed that type V collagen was important for fibril formation and is present as part of an initial core filament (Wenstrup et al., 2004). The subsequent steps in fibril aggregation and growth to give the larger fibrils found in mature tissues are poorly understood.

In vitro studies, often using only purified type I collagen, have provided additional data on fibrillogenesis. For example, in vitro studies on purified type I collagen, with

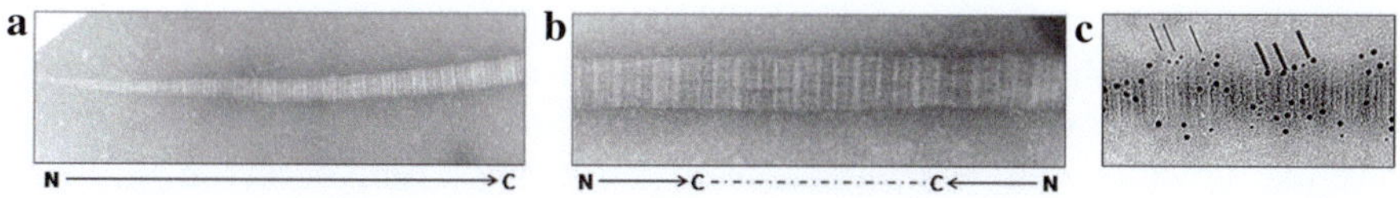

Fig. 8 TEM images of newly formed, 14 day, collagen fibrils from a wound healing model. **a** the tip of a new fibril, **b** a region of polarity reversal, and **c** dual immuno-labelling of a 28-day fibril showing the co-localisation of collagen types I (large gold label and wide arrows) and III (small gold label and thin arrows) in the same fibril (Adapted from Ramshaw & Glattauer, 2020. © IOP Publishing. Reproduced with permission. All rights reserved. https://doi.org/10.1088/978-0-7503-2096-2)

telopeptides present, showed that solutions at neutral pH when heated to 37 °C could readily reconstitute into D-periodic fibrils, without any templates or other added factors, that then entangle to form a hydrogel (Gross & Kirk, 1958).

Tracking this process by turbidity or rheology showed that this was not instantaneous, but rather starts after a lag phase, generally of about 8–12 min, followed by rapid fibril accumulation until an end point. These data suggest that there is a nucleation phase, in which small fibrillar aggregates form and grow to a stage where lateral association leads to rapid growth to form mature fibrils. When collagen lacking telopeptides is used, the lag phase is extended and the rate and extent of fibril formation was much reduced, demonstrating the importance of the telopeptides in fibril formation and growth (Gelman & Piez, 1980), and suggesting that a specific mechanism is used to add additional single collagen molecules onto a developing fibril (Prockop & Fertala, 1998). However, the polarity reversal seen in tissue-derived fibrils is generally absent from these biochemical fibrils.

In vitro studies, using enzyme initiation of fibrillogenesis have provided further data. The availability of collagen precursors with attached propeptides, enables fibrillogenesis to be initiated by addition of purified N- and C-terminal propeptidases. This led to the formation of distinct, needle-like entities, that were distinct from fibrils from tissues (Kadler et al., 1990). However, TEM examination showed that these structures had the characteristic D-periodic banding pattern, as well as the polarity reversal observed in tissue-derived fibrils (Fig. 8). They had one relatively blunt end and one sharp end, with fibril growth, away from the reversal point, happening at the sharper end (Kadler et al., 1990).

While in vitro studies have shown uncontrolled fibril growth, this would not be appropriate in vivo where different tissues have different requirements for fibril size and organisation at different times during growth and aging. Hence a complex set of regulatory steps are involved during fibrillogenesis. The processes involved are not well understood and a variety of mechanisms and active molecules have been proposed. They include feedback mechanisms based on release of propeptides, and by control using various binding molecules, particularly the small interstitial leucine-rich repeat proteoglycans including decorin, biglycan, lumican and fibromodulin (Chen & Birk, 2013), which themselves are subject to further control pathways. The actual mechanism may also depend on the particular tissue environment, with tendon, for example, showing differences to cornea and dermis (Birk et al., 1996).

The final structural element of fibre-forming fibrils is crimp, which is widespread in various native tissues. Crimp develops during embryonic growth and is clear in young animals, with cell contact seeming to play a central role in crimp formation. It is observed, for example by polarised light microscopy, as a zig-zag rather than wave-like structure, with wide variations in periodicity. Crimp occurs in all tissues that experience tensile forces, where the fibres and crimp are organised to minimise the shock of a load.

Although fibrillogenesis models have focused almost exclusively on type I collagen, many tissues are a complex mixture of several collagen types, some of which may exist in organised structures. Other collagen types in skin may also be in ordered aggregates or are associated directly with the fibril structure. For example,

Type VI collagen also forms into fibres, built from disulfide-bonded antiparallel dimers and occasionally into larger sheet-like structures, while type VII is also aggregated into a possible fibrous form, which attaches the dermis to the basement membrane. On the other hand, the main collagen of the basement membrane, type IV forms an essentially 2-dimensional irregular net-like arrangement as the underlying structure. However, unlike the interstitial collagens, the processes by which these other ordered structures are formed are not yet well established.

The knowledge concerning the formation, control and structure of larger features including fibre bundles and tissues is considerably less, despite the importance of these steps in forming functional tissues. At this complex, high level of assembly, the collagen structures need to be considered in the context of the many other molecules, cell types and environmental factors (such as mechanical forces) that are integral aspects during tissue formation. Further, these additional molecular components vary with the developmental stage and maturity of the tissue. Of these the proteoglycans, with their glycosaminoglycan chains, which are present in higher concentration when skin is load bearing (Gillard et al., 1977). While there has been extensive research on the cellular components and changes, there is still much to understand, as little is known about how these changes translate into an organised tissue (Lecuit & Lenne, 2007).

The structure and composition of skin varies with age and location. Initial studies were on human skin and indicated that type III collagen was more abundant in younger, particularly foetal skin tissue, with similar amounts at all depths in abdominal tissue (Epstein, 1974; Epstein & Munderloh, 1978). Subsequent studies of bovine hides, using samples consistently from the rear flank of the animals, showed the presence of type III collagen in the skin of animals of up to 10 years (Ramshaw, 1986). These data confirmed that type III collagen was more abundant in young tissues, and in the papillary dermis, associated with the various appendages. The proportion in this zone declined until about 18 months after which it remained essentially constant. On the other hand, the proportion of type III collagen in the lower, reticular dermis continued to decline with age, up to around 10 years (Ramshaw, 1986) (Fig. 9).

A normal hide can be the source of different types of leather materials, these being full-grain leather, top-grain leather, genuine leather, and corrected-grain leather, in decreasing value. 'Full grain leather' is a premium product, which is the best quality, strongest and most dependable product. It is unlike the other formats in that it has the finest upper, natural grain surface still linked in an integral structure with part of the lower corium, the amount of corium depending on the type and thickness of the required product. The best leathers are fully natural, or with very minimal protective coating. The next highest quality product is 'top-grain leather'. It is derived from the upper portion of the hide, with a finish coating that increases durability while losing some breathability (Fig. 10). 'Genuine leather' is produced from the central and lower parts of the corium. Genuine leather has less durability and lacks the appearance and feel of premium materials. Suede leathers are made from the underside of the corium of the animal hide (Fig. 10). Finally, corrected grain leather material is made from hides that do not meet the standards for premium products and is modified by removal

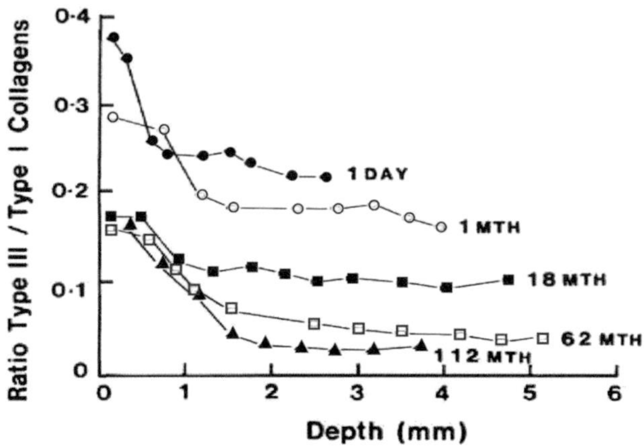

Fig. 9 The ratio of type III to type I collagen at different depths in bovine skin from animals of various ages. The papillary dermis was between 1.0 and 1.3 mm deep (Redrawn from Ramshaw, 1986)

of imperfections followed by coatings and embossing to give the appearance of better products.

The direction of collagen fibres in skin is important as it impacts the strength of the products (Sizeland et al., 2013) (see later). A heritable feature in skin has been identified in Hereford cattle which has no obvious impact on animal well-being but can be detrimental to the resultant leather (Amos, 1958). Termed the 'vertical fibre defect' it is present in the main butt area of some hides (Fig. 10b). In affected hides the collagen bundles at the top of the reticular dermis are aligned vertically, perpendicular to the surface, in contrast to normal hides, and the lower dermis of affected hides, where a complex interwoven fibre network is found. In the leather

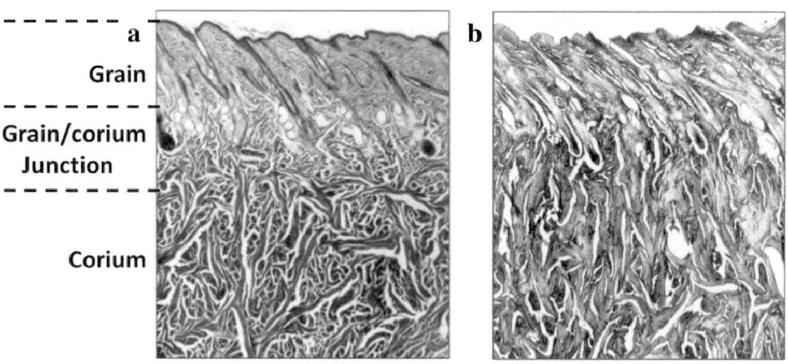

Fig. 10 A comparison between the collagen organisation in **a** normal hide and **b** vertical fibre hide, from Hereford cattle (Adapted from Ramshaw & Glattauer, 2020. © IOP Publishing. Reproduced with permission. All rights reserved. https://doi.org/10.1088/978-0-7503-2096-2)

industry, if an affected hide is split to give a 1.5–2.5 mm thick leather the resulting materials can be weak and can tear readily as no interwoven fibres are present (Dufty et al., 1983). This structural variation in the arrangement of the collagen fibre bundles is inherited as a recessive trait (Cundiff et al., 1987). It probably does not reflect any changes in the collagen (Ramshaw, 1985), but rather changes in the various other extracellular matrix molecules and cellular interactions.

One key process in the formation of mature, covalent tissues that has been extensively studied is the formation of the various intra- and intermolecular crosslinks that are critical to giving mature tissues their strength, stability and functional integrity. The initial crosslinks that form in the extracellular space, connecting at least types I and III collagens, are enzyme-mediated and involve specific amino acids in the polypeptide chains.

Lysyl oxidase catalyses the oxidation of the Lys and hydroxylysine side chains in the telopeptide regions of fibrillar collagens, (α1-9^{N}, α1-16^{C}, α2-5^{N}), giving peptidyl α-aminoadipic-δ-semialdehyde (allysine) and peptidyl-δ-hydroxy-α-aminoadipic-δ-semialdehyde (hydroxyallysine) residues respectively. These highly reactive aldehydes can undergo an aldol condensation reaction with an adjacent aldehyde forming an intramolecular crosslink. However, it is the formation of intermolecular crosslinks that is functionally important. These form initially from a condensation reaction of an aldehyde with a specific amino group (α1-87, α1-930, α2-87, α2-933), on an adjacent collagen molecule (Yamauchi & Sricholpech, 2012). They have unsaturated structures that can be reduced, normally using sodium borohydride. The initial crosslinks that form in skin are predominantly aldimines, which are reversible under acid conditions, although some non-reversible keto-imines may be present (Robins, 2007). These initial crosslinks subsequently go through a series of non-enzymatic, spontaneous reactions, to give the mature crosslinks, many of which have had structures established.

As tissues age, a range of non-enzymatic covalent crosslinks accumulate in tissues. These are advanced glycosylation end-products (AGE) and result from a body-temperature Maillard-type reaction using sugars (such as glucose and ribose) in the body and the ε-amino groups of Lys residues in the triple-helix. These crosslinks lead to a stiffening of critical tissues, such as blood vessels. Some of these crosslinks, both bifunctional and trifunctional, have been identified (Monnier et al., 2016).

1.4 Factors Contributing to the Stability of Collagen

For any given species the key aspect of collagen stability is that the collagen and collagenous structures should be stable at the environment temperature of the species, so that newly synthesised collagen is stable prior to incorporation in the ECM. For warm-blooded animals, homeotherms, this becomes the body temperature, while for cold-blooded animals, poikilotherms, it is the maximum environmental temperature. Thus, for example, for pelagic fish it is the ocean temperature, and if they typically

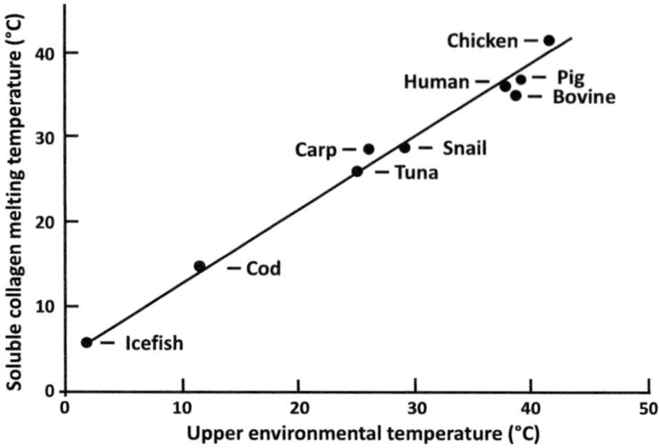

Fig. 11 Thermal stabilities of soluble collagen from various species compared to their upper environmental temperature. Data redrawn from (Mathews, 1975; Rigby, 1968)

live only in deeper waters, it is that low temperature, whereas if these fish were to come up into surface waters for feeding it is the higher surface temperature.

The stability of the collagen for a species is governed primarily by the extent of Hyp incorporation. In higher mammals, including humans, almost every Pro in the Yaa position is fully modified to Hyp, whereas in fish the content of Hyp is incomplete and declines to match but seemingly does not exceed the required stability (Rigby & Robinson, 1975) (Fig. 11). The importance of Hyp for stability has been shown by cell culture approaches where reducing the extent of hydroxylation reduced the collagen stability, while full cessation of hydroxylation reduced stability by some 15 °C (Rosenbloom et al., 1973), which was also confirmed by an un-hydroxylated recombinant collagen, which also showed poor fibril forming capacity (Perret et al., 2001). The amino acid sequence of the collagen can also provide variation to the collagen stability (Persikov et al., 2005).

Studies using collagen molecules only allow the aggregate effect of sequence changes, including hydroxylation, to be studied. The size of type I collagen, having 3 chains each of more than 1000 amino acid residues, makes finer resolution difficult. Initially, poly-hexapeptide structures gave limited opportunities for examining the role of certain amino acids in providing triple-helical stability, but this approach was limited, and models based on shorter synthetic collagen-peptides of defined length became available as useful tools. One example, which focussed on the roles of individual amino acids on collagen stability, was using host-guest peptides (Shah et al., 1996). In this approach, the differences in stability arising from single (or a small number) change in a constant host system, acetyl-$(Gly\text{-}Pro\text{-}Hyp)_8$-amide, could be readily examined using CD studies. Overall, studies on all the single, central triple substitutions of Gly-Xaa-Hyp and Gly-Pro-Yaa and an equal number of Gly-Xaa-Yaa substitutions led to a description of how each amino acid in both positions affected collagen stability (Persikov et al., 2000), and allowed algorithms to predict peptide stability and changes in stability along the length of the triple-helix (Persikov et al., 2005). These data also showed for the leather and gelatin industries, that decreases in stability arising from the loss of amide groups from Gln and Asn residues by alkaline treatment, were greatest from deamidation of Gln residues in the Yaa position, while the racemisation of amino acids was particularly destabilising (see: Ramshaw & Glattauer, 2020).

Overall, beyond the stability of individual molecules, the other important stability concerns are those of collagenous tissues, and of collagenous tissues after stabilisation (crosslinking), as in the biomedical and leather industries.

1.4.1 Stability of Individual Collagen Molecules

A wide variety of methods have been used to examine the stability of soluble collagen molecules. These include circular dichroism spectroscopy (CD), optical rotary dispersion (ORD), calorimetry, viscosity, protease susceptibility and hydrogen exchange (Ramshaw & Glattauer, 2020). These data are reported as the denaturation temperature (T_d), otherwise known as the melting temperature (T_m). The method of choice is CD. The CD spectrum for collagens has a maximum around 222–244 nm, and this signal is lost with increasing temperature as the collagen denatures, with the temperature of the midpoint of the transition, where the fraction of molecules folded is half, given as the T_m.

Stability data have shown the effects of pH on stability, where the greatest stability is found at neutral pH, while lower stability is found at acidic and basic conditions. Different salts can also affect stability, where some, such as KSCN and LiBr, destabilise the triple-helix, while others, such as Na_2SO_4 and KCl, are only slightly destabilising. Others, such as NaCl and $(NH_4)_2SO_4$, are used without any stability issues for collagen precipitation during preparation (Ramshaw & Glattauer, 2020). The effects of various modification reactions, without any crosslinking, on collagen stability have also been examined. Generally, such modifications, like succinylation, and methylation, lead to loss of stability for the triple-helix. Similarly, modification by alkali treatment, which can happen in both the leather and gelatin industries, leads to modifications, with the amide groups of Asn and Gln being lost, giving the corresponding acidic residues and reducing the collagen stability by a few degrees (Ramshaw & Glattauer, 2020).

1.4.2 Stability of Collagenous Tissues

A key process in the development of skin and other vertebrate tissues is the intermolecular crosslinking that stabilises the tissues, especially those containing a significant amount of types I and III collagens. A consequence of the cross-linking is that the tissues, in an aqueous environment, show an enhanced stability compared to individual molecules. When a tissue is heated, it reaches a denaturation point that is accompanied by shrinkage. This temperature is referred to as the shrinkage temperature, T_s, and is frequently measured by increasing the temperature in a water bath until shrinkage occurs.

1.4.3 Additional Crosslinking

In the leather industry, and also for some collagenous biomedical products, the incorporation of further chemical or physical crosslinking is an essential process to produce a durable product, which generally will lead to a further increase in collagen stability. For any given crosslinking strategy, the T_s will increase with the extent of reaction up to a maximum value, with these maximum values depending on the choice of reagent. In some cases, particularly in the leather industry, stability in boiling water of tanned collagen is seen as a satisfactory minimum.

A wide range of tanning agents, including basic chromium salts, aldehydes, vegetable tannins, etc., have been used in the stabilisation of collagen in leather manufacture. The structure of collagen is such that all the functional side chain groups of the triple helical chains are exposed outward. There are several functional groups available in collagen for the interaction with tanning agents and they are, backbone peptide group, hydroxy, carboxy, amino and amide functionality. The crosslinking interaction of tanning agent with the functional groups in collagen varies from covalent to non-covalent interactions depending on the nature of tanning agent. In the fibrillar collagen, though there are intermolecular side chain interactions that provide long range-ordered arrangement. The nature, including size and functionality, of a tanning agent through crosslinking further enhances the ordered arrangement and stabilises the collagen matrix against enzymatic and hydrothermal stress.

1.4.4 The Effects of Solvents on Collagen and Tissue Stability

A summary on a wide variety of possible solvent variations on the stability of collagens has been presented (Ramshaw & Glattauer, 2020). The effects of solvents that have been reported include organic solvents, salts and pH, and initially involved tissues samples. Tissue shrinkage temperature was reduced significantly by a wide range of compounds, including KSCN, $NaClO_4$, KI, salicylate, guanidine and urea. Also, addition at 1 M of various primary amines all showed a decrease in stability, with the effect increasing with the number of carbons in the compounds, whereas a limited range of 1 M solutions of non-ionised organic solvents, mostly alcohols, which showed little effect on stability. The effect of pH showed that there was little change in stability between pH 4 and pH 10, but outside these values, stability decreased substantially. Overall, the effects of more than 70 different compounds were evaluated in this initial detailed study (Lennox, 1949), with many subsequent studies confirming and adding to these findings.

The emerging availability of characterised soluble collagen led to many similar studies (see: Ramshaw & Glattauer, 2020), generally showing comparable results. There was a growing interest in various organic solvents over a range of concentrations. Generally, solutions of monohydric alcohols lead to loss of stability (Gopinath et al., 2014, 2017). In contrast, some diols, including ethylene glycols were stabilising, whereas others caused destabilisation. Some polyols, as sugars, can also provide a stabilising effect. Various proposals to use fluorinated solvents in electrospinning, especially hexafluoro-iso-propanol, has led to studies showing the notable destabilising effects of this group of solvents (Zeogolis et al., 2008). On the other hand, trimethylamine N-oxide (TMAO) solutions were found to be stabilising when studied using collagen peptide models (Beck et al., 2000), consistent with a role in nature where it counteracts the destabilising effects of urea.

1.5 Physico-Chemical Characteristics of Collagen Fibres

The overall physical properties of leather result from the original properties of the collagen fibres in the skin, and on the changes that emerge during processing. These natural features and the effects of processing can be observed from macroscopic down to microscopic levels.

Although skin is generally seen as an anisotropic material, it has long been known that there are differences arising from variation in fibre orientation at different locations. This was first noted by Dupuytren, who noted that a round puncture wound could resolve into an ellipsoidal shape (see: Cox, 1941). Subsequent studies by Langer using circular punctures on cadavers confirmed the previous observation, and lines linking the long axes of the ellipses are now termed Langer lines (Langer, 1862), or skin tension lines (Corps, 1969). Studies on various commercial, experimental and companion animal hides and skins have followed (e.g.: Irwin, 1966; Osaki, 1999). It is now recognised that the skin tension lines are not necessarily the lines of true skin tension, as in some species the skin is more relaxed, and also that living skin may differ from that of a cadaver.

Various approaches have been used to provide quantitative measures for the collagen orientation in skin and to correlate these measurements with tensile strength measurements of leather and observed tension lines. These include traditional microscopy, but also including confocal microscopy and second-harmonic-generation microscopy, as well as light scattering, together with automated quantification programs. Most recently, X-ray diffraction using synchrotron irradiation has provided a method to assess collagen orientation in numerous samples (Sizeland et al., 2013). This approach allows analysis of many side-on samples and their rotations, as opposed to flat analysis through the thickness of the sample, thus giving a better data set. These data, collected from 7 mammalian species, showed that considerable variation in collagen organisation exists between different locations on an animal's skin and between species (Sizeland et al., 2013).

Some studies have been reported on the fibril diameters found in skin, using transmission electron microscopy. While the distributions in well-ordered systems with a narrow size range can potentially be obtained from measurements on just a few hundred fibrils, in a complex tissue like skin where there is a wide range of fibril size and orientation several thousand measurements could be required (Parry et al., 1978). Again, development of synchrotron-based X-ray diffraction has the potential to provide new data on this and on other collagen properties that determine leather characteristics (Sizeland et al., 2015, 2017; Wells et al., 2013, 2017).

1.5.1 Viscoelasticity

Collagenous fibres in skin are known to exhibit a strain-stiffening behaviour. The stiffening mechanism becomes prominent at higher strain levels because of recruitment and elongation of the aligned collagen fibres. The stages of tensile response of collagen fibres include: (i) random orientation of collagen fibres when the applied stress is low; (ii) recruitment and straightening of collagen fibres along the tension direction; (iii) sliding, stretching, and delaminating of fibres and their alignment along the tension direction (strain stiffening); and (iv) fracture of the fibres when the applied stress is high (Yang et al., 2015). Collagen fibres have shown time-dependent viscoelastic behaviour which is quantified using cyclic loading, creep, and stress-relaxation behaviours. During creep, the stress is kept constant while the increase in the strain is recorded; for stress-relaxation, the strain is kept constant, and the decreasing stress is noted over time. In cyclic loading, the energy dissipation is estimated from the area under the loading–unloading curve. It is well established that collagenous tissues accumulated plastic strain (ratchetting behaviour) with increasing cycles (Remache et al., 2018). The increase in creep strain and the exponential decay of stress during relaxation tests indicated the viscous nature of collagen (Shen et al., 2011; Remache et al., 2018).

1.5.2 Shrinkage Temperature

It was noted above, that the melting temperatures of soluble collagens depend on the environmental temperature of the species (Rigby & Robinson, 1975), and that the denaturation temperature (shrinkage temperature) of tissues typical fall above that of the soluble collagen (Mathews, 1975). Melting temperature (T_m) is a molecular phenomenon when native collagen triple-helical structures transition to a denatured, random form. T_m is influenced by the factors that stabilise and hold the triple-helical chains together, whereas shrinkage temperature is macroscopic phenomenon at which the tissue denatures and is influenced by the crosslinking and intermolecular forces between supramolecular assemblies of collagenous tissues. Mammalian soluble type I collagen has a melting temperature around 39 °C, whereas the temperature at which the fibrous collagenous skin/hide tissue shrink to one-third of its original length is referred to as shrinkage temperature usually observed around 60 °C in its native form (Mathews, 1975). Shrinkage temperature (T_s) is also referred to as hydrothermal stability, as the thermal stability determination of collagen fibre happens in the presence of water. With every degree rise in temperature during the determination of T_s, the volume fraction expansion coefficient of both free and bound water in collagenous fibre and their effects on the fibres needs to be considered during their transition to denatured form. Both the structure and dynamics of water influences the hydrothermal stability of collagen fibres. Enhancement of stability of the collagenous fibre against the hydrothermal stress is one of the desired effects brought about by additional crosslinking and long range interactions that are formed in tanning (Madhan et al., 2003). Chrome tanning increases the point at which shrinkage occurs to temperatures up to a maximum of around 120ºC. The increased resistance to wet heat is an important requirement for leather while making footwear, wherein the leather is subjected to moisture and high temperatures as part of the manufacturing process.

The 'boil-test' for tanned material, first discussed over a century ago (Fahrion, 1908), has proven to be a valuable quality control parameter followed by leather manufacturers even today, as properly chrome-tanned leather does not shrink in boiling water. On the other hand, materials tanned by aldehydes and vegetable reagents will generally shrink during the boil test. In these tanning systems specific shrinkage temperatures can be determined by shrinkage testing devices (Allen, 2015). The shrinkage temperature is characteristic of the tannage of the leather. These values reflect on the chemical nature and thermal stability of the additional crosslinking and stabilization introduced to collagenous fibres by different methods.

1.5.3 Swelling

A critical property of hides and skins that is essential to effective tanning and leather production is the swelling of the tissue that occurs at high and low pH values. Briefly, a typical approach will be to clean the hides, first by treatment in lime (alkali) up to a pH of around 12, then to de-lime the material, and then adjusting to low pH, around a pH of 2, in the pickling step. Broadly, the extent of swelling is greatest at these extremes of pH, but the extent can vary with process selections, including the extent of use of sulphide during liming to eliminate hair, the temperature of the process, along with the choice of the reagents used and their concentrations (Liu et al., 2019). This includes, for example, the inclusion of salt during pickling to limit of control the extent of swelling. When the skins or hides are processed, the collagen matrix in the skin is surrounded by water and hence are subject to dimensional changes caused by absorption of water. Any increase in volume of the skin due to absorption of water is associated with increase in thickness of skin since the resistance offered by the skin against the hydrostatic pressure is least in this direction. Skin being anisotropic in nature undergoes differential swelling owing to the differences in fibre structure and compaction amidst the various layers of skin like grain, corium and flesh. There will be a difference in swelling at different regions of skin such as the butt, belly and neck due to the same reason.

The empirical knowledge of the importance of swelling steps goes back to time immemorial (CE 1189) and before, while quantitative examination of the process extends back a century (Kaye & Lloyd, 1924), with various studies subsequently reporting on the extent of swelling of native, limed and pickled tissues. However, in moving between these states, there are both reversible and irreversible processes (Rao et al., 1994). The swelling observed on moving to either pH 2 or pH 12 from neutral pH is irreversible and subsequent neutralisation does not return the material to its native state, but rather it remains partly swollen. However, for a pre-swollen tissue, the changes from acid to alkali and back are reversible (Rao et al., 1994).

The swelling can be seen as arising from the changes in ionisation of changed side chains, such that at extremes of pH suppression of the ionisation of carboxyl or amino groups leads to imbalance of change and repulsion between collagen structures. The important aspect of this resulting swelling is that the non-collagenous materials can be washed free of the collagen matrix. For the proteoglycans in the tissue the use of a high pH leads to the cleavage of the many glycosaminoglycan chains from the protein core, significantly improving the removal of otherwise bulky components. The effective removal of the non-collagenous components is essential for the effectiveness of the subsequent tanning and to the eventual feel and handle of the leather product.

1.6 Uniqueness of Leather Derived from the Properties of Hides and Skins

Leather has a range of special, unique properties that are derived, to a great or lesser extent, from the properties of hides and skins, and particularly the collagen content. Further, the range of properties is extended by the range of different processing strategies that can be used. Thus, in the hands of a skilled leather maker, the same basic raw material can be turned into many different materials that have different qualities and benefits. And as a naturally derived material the wide range of different raw materials that are available further extends the range of products and properties that can be provided, so a match to a particular requirement can normally be found. And for certain applications, such as upholstery, the natural uniqueness of each piece is a further attraction.

1.6.1 Durability

Animal hides and skins have layers with distinct components that vary widely among various species, and which have different fibre-packing density across each layer. These differences dictate the overall strength and other mechano-physical properties contributed by the skin that get translated to the final leather post processing (Dutta, 1999). Generally, the skin has stratified layers classified as grain, grain-corium junction, corium and the adipose (fat and flesh) layer (Fig. 10). The collagen fibre size and packing density varies widely across these layers and the response to mechanical stress also varies with the packing density (Kelly et al., 2019). Other key factors that contribute to biomechanical durability of skin include collagen structure (arrangement and type of collagen), nature of crosslinking present between the fibres, ageing, collagen fibril diameter and orientation (Naffa et al., 2019). In the corium, fibres are largest with diameters of approx. 0.1 mm whereas fibre diameter is around 0.001 mm at the grain layer, which contributes to smoother appearance on the surface owing to finer interweaving (Haines & Barlow, 1975). Larger fibres in the corium interweave more loosely than in grain allowing for fibril movement for stretching and resistance to tear propagation (O'Leary & Attenburrow, 1996). Fibril diameter changes with age and species; typically, a bovine leather exhibiting higher strength has larger fibril diameters, as evaluated by small angle X-ray scattering (Wells et al., 2013). The development of covalent crosslinks within collagen fibrils is essential for stabilisation of the tissue structure and is therefore critical to the biomechanical properties of specific tissues which are defined by the alignment of their collagen fibrils (Naffa et al., 2019). The anisotropic nature of skin is associated with random organisation of fibres which is crucial to impart superior tear strength (Kelly et al., 2018). Tear strength of leather has been correlated with collagen fibril orientation which suggested leather was stronger when the collagen fibrils were arranged in parallel orientation (Wells et al., 2015). Besides these natural factors, the physical stability of leather is also influenced by the chemical composition and hydration of skin. The chemical constituents of bovine and ovine raw material comprise of varying percentages of protein, interstitial cementing substances such as proteoglycans with their glycosaminoglycan side chain, and lipids. Selective removal of the cementing substances and fat from certain type of raw material reflects in the feel and handle of the leather (Wise & Covington, 2019). Moisture affects both flexibility and strength of a material; dry leathers are stiffer and brittle whereas stronger leathers typically have higher moisture contents (Evans & Critchfield, 1933).

1.6.2 Breathability

The skin matrix constitutes three-dimensional weave of collagen fibre bundles, leading to numerous pores and paths in its structure, ranging from macro to nanopores. All these pores and paths in hides and skins are inter-connected, providing a long-range ordering that supports the breathability properties of the leather, and makes it comfortable when compared with the synthetic substitutes. The porosity of the leather material is the structural advantage of leather, allowing its permeability to the water vapour that makes it breathable. The pores in the skin and their connective network, facilitates the two-way flow of the body fluids in the living animals, and this helps the body in thermoregulation with the fluctuation in the environmental temperature. The uniqueness of leather with respect to the water vapour permeability is not readily mimicked by the synthetic fabrics as this characteristic is particularly difficult to produce (Sudha et al., 2009).

Leather processing has four process steps, namely pre-tanning, tanning, post-tanning and finishing (Ramasami & Prasad, 1991). Pre-tanning is a set of operations which are likely a 'do and undo processes', where a process is done to achieve physical/chemical characteristics upon the pelt and then it is removed out of the processed pelt. Mostly pre-tanning focuses on fibre opening and preparative operation of the leather for the tanning process. Fibre opening is the main part of concern for water vapour permeability as it facilitates to retain or widen the pore and its network size. Fibre opening is achieved through swelling process in liming operation and unwanted fibre removal in bating operation (Dutta, 1999). An adult human body can produce up to 10 L of sweat per day (Jessen, 2000), depending on the level of metabolic activity, which is equivalent to about 3–4 g/min per m^2 of the skin. The water vapour permeability of the leather has been quantified as up to 4.93 g/10 cm^2/24 h (Smiechowski et al., 2014), which is about 3.4 g/min per m^2. Thus, leather has a good amount of capacity, with benefits from the skin structure facilitating the perspiration absorption and removal from the foot, making it the best-suited material for footwear manufacture.

Leather is a bad conductor of heat as it has a great number of air pockets. Hence, ancient human populations used animal hide products as thermal insulation material. Leather has an extremely low thermal conductivity of and can reflect near infrared waves due to its hierarchical structure (Wang et al., 2019). Due to this thermal insulation ability, it has been used alone, and along with the carpentry waste, for building heat and sound insulation purposes (Lakrafli et al., 2017).

1.6.3 Shape Adaptation

A significant property of leather, which is particularly important in the clothing and footwear sectors, is its ability to adapt to and retain a particular shape. For example, a leather shoe will mould to the shape of an individual's foot and will retain this adaptation. In contrast, a plastic or other synthetic shoes typically will not change to accommodate the shape of the foot, and if there are any changes, these are lost when the shoe is removed whence it returns to the original state. This moulding of leather to the individual provides comfort, whether for shoes, gloves, jackets and other clothing and consumer items. Leather has an ability to get modified to the shape to which it is applied (Dutta, 1999). The skin itself in living animals gets adjusted to the structure of the internal organ arrangements. Thus, skin takes the shape of (i) the bone which gives the internal outline for the structure of the body and (ii) muscles which cover the bone. This ability of the skin, which is primarily derived from the visco-elastic characteristics of the collagen fibres to take the shape of the internal structure is transmitted to the leather where the leather also takes the shape of the internal body that it covers. The malleability property of the leather footwear makes it more a much more comfortable product than synthetic materials where an equivalent malleability is very difficult to achieve. In leather, both the strength and the flexibility properties are achieved from the interweaving of the collagen fibrils in three-dimensional matrixes, thus making a leather product with good malleability without compromising its strength characteristics (Sudha et al., 2009).

1.6.4 Aesthetics

Undoubtedly, leather is a beautiful and luxurious material. But another important feature for many people is that it is hypoallergenic, and lint and dust-free. Nor does it gather or retain dust, pollen, hair, and other allergenic materials. Also, leather is kind to the body, while synthetic fibres (and even some natural fibres such as wool) can lead to prickle and skin itch, and occasionally skin rashes.

The handle of leather, its individual and distinctive touch and wear qualities, never fails to impress, nor does it lose these qualities over time. Leather can be unique through its natural colour but is also readily adapted to consumer demands through dying. Its surface texture can be manipulated, although the natural textures obtained through use of species such as crocodile and ostrich are preferable. In addition, leather typically has a characteristic, yet subtle smell that is appealing to most, and distinct from the often-unpleasant chemical smell of synthetic materials.

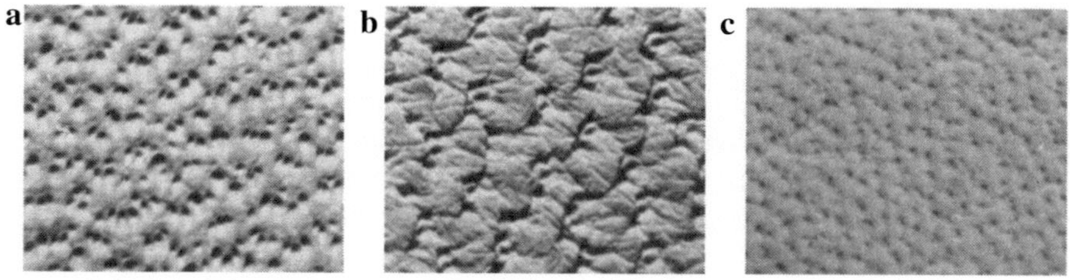

Fig. 12 The grain layer of leather from different species, showing the different characteristic follicle patterns. **a** Bovine, **b** Caprine, **c** Ovine, each at the same magnification

The skin from which the leather is made provides a unique pattern to the grain, usually called the grain pattern of the skin. This pattern varies from species to species (Fig. 12). The hair follicle extension for the epidermis to the corium is the cause of these pores. The hair density and thickness will directly determine the structure of gain pattern with respect to dense nature and the size of the pores. As well as contributing to aesthetics, the grain pattern can be used for the identification of the skin species that was used for leather processing. This can be quite valuable when assessing archaeological or historic samples.

1.7 Global Trends in Production of Hides and Skins

The production and future trends in production of hides and skins are intimately linked to the demand for meat. Thus, leather is almost entirely (>99%) obtained as a by-product of the food (meat) processing industry, with the residual niche area (<1%) being exotic skins for top-end fashion accessories (Anon, 2020). Thus the approximately 270 million cattle hides, as well as other skins, are by-products of the food industry, and by transforming them into a valuable material the tanning industry prevents many millions of tonnes of biological waste going to landfill, with a large environmental impact. Even though leather production has its own waste and uses chemicals the net benefit is positive, and the chemical use is less than that in synthetic materials.

Over the last 3 decades global meat production has more than doubled (Shahbandeh, 2021) and while this has included significant increases in poultry consumption, the meat production that provides hides and skins as by-products for leather manufacture remains strong. It was estimated that in 2020, 324 million metric tons (MMT) of meat were consumed worldwide. Of this beef and veal accounted for 71 MMT, while sheep and goat accounted for 14 MMT. Other species, such as pigs where skin on meat products are common in some regions, accounted for 106 MMT, while poultry meat, now the largest part of this sector, accounted for 111 MMT (Shahbandeh, 2021). The Food and Agriculture Organization (FAO) of United Nations has prepared a compendium on the production of raw hides and skins, leather and leather footwear for 1999–2015, which provides data on the availability of hides and skins.

In order to determine the updated status of raw hides and skins, data for some of the major raw materials (hides/skins) are projected and provided in Table 3.

A variety of proprietary reports on the leather industry and allied sectors suggest strong ongoing growth in these activities, perhaps of with a cumulative annual growth rate (CAGR) of 5 to 6% for the industry (Pulidindi & Mukherjee, 2019). Estimates of current market value range mostly from USD 350–400 billion per annum, with the added value of the secondary product manufacturing having added 15–20 fold to the base value of the tannery leather production (Anon, 2021a, 2021b, 2022a). In order to enhance the sustainability of leather manufacturing there are a variety of by-products, including off-cuts, splits and shavings during processing and off-cuts and trimmings from product manufacturing, all of which can be recycled. Also, end of use material waste is also available to recycle.

Some of the solid waste materials that have not been treated with tanning agents, or have been subject to eco-benign processing (Sathish et al., 2019), allow extraction of collagen (Masilamani et al., 2016) and gelatin for other applications, such as packaging (Masilamani et al., 2017), or production of collagen hydrolysate (Hussien et al., 2017).

Other solid wastes are processed, including using chrome shavings and splits, where there can be a limit on utility even when the chrome content is minimised. However, in these cases, and with end of use materials, the materials can be repurposed as stuffing, or combined with other materials to make composite products such as leatherboard, which can be reused, for example in footwear, and other innovative and artistic applications.

One aspect of these reports is what is included in the estimates. So synthetic leather is often included, distorting the figures. Similarly, the newly, named 'vegan leather' can also be included, even though it also is not a natural leather and is presently experiencing high research and development costs (Anon, 2022d). Overall, it is estimated that natural, real leather accounts for about 55% of the total market (Pulidindi & Mukherjee, 2019). Whether regulations will tackle the tricky problem of misleading nomenclature, is uncertain. For example, it remains to be addressed, and is becoming more prevalent, in other agriculture-derived products, including plant-derived 'milk' products that never see a mammal (plant homogenates), and non-meat 'meat' products that have never seen an animal (extruded/coagulated plant protein pastes).

Table 3 Global annual production of hides/skins (2021)

Hides/skins	Numbers (million pieces)*	Weight (thousand tonnes)*
Bovine	395	7090[#]
Sheep	564	423[$]
Goat	555	394[$]

Note All values (*, #, $) are extrapolated based on the projection from average Hides/Skins production data available in FAO World statistical compendium for raw hides and skins, leather and leather footwear. [#]Values based on wet salted weight of Hides. [$]Values based on dry weight of skins

Synthetic leathers are frequently used in low-cost, high-volume products. It has been suggested that they are more eco-friendly, but this is not necessarily the case. Most synthetic materials are polymers derived from petroleum products, and from manufacture may still contain harmful substances, such as unreacted monomers, as well as phthalates or other catalysts and plasticisers. Further, on disposal biodegradation is limited and incineration or other degradation can lead to dioxin by-products. Leather substitutes, often made by cell-culture approaches, may seem preferable via targeted marketing to some, even though foetal calf serum may be used in manufacturing. And although so-called 'vegan leathers' are broadly bio-based, they too have added chemicals for durability and for aesthetic considerations.

Equally, future estimates can be affected by unforeseen events. Those talking about the anticipated growth in 2018 could not have expected the negative impact that emerged with the emergence of SARS-CoV-2 (Covid-19) as an international concern. There was a weakening in the demand for animal hides and skins for utilisation in the leather industry, which had a significant impact on many abattoirs. This forced some to resort to the option of dumping to landfills (Patel et al., 2021), as a result of sales dropping in some sub-sectors of the market (Anon, 2021a), with recovery still in progress as trade restrictions and supply chain issues slowly resolve (Anon, 2022d). On the other hand, a possible market benefit that has emerged during Covid-19 has been the increased use of on-line purchasing and this trend is expected to continue with positive effects for growth.

Currently, aggression and conflict have again arisen as another blow to trade, with sanctions and other restrictions changing once again. Nevertheless, while there will always be fluctuations in the global economy, there are no fundamental weaknesses in the leather and leather goods markets (Anon, 2022a).

Leather, derived from hides and skins, is synonymous with quality, and this will be the main factor in future growth. The predominant application for leather at present is in footwear, just under half of total use, followed by apparel, fashion accessories and then upholstery (Anon, 2021a). Anticipated growth varies between regions and between product types with a range of perspectives presented. For example, although North America is the main market at present, about one-third, it is anticipated that better growth will occur in the Asia Pacific region (Anon, 2021a). A general sense is that the unfortunate widening of the gap in incomes between rich and poor, worldwide, and the growing number of high-net-worth individuals will drive the high-end markets in all sectors. This is particularly the case for fashion, especially for trendy and for well-recognised international brands, but also includes the upholstery for luxury and top-end automobiles (Anon, 2021a, 2021b). Other topics suggested for better than average growth include sporting footwear (Anon, 2021a) while an increase in the apparel sector may come from cold weather safety products.

Additional growth is anticipated from structural changes within the industry, and consideration of the value chain (Memedovic & Mattila, 2008). For example, increased education and automation in developing countries through better capital availability, especially in the least developed countries, will reduce damage and loss in value for hides, while also leading to processing quality improvements. Reduction in trade barriers will also benefit these countries (Memedovic & Mattila, 2008). There could also be structural changes to the industry. For example, in southern Europe, particularly Italy, the industry is mostly based on many small to medium enterprises, focused often on fashion. Consolidation may occur, as it has in Northern Europe, where there are fewer larger enterprises that are competitive through economies of scale (Anon, 2020).

Certain initiatives by international organisations could also bring benefits to legitimate enterprises and their staff. For example, the United Nations Economic Commission for Europe maintains its pressure to seek sustainable production and to target labour and human rights violations by rogue manufacturers (UNECE, 2022). Losses to legitimate enterprises through trade in fake goods are also being addressed by the members of the Organisation for Economic Co-operation and Development (OECD). It is estimated, based on seizures, that around 3.3% of world trade comprises counterfeited or pirated goods, of which leather goods are a third of these seizures. About two-thirds of these seizures are small parcels, and this proportion is increasing, perhaps with more on-line commerce (OECD/EUIPO, 2019). The OECD is also addressing responsible supply chains, where a neutral environment is targeted through proper controls on transparency and traceability, which will again bring benefits to legitimate enterprises (OECD, 2018).

References

Abreu-Velez, A. M., & Howard, M. S. (2012). Collagen IV in normal skin and in pathological processes. *North American Journal of Medicine and Science, 4*(1), 1.

Akagi, A., Tajima, S., Nagai, Y., Ishibashi, A., & Yamaguchi, N. (1999). Expression of type XVI collagen in human skin fibroblasts: Enhanced expression in fibrotic skin diseases. *The Journal of Investigative Dermatology, 113*(2), 246–250.

Alibardi, L. (2003). Adaptation to the land: The skin of reptiles in comparison to that of amphibians and endotherm amniotes. *Journal of Experimental Zoology Part B. Molecular and Developmental Evolution, 298*(1), 12–41.

Allen, P. (2015). Determining leather shrinkage temperature. The use of SATRA's STD 114 test device. *SATRA Bulletin, 48*.

Amos, G. L. (1958). Vertical fibre in relation to the properties of chrome side leather. *Journal of the Society of Leather Trades and Chemists, 42*, 79–90.

Anon. (2020). European leather industry. EURATEX: The European apparel and textile confederation. Brussels, Belgium. https://s4tclfblueprint.eu/project/tclf-sectors/european-leather-industry/. Accessed 25 Jun 2022.

Anon. (2021a). *Leather goods market size*. Grand View Research, San Francisco, CA. https://www.grandviewresearch.com/industry-analysis/leather-goods-market. Accessed 25 Jun 2022.

Anon. (2021b). *Leather goods market*. Research and Markets, Dublin, Ireland. https://www.researchandmarkets.com/reports/5441854/leather-goods-market-global-industry-trends. Accessed 25 Jun 2022.

Anon. (2022a). *Leather and allied products global market report 2022*. The Business Research Company. https://www.thebusinessresearchcompany.com/report/leather-and-allied-products-global-market-report. Accessed 25 Jun 2022.

Anon. (2022b). *History of leather in India*. https://saddlesindia.com/history-of-leather-in-india/. Accessed 25 Jun 2022.

Anon. (2022c). *The incredible history of leather*. https://www.libertyleathergoods.com/history-of-leather/. Accessed 25 Jun 2022

Anon. (2022d). *Leather and leather substitute product manufacturing in Australia—Market research report*. IbisWorld. https://www.ibisworld.com/au/industry/leather-substitute-product-manufacturing/1864/. Accessed 25 Jun 2022.

Arumugasaamy, N., Navarro, J., Leach, J. K., Kim, P. C., & Fisher, J. P. (2019). In vitro models for studying transport across epithelial tissue barriers. *Annals of Biomedical Engineering, 47*(1), 1–21.

Astbury, W. T. (1933). Some problems in the X-ray analysis of the structure of animal hairs and other protein fibres. *Transactions of the Faraday Society, 29*(140), 193–205.

Bagnara, J. T., Taylor, J. D., & Hadley, M. E. (1968). The dermal chromatophore unit. *Journal of Cell Biology, 38*(1), 67–79.

Barnes, M. J., Morton, L. F., Bennett, R. C., Bailey, A. J., & Sims, T. J. (1976). Presence of type III collagen in guinea-pig dermal scar. *The Biochemical Journal, 157*(1), 263–266.

Bar-On, Y. M., Phillips, R., & Milo, R. (2018). The biomass distribution on Earth. *Proceedings of the National Academy of Sciences of the United States of America, 115*(25), 6506–6511.

Baroni, A., Buommino, E., De Gregorio, V., Ruocco, E., Ruocco, V., & Wolf, R. (2012). Structure and function of the epidermis related to barrier properties. *Clinics in Dermatology, 30*(3), 257–262.

Beck, K., Chan, V. C., Shenoy, N., Kirkpatrick, A., Ramshaw, J. A. M., & Brodsky, B. (2000). Destabilization of osteogenesis imperfecta collagen-like model peptides correlates with the identity of the residue replacing glycine. *Proceedings of the National Academy of Sciences of the United States of America, 97*(8), 4273–4278.

Bella, J. (2016). Collagen structure: New tricks from a very old dog. *The Biochemical Journal, 473*(8), 1001–1025.

Bella, J., Brodsky, B., & Berman, H. M. (1995). Hydration structure of a collagen peptide. *Structure., 3*(9), 893–906.

Bella, J., Eaton, M., Brodsky, B., & Berman, H. M. (1994). Crystal and molecular structure of a collagen-like peptide at 1.9 Å resolution. *Science, 266*(5182), 75–81.

Betts, J. G., Desaix, P., & Johnson, E. (2013). Anatomy & physiology, connexions web site. http://cnx.org/content/col11496/1.6/Wikimedia Commons contributors, "File:501 Structure of the skin.jpg," *Wikimedia Commons, the free media repository,* https://commons.wikimedia.org/w/index.php?title=File:501_Structure_of_the_skin.jpg&oldid=548453610. Accessed 25 Jun 2022.

Birk, D. E., Fitch, J. M., Babiarz, J. P., Doane, K. J., & Linsenmayer, T. F. (1990). Collagen fibrillogenesis in vitro: Interaction of types I and V collagen regulates fibril diameter. *Journal of Cell Science, 195*(4), 649–657.

Birk, D. E., Hahn, R. A., Linsenmayer, C. Y., & Zycband, E. I. (1996). Characterization of collagen fibril segments from chicken embryo cornea, dermis and tendon. *Matrix Biology, 15*(2), 111–118.

Boedtker, H., & Doty, P. (1956). The native and denatured states of soluble collagen. *Journal of American Chemical Society, 78*(17), 4267–4280.

Brodsky, B. (1999). Hydrogen bonding in the triple-helix. *Proceedings of Indian Academy of Science, (Chem. Sci.) 111*, 13–18.

Brodsky, B., Eikenberry, E. F., & Cassidy, K. (1980). An unusual collagen periodicity in skin. *Biochimica Et Biophysica Acta, 621*(1), 162–166.

Brodsky, B., & Ramshaw, J. A. M. (1997). The collagen triple-helix structure. *Matrix Biology, 15*(8–9), 545–554.

Byers, P. H., McKenney, K. H., Lichtenstein, J. R., & Martin, G. R. (1974). Preparation of type III procollagen and collagen from rat skin. *Biochemistry, 13*(25), 5243–5248.

Chan, F. L., & Inoue, S. (1994). Lamina lucida of basement membrane: An artefact. *Microscopy Research and Technique, 28*(1), 48–59.

Chen, S., & Birk, D. E. (2013). The regulatory roles of small leucine-rich proteoglycans in extracellular matrix assembly. *FEBS Journal, 280*(10), 2120–2137.

Corps, B. V. M. (1969). Wound contracture in the hooded rat in relation to skin tension lines and depth of injury. *British Journal of Plastic Surgery, 22*(1), 44–47.

Cowan, P. M., North, A. C. T., & Randall, J. T. (1953). High-angle X-ray diffraction of collagen fibres. In J.T. Randall (Ed.), *Nature and structure of collagen* (pp. 241–249). Butterworths Scientific.
Cox, H. T. (1941). The cleavage lines of the skin. *Journal of British Surgery, 29*(114), 234–240.
Cundiff, L. V., Buechler, P. R., Hannigan, M. V., Everett, A. L., & Dahms, M. P. (1987). Inheritance of vertical fiber hide defect in cattle. *Journal of Heredity, 78*(1), 24–28.
Darmon, M., & Blumenberg, M. (Eds.). (2012). *Molecular biology of the skin: The keratinocyte.* Academic Press.
Dennis, M. (1893). *Tanning leather*. US Patent 495,028.
Dufty, J. H., Peters, D. E., & Bavinton, J. H. (1983). Studies of the cause of the vertical fibre hide structure in Hereford cattle. *Journal of the Society of Leather Technologists and Chemists, 67*, 70–73.
Dutta, S. S. (1999). *An introduction to the principles of leather manufacture* (4th ed.). Indian Leather Technologists' Association.
Dumay, O., Karam, A., Vian, L., Moyal, D., Hourseau, C., et al. (2001). Ultraviolet AI exposure of human skin results in Langerhans cell depletion and reduction of epidermal antigen-presenting cell function: Partial protection by a broad-spectrum sunscreen. *British Journal of Dermatology, 144*(6), 1161–1168.
Edman, P. (1970). Sequence determination. *Molecular Biology, Biochemistry, and Biophysics, 8*, 211–255.
Epstein, E. H. (1974). [α1 (III)] 3 human skin collagen: Release by pepsin digestion and preponderance in fetal life. *Journal of Biological Chemistry, 249*(10), 3225–3231.
Epstein, E. H., & Munderloh, N. H. (1978). Human skin collagen. Presence of type I and type III at all levels of the dermis. *Journal of Biological Chemistry, 253*(5), 1336–7.
Epstein, H. T. (2002). Evolution of the reasoning hominid brain. *The Behavioral and Brain Sciences, 25*(3), 408–409.
Evans, W. D., & Critchfield, C. L. (1933). The effects of atmospheric moisture on the physical properties of vegetable and chrome tanned calf leathers. *Journal of Research of the National Bureau of Standards, 11*, 147–162.
Fahrion, W. (1908). Über eine neue Methode der Lederprüfung. *Collegium, 338–339*, 495–502.
Fietzek, P. P., & Kühn, K. (1975). Information contained in the amino acid sequence of the α1(I)-chain of collagen and its consequences upon the formation of the triple helix, of fibrils and crosslinks. *Molecular and Cellular Biochemistry, 8*(3), 141–157.
Fietzek, P. P., & Kühn, K. (1976). The primary structure of collagen. *International Review of Connective Tissue Research, 7*, 1–60.
Fraser, R. D. B., MacRae, T. P., & Suzuki, E. (1979). Chain conformation in the collagen molecule. *Journal of Molecular Biology, 129*(3), 463–481.
Fraser, R. D. B., MacRae, T. P., Miller, A., & Suzuki, E. (1983). Molecular conformation and packing in collagen fibrils. *Journal of Molecular Biology, 167*(2), 497–521.
Fukushige, T., Kanekura, T., Ohuchi, E., Shinya, T., & Kanzaki, T. (2005). Immunohistochemical studies comparing the localization of type XV collagen in normal human skin and skin tumors with that of type IV collagen. *Journal of Dermatology, 32*(2), 74–83.
Gelman, R. A., & Piez, K. A. (1980). Collagen fibril formation in vitro. A quasielastic light-scattering study of early stages. *Journal of Biological Chemistry, 255*(17), 8098–102.
Gillard, G. C., Reilly, H. C., Bell-Booth, P. G., & Flint, M. H. (1977). A comparison of the glycosaminoglycans of weight-bearing and non-weight bearing human dermis. *The Journal of Investigative Dermatology, 69*(2), 257–261.
Gilligan, I. (2010). The prehistoric development of clothing: archaeological implications of a thermal model. *Journal of Archaeological Methods and Theory, 17*(1), 15–80.
Goldsmith, L. A. (1990). My organ is bigger than your organ. *Archives of Dermatology, 126*(3), 301–302.
Gopinath, A., Reddy, S. M., Madhan, B., Shanmguam, G., & Rao, J. R. (2014). Effect of aqueous ethanol on the triple helical structure of collagen. *European Biophysics Journal, 43*, 643–652.

Gopinath, A., Shanmugam, G., Madhan, B., & Rao, J. R. (2017). Differential behavior of native and denatured collagen in the presence of alcoholic solvents: A gateway to instant structural analysis. *International Journal of Biological Macromolecules, 102*, 1156–1165.

Graham, L. D., Glattauer, V., Huson, M. G., Maxwell, J. M., Knott, R. B., et al. (2005). Characterization of a protein-based adhesive elastomer secreted by the Australian frog *Notaden bennetti. Biomacromolecules, 6*, 3300–3312.

Gross, J., & Kirk, D. (1958). The heat precipitation of collagen from neutral salt solutions: some rate-regulating factors. *Journal of Biological Chemistry, 233*(2), 355–360.

Haines, B. M., & Barlow, J. R. (1975). The anatomy of leather. *Journal of Materials Science, 10*, 525–538.

Hall, C. E., Jakus, M. A., & Schmitt, F. O. (1942). Electron microscope observations of collagen. *Journal of the American Chemical Society, 64*(5), 1234.

Hallett, E. Y., Marean, C. W., Steele, T. E., Álvarez-Fernández, E., Jacobs, Z., et al. (2021). A worked bone assemblage from 120,000–90,000 year old deposits at Contrebandiers Cave, Atlantic Coast, Morocco. *Iscience., 24*(9), 102988.

Hawkes, J. W. (1974). The structure of fish skin. *Cell and Tissue Research, 149*(2), 147–158.

Herzog, R. O., & Jancke, W. (1921). *Festschrift der Kaiser Wilhelm Gesellschaft zur Forderung der Wissenschaften* (pp. 118–121). Springer.

Hodge, A. J. (1967). Structure at the electron microscopic level. In G.N. Ramachandran (Ed.), *Treatise on collagen* (Vol. 1, pp. 185–205). Academic.

Hussien, M., Kumar-Ramadass, S., Madhan, B., & Rao, J. R. (2017). Enzymatic hydrolysis of limed trimmings: Preparation, characterization and application of collagen hydrolysate. *Journal of American Leather Chemists Association, 112*, 44–51.

Irwin, D. H. G. (1966). Tension lines in the skin of the dog. *Journal of Small Animal Practice, 7*(9), 593–598.

James, W. D., Berger, T. G., & Elston, D. M. (2006). *Andrews' diseases of the skin: Clinical dermatology* (10th ed.). Elsevier Saunders.

Jeon, Y., Park, S. G., Lee, N., Weber, J. A., Kim, H. S., et al. (2019). The draft genome of an octocoral, *Dendronephthya Gigantea. Genome Biological Evolution, 11*(3), 949–953.

Jessen, C. (2000). *Temperature regulation in humans and other mammals*. Springer.

Kadler, K. E., Hojima, Y., & Prockop, D. J. (1990). Collagen fibrils in vitro grow from pointed tips in the C-to N-terminal direction. *The Biochemical Journal, 268*(2), 339–343.

Kaye, M., & Lloyd, D. J. (1924). A histological and physico-chemical investigation of the swelling of a fibrous tissue. *Proceedings of Royal Society of London Series B, 96*(677), 293–316.

Keeley, L. H. (1977). The functions of paleolithic flint tools. *Scientific American, 237*(5), 108–27.

Kelly, S. J., Eldmonds, R. L., Cooper, S., Sizeland, K. H., Wells, H. C., Ryan, T., Kirby, N., Hawley, A., Mudie, S., & Haverkamp, R. G. (2018). Mapping tear strength and collagen fibril orientation in bovine, ovine and cervine hides and skins. *Journal of American Leather Chemists Association, 113*, 1–11.

Kelly, S. J., Weinkamer, R., Bertinetti, L., Edmonds, R. L., Sizeland, K. H., Wells, H. C., Fratzl, P., & Haverkamp, R. G. (2019). Effect of collagen packing and moisture content on leather stiffness. *Journal of the Mechanical Behavior of Biomedical Materials, 90*, 1–10.

Kipp, D. E., McElvain, M., Kimmel, D. B., Akhter, M. P., Robinson, R. G., & Lukert, B. P. (1996). Scurvy results in decreased collagen synthesis and bone density in the guinea pig animal model. *Bone, 18*(3), 281–288.

Koch, M., Foley, J. E., Hahn, R., Zhou, P., Burgeson, R. E., et al. (2001). α1(XX) collagen, a new member of the collagen subfamily, fibril-associated collagens with interrupted triple helices. *Journal of Biological Chemistry, 276*(25), 23120–23126.

Koch, M., Schulze, J., Hansen, U., Ashwodt, T., Keene, D. R., et al. (2004). A novel marker of tissue junctions, collagen XXII. *Journal of Biological Chemistry, 279*(21), 22514–22521.

Koch, M., Veit, G., Stricker, S., Bhatt, P., Kutsch, S., et al. (2006). Expression of type XXIII collagen mRNA and protein. *Journal of Biological Chemistry, 281*(30), 21546–21557.

Kolarsick, P. A., Kolarsick, M. A., & Goodwin, C. (2011). Anatomy and physiology of the skin. *Journal of Dermatology Nurses' Association, 3*(4), 203–213.

Kramer, R. Z., Bella, J., Brodsky, B., & Berman, H. M. (2001). The crystal and molecular structure of a collagen-like peptide with a biologically relevant sequence. *Journal of Molecular Biology, 311*(1), 131–147.

Kronick, P. L., & Iandola, S. (1997). Persistence of minority macromolecules of hide through the beamhouse. II. Removal of collagen type XIV. *Journal of American Leather Chemists Association, 92*(7), 172–8.

Kronick, P. L., Maleeff, B. E., & Dahms, M. P. (1991). Removal of collagen-VI and collagen-XII by beamhouse chemistry. *Journal American Leather Chemists Association, 86*(6), 209–224.

Lakrafli, H., Tahiri, S., Albizane, A., El Houssaini, S., & Bouhria, M. (2017). Effect of thermal insulation using leather and carpentry wastes on thermal comfort and energy consumption in a residential building. *Energy Efficiency, 10*, 1189–1199.

Langer, K. (1862). On the anatomy and physiology of the skin: II. Skin tension (with 1 figure). *British Journal of Plastic Surgery, 31*(2), 93–106.

Lecuit, T., & Lenne, P. F. (2007). Cell surface mechanics and the control of cell shape, tissue patterns and morphogenesis. *Nature Reviews Molecular Cell Biology, 8*(8), 633–644.

Lennox, F. G. (1949). Shrinkage of Collagen. *Biochim Biophy Acta, 3*, 170–187.

Lim, K. M. (2021). Skin epidermis and barrier function. *International Journal of Molecular Sciences, 22*(6), 3035.

Liu, H., Yin, Z., Zhang, Q., Li, X., Tang, K., et al. (2019). Mathematical modeling of bovine hides swelling behavior by response surface methodology for minimization of sulfide pollution in leather manufacture. *Journal of Cleaner Production, 237*, 117800.

Looney, M., Kyratzis, I., Truong, Y., & Wassenberg, J. (2002). Enhancing the unique properties of kangaroo leather. Rural Industries Research and Development Corporation. Publication No. 02/105.

Madhan, B., Fathima, N. N., Rao, J. R., Subramanian, V., Nair, B. U., & Ramasami, T. (2003). Hydration in matrix stabilization of collagen fibres: Relevance to mechanistic insight into tanning. *Journal of American Leather Chemists Association, 98*, 263–272.

Madison, K. C. (2003). Barrier function of the skin:"la raison d'etre" of the epidermis. *The Journal of Investigative Dermatology, 121*(2), 231–241.

Mathews, M. B. (1975). *Connective tissue: Macromolecular structure and evolution*. Springer.

Masilamani, D., Madhan, B., Shanmugam, G., Palanivel, S., & Narayan, B. (2016). Extraction of collagen from raw trimming wastes of tannery: A waste to wealth approach. *Journal of Cleaner Production, 113*, 338–344.

Masilamani, D., Srinivasan, V., Ramachandran, R. K., Gopinath, A., Madhan, B., & Saravanan, P. (2017). Sustainable packaging materials from tannery trimming solid waste: A new paradigm in wealth from waste approaches. *Journal of Cleaner Production, 164*, 885–891.

Memedovic, O., & Mattila, H. (2008). The global leather value chain: The industries, the main factors and prospects for upgrading in LDCs. *International Journal of Technological Learning and Innovation Development, 1*(4), 482519.

Michod, R. E., & Roze, D. (2001). Cooperation and conflict in the evolution of multicellularity. *Heredity, 86*(1), 1–7.

Miller, E. J., & Matukas, V. J. (1969). Chick cartilage collagen: A new type of $\alpha 1$ chain not present in bone or skin of the species. *Proceedings of the National Academy of Sciences of the United States of America, 64*(4), 1264–1268.

Monnier, V. M., Genuth, S., & Sell, D. R. (2016). The pecking order of skin Advanced Glycation End products (AGEs) as long-term markers of glycemic damage and risk factors for micro- and subclinical macrovascular disease progression in Type 1 diabetes. *Glycoconjugate Journal, 33*(4), 569–579.

Montagna, W. (2012). *The structure and function of skin*. Elsevier.

Müller, W. E. G. (1995). Molecular phylogeny of Metazoa (animals): Monophyletic origin. *Naturwissenschaften, 82*(7), 321–329.

Myers, J. C., Li, D., Bageris, A., Abraham, V., Dion, A. S., & Amenta, P. S. (1997). Biochemical and immunohistochemical characterization of human type XIX defines a novel class of basement membrane zone collagens. *American Journal of Pathology, 151*(6), 1729–1740.

Naffa, R., Maidment, C., Ahn, M., Ingham, B., Hinkley, S., & Norris, G. (2019). Molecular and structural insights into skin collagen reveals several factors that influence its architecture. *International Journal of Biological Macromolecules, 128*, 509–520.

Niklas, K. J. (2014). The evolutionary-developmental origins of multicellularity. *American Journal of Botany, 101*(1), 6–25.

Nriago, E. N., & Nieboer, E. (1988). *Chromium in the natural and human environments*. Wiley.

Nyström, A. (2016). Collagens in wound healing Ch 9, Wound Healing Biomaterials (Vol. 2, pp. 171–201). Functional Biomaterials. https://doi.org/10.1016/B978-1-78242-456-7.00009-X

OECD. (2018). *OECD due diligence guidance for responsible supply chains in the garment and footwear sector*. OECD Publishing. https://doi.org/10.1787/9789264290587-en. Accessed 25 Jun 2022.

OECD, EUIPO. (2019). Trends in trade in counterfeit and pirated goods. *OECD Publishing, 2022*,. https://doi.org/10.1787/g2g9f533-en. Accessed 25 Jun.

O'Leary, D. N., & Attenburrow, G. E. (1996). Differences in strength between the grain and corium layers of leather. *Journal of Materials Science, 31*, 5677–5682.

Osaki, S. (1999). Distribution map of collagen fiber orientation in a whole calf skin. *Anatomical Record, 254*(1), 147–152.

O'Sullivan, N. J., Teasdale, M. D., Mattiangeli, V., Maixner, F., & Pinhasi, R., et al. (2016). A whole mitochondria analysis of the Tyrolean Iceman's leather provides insights into the animal sources of Copper Age clothing. *Science Report, 6*(1), 1–9.

Patel, K., Munir, D., & Santos, R. M. (2021). Beneficial use of animal hides for abattoir and tannery waste management: A review of unconventional, innovative, and sustainable approaches. *Environmental Science and Pollution Research, 29*, 1–7.

Parfrey, L. W., & Lahr, D. J. (2013). Multicellularity arose several times in the evolution of eukaryotes. *BioEssays, 35*(4), 339–347.

Parry, D. A. D., Barnes, G. R. G., & Craig, A. S. (1978). A comparison of the size distribution of collagen fibrils in connective tissues as a function of age and a possible relation between fibril size distribution and mechanical properties. In *Proceedings of the Royal Society of London. Series B. Biological Sciences, 203*(1152), 305–321.

Peltonen, S., Hentula, M., Hägg, P., Ylä-Outinen, H., & Tuukkanen, J., et al. (1999). A novel component of epidermal cell–matrix and cell–cell contacts: transmembrane protein type XIII collagen. *Journal of Investigative Dermatology, 113*(4), 635–42.

Perret, S., Merle, C., Bernocco, S., Berland, P., Garrone, R., et al. (2001). Unhydroxylated triple helical collagen I produced in transgenic plants provides new clues on the role of hydroxyproline in collagen folding and fibril formation. *Journal of Biological Chemistry, 276*(47), 43693–43698.

Persikov, A. V., Ramshaw, J. A. M., & Brodsky, B. (2005). Prediction of collagen stability from amino acid sequence. *Journal of Biological Chemistry, 280*(19), 19343–19349.

Persikov, A. V., Ramshaw, J. A. M., Kirkpatrick, A., & Brodsky, B. (2000). Amino acid propensities for the collagen triple-helix. *Biochemistry, 39*(48), 14960–14967.

Pihlajaniemi, T., Myllylä, R., & Kivirikko, K. I. (1991). Prolyl 4-hydroxylase and its role in collagen synthesis. *Journal of Hepatology, 13*, S2–S7.

Pinhasi, R., Gasparian, B., Areshian, G., Zardaryan, D., Smith, A., et al. (2010). First direct evidence of chalcolithic footwear from the near eastern highlands. *PLoS ONE, 5*(6), e10984.

Plenz, G. A., Deng, M. C., Robenek, H., & Völker, W. (2003). Vascular collagens: Spotlight on the role of type VIII collagen in atherogenesis. *Atherosclerosis, 166*(1), 1–11.

Plumb, D. A., Dhir, V., Mironov, A., Ferrara, L., Poulsom, R., et al. (2007). Collagen XXVII is developmentally regulated and forms thin fibrillar structures distinct from those of classical vertebrate fibrillar collagens. *Journal of Biological Chemistry, 282*(17), 12791–12795.

Pulidindi, K., & Mukherjee, S. (2019). *Leather chemicals market size*. Global Market Insights, Delaware. https://www.gminsights.com/industry-analysis/leather-chemicals-market. Accessed 25 Jun 2022.

Prockop, D. J., & Fertala, A. (1998). Inhibition of the self-assembly of collagen I into fibrils with synthetic peptides: Demonstration that assembly is driven by specific binding sites on the monomers. *Journal of Biological Chemistry, 273*(25), 15598–15604.

Ramachandran, G. N., & Kartha, G. (1954). Structure of collagen. *Nature, 174*(4423), 269–270.

Ramasami, T., & Prasad, B. (1991). Environmental aspects of leather processing. Journal of Indian Leather Technologists Association, 2, 43-71.

Ramshaw, J. A. M. (1985). Examination of the collagen from normal and vertical fibre hides. *Journal of American Leather Chemists Association, 80*, 219–225.

Ramshaw, J. A. M. (1986). Distribution of type III collagen in bovine skin of various ages. *Connective Tissue Research, 14*, 307–314.

Ramshaw, J. A. M., & Glattauer, V. (2020). *Biophysical and chemical properties of collagen: Biomedical applications*. IOP Publishing.

Ramshaw, J. A. M., Shah, N. K., & Brodsky, B. (1998). Gly-XY tripeptide frequencies in collagen: A context for host–guest triple-helical peptides. *Journal of Structural Biology, 122*(1–2), 86–91.

Randles, M. J., Humphries, M. J., & Lennon, R. (2017). Proteomic definitions of basement membrane composition in health and disease. *Matrix Biology, 57*, 12–28.

Rao, S. P., Murugesan, T., & Raghavan, K. V. (1994). Hysteresis phenomena in swelling of collagen biomatrix in animal hides. *Journal of Applied Polymer Science, 51*(4), 765–67.

Rasmussen, M., Jacobsson, M., & Björck, L. (2003). Genome-based identification and analysis of collagen-related structural motifs in bacterial and viral proteins. *Journal of Biological Chemistry, 278*(34), 32313–32316.

Rattenholl, A., Pappano, W. N., Koch, M., Keene, D. R., Kadler, K. E., et al. (2002). Proteinases of the bone morphogenetic protein-1 family convert procollagen VII to mature anchoring fibril collagen. *Journal of Biological Chemistry, 277*(29), 26372–26378.

Remache, D., Caliez, M., Gratton, M., & Dos Santos, S. (2018). The effects of cyclic tensile and stress-relaxation tests on porcine skin. *Journal of the Mechanical Behavior of Biomedical Materials, 77*, 242–249.

Ricard-Blum, S. (2011). The collagen family. *Cold Spring Harbor Perspectives in Biology, 3*(1), a004978.

Rich, A., & Crick, F. H. C. (1961). The molecular structure of collagen. *Journal of Molecular Biology, 3*(5), 483–506.

Rigby, B. J. (1968). Amino-acid composition and thermal stability of the skin collagen of the Antarctic ice-fish. *Nature, 219*, 166–167.

Rigby, B. J., & Robinson, M. S. (1975). Thermal transitions in collagen and the preferred temperature range of animals. *Nature, 253*(5489), 277–279.

Ringrose, J. H., Van Den Toorn, H. W., Eitel, M., Post, H., & Neerincx, P. (2013). Deep proteome profiling of *Trichoplax adhaerens* reveals remarkable features at the origin of metazoan multicellularity. *Nature Communication, 4*(1), 1–7.

Robins, S. P. (2007). Biochemistry and functional significance of collagen cross-linking. *Biochemical Society Transactions, 35*(5), 849–852.

Rosenbloom, J., Harsch, M., & Jimenez, S. (1973). Hydroxyproline content determines the denaturation temperature of chick tendon collagen. *Archives of Biochemistry and Biophysics, 158*(2), 478–484.

Saikia, P., Medeiros, C. S., Thangavadivel, S., & Wilson, S. E. (2018). Basement membranes in the cornea and other organs that commonly develop fibrosis. *Cell and Tissue Research, 374*, 439–453.

Saporito, R. A., Donnelly, M. A., Spande, T. F., & Garraffo, H. M. (2012). A review of chemical ecology in poison frogs. *Chemoecology, 22*(3), 159–168.

Sathish, M., Madhan, B., & Rao, J. R. (2019). Leather solid waste: An eco-benign raw material for leather chemical preparation. A circular economy example. *Waste Management, 87*, 357–367.

Schierwater, B. (2005). My favorite animal, *Trichoplax adhaerens*. *BioEssays, 27*(12), 1294–1302.
Schultz, A. (1884). *Tawing hides and skins*. US Patent 291,784.
Seppinen, L., & Pihlajaniemi, T. (2011). The multiple functions of collagen XVIII in development and disease. *Matrix Biology, 30*(2), 83–92.
Smiechowski, K., Zarlok, J., & Kowalska, M. (2014). The relationship between water vapour permeability and softness for leathers produced in Poland. *Journal of the Society of Leather Technologies and Chemists, 98*, 259–263.
Shahbandeh, M. (2021). *Meat consumption worldwide from 1990 to 2021, by meat type*. https://www.statista.com/statistics/274522/global-per-capita-consumption-of-meat/. Accessed 25 Jun 2022.
Shah, N. K., Ramshaw, J. A. M., Kirkpatrick, A., Shah, C., & Brodsky, B. (1996). A host−guest set of triple-helical peptides: Stability of Gly-XY triplets containing common nonpolar residues. *Biochemistry, 35*(32), 10262–10268.
Shen, Z. L., Kahn, H., Ballarini, R., & Eppell, S. J. (2011). Viscoelastic properties of isolated collagen fibrils. *Biophysical Journal, 100*, 3008–3015.
Shin, H., Ma, Y., Chanturiya, T., Cao, Q., Wang, Y., Kadegowda, A. K. G., Jackson, R., Rumore, D., Xue, B., Shi, H., Gavrilova, O., & Yu, L. (2017). Lipolysis in brown adipocytes is not essential for cold-induced thermogenesis in mice. *Cell Metabolism, 26*(5), 764–777.
Sizeland, K. H., Basil-Jones, M. M., Edmonds, R. L., Cooper, S. M., Kirby, N., Hawley, A., & Haverkamp, R. G. (2013). Collagen orientation and leather strength for selected mammals. *Journal of Agriculture and Food Chemistry, 61*(4), 887–892.
Sizeland, K. H., Edmonds, R. L., Basil-Jones, M. M., Kirby, N., Hawley, A., Mudie, S., & Haverkamp, R. G. (2015). Changes to collagen structure during leather processing. *Journal of Agriculture and Food Chemistry, 63*(9), 2499–2505.
Sizeland, K. H., Wells, H. C., Kelly, S. J. R., Edmonds, R. L., Kirby, N. M., Hawley, A., Mudie, S. T., Ryan, T. M., & Haverkamp, R. G. (2017). The influence of water, lanolin, urea, proline, paraffin and fatliquor on collagen D-spacing in leather. *RSC Advances, 7*(64), 40658–63.
Sontheimer, R. D. (2014). Skin is not the largest organ. *The Journal of Investigative Dermatology, 134*(2), 581.
Straits Research, Leather Goods Market: Information by Product (Footwear, Luggage, Accessories, Clothing, Apparel), Distribution Channel (Store-Based Retail), and Region—Forecast till 2030. https://straitsresearch.com/report/leather-goods-market. Accessed 5 Jan 2023.
Srivastava, M., Begovic, E., Chapman, J., Putnam, N. H., Hellsten, U., et al. (2008). The Trichoplax genome and the nature of placozoans. *Nature, 454*(7207), 955–960.
Srivastava, M., Simakov, O., Chapman, J., Fahey, B., & Gauthier, M. E. (2010). The *Amphimedon queenslandica* genome and the evolution of animal complexity. *Nature, 466*(7307), 720–726.
Stephens, L. J., Werkmeister, J. A., Caine, J. M., & Ramshaw, J. A. M. (1988). Pepsin soluble collagens from kangaroo skin. *Das Leder, 39*, 88–93.
Stephens, L. J., Werkmeister, J. A., & Ramshaw, J. A. M. (1993). Changes to bovine hides during leather processing. *Journal of the Society of Leather Technologists and Chemists, 77*, 71–74.
Stephens, L. J., Werkmeister, J. A., Tebb, T. A., & Ramshaw, J. A. M. (1991). Identification of type III collagen from kangaroo skin. *Das Leder, 42*, 41–44.
Sudha, T. B., Thanikaivelan, P., Aaron, K. P., Krishnaraj, K., & Chandrasekaran, B. (2009). Comfort, chemical, mechanical, and structural properties of natural and synthetic leathers used for apparel. *Journal of Applied Polymer Science, 114*(3), 1761–1767.
Tamai, K., Kaneda, Y., & Uitto, J. (2009). Molecular therapies for heritable blistering diseases. *Trends in Molecular Medicine, 15*(7), 285–292.
Tong, K., Bozdag, G. O., & Ratcliff, W. C. (2022). Selective drivers of simple multicellularity. *Current Opinion in Microbiology, 67*, 102141.
UNECE. (2022). *Enhancing traceability and transparency of sustainable value chains in the garment and footwear sector (ECE/TRADE/463)*. United Nations.

Veit, G., Kobbe, B., Keene, D. R., Paulsson, M., Koch, M., & Wagener, R. (2006). Collagen XXVIII, a novel von Willebrand factor A domain-containing protein with many imperfections in the collagenous domain. *Journal of Biological Chemistry, 281*(6), 3494–3504.

Wang, X., Tang, Y., Wang, Y., Ke, L., Ye, X., Huang, X., & Shi, B. (2019). Leather enabled multifunctional thermal camouflage armor. *Chemical Engineering and Science, 196*, 64–71.

Wells, H. C., Edmonds, R. L., Kirby, N., Hawley, A., Mudie, S. T., & Haverkamp, R. G. (2013). Collagen fibril diameter and leather strength. *Journal of Agriculture and Food Chemistry, 61*(47), 11524–11531.

Wells, H. C., Sizeland, K. H., Kirby, N., Hawley, A., Mudie, S., & Haverkamp, R. G. (2015). Collagen fibril structure and strength in acellular dermal matrix materials of bovine, porcine and human origin. *ACS Biomaterials Science & Engineering, 1*, 1026–1038.

Wells, H. C., Holmes, G., Jeng, U. S., Wu, W. R., Kirby, N., Hawley, A., Mudie, S., & Haverkamp, R. G. (2017). A small angle X-ray scattering study of the structure and development of looseness in bovine hides and leather. *Journal of the Science of Food and Agriculture, 97*(5), 1543–51.

Wenstrup, R. J., Florer, J. B., Brunskill, E. W., Bell, S. M., Chervoneva, I., & Birk, D. E. (2004). Type V collagen controls the initiation of collagen fibril assembly. *Journal of Biological Chemistry, 279*(51), 53331–53337.

Werkmeister, J. A., Tebb, T. A., White, J. F., & Ramshaw, J. A. M. (1993). Monoclonal antibodies to type VI collagen demonstrate new tissue augmentation of a collagen-based biomaterial implant. *Journal of Histochemistry and Cytochemistry, 41*(11), 1701–1706.

White, J. F., Werkmeister, J. A., Bisucci, T., Darby, I. A., & Ramshaw, J. A. M. (2014). Temporal variation in the deposition of different types of collagen within a porous biomaterial implant. *Journal of Biomedical Materials Research. Part A, 102*(10), 3550–3555.

White, J. F., Werkmeister, J. A., Darby, I. A., Bisucci, T., Birk, D. E., & Ramshaw, J. A. M. (2002). Collagen fibril formation in a wound healing model. *Journal of Structural Biology, 137*(1–2), 23–30.

White, S. D., & Yager, J. A. (1995). Resident dendritic cells in the epidermis: Langerhans cells, Merkel cells and melanocytes. *Veterinary Dermatology, 6*(1), 1–8.

Wise, W., & Covington, A. D. (2019). *Tanning chemistry: The science of leather* (2nd ed.). RSC publishing.

Yamauchi, M., & Sricholpech, M. (2012). Lysine post-translational modifications of collagen. *Essays in Biochemistry, 52*, 113–133.

Yang, W., Chan, V. C., Kirkpatrick, A., Ramshaw, J. A. M., & Brodsky, B. (1997). Gly-Pro-Arg confers stability similar to Gly-Pro-Hyp in the collagen triple-helix of host-guest peptides. *Journal of Biological Chemistry, 272*(46), 28837–28840.

Yang, W., Sherman, V. R., Gludovatz, B., Schaible, E., Stewart, P., Ritchie, R. O., & Meyers, M. A. (2015). On the tear resistance of skin. *Nature Communication, 6*, 1–10.

Yaoita, H., Foidart, J. M., & Katz, S. I. (1978). Localization of the collagenous component in skin basement membrane. *The Journal of Investigative Dermatology, 70*(4), 191–193.

Yu, Z., An, B., Ramshaw, J. A. M., & Brodsky, B. (2014). Bacterial collagen-like proteins that form triple-helical structures. *Journal of Structural Biology, 186*(3), 451–461.

Zafalon, R. V., Ruberti, B., Rentas, M. F., Amaral, A. R., Vendramini, T. H., et al. (2020). The role of vitamin D in small animal bone metabolism. *Metabolites, 10*(12), 496.

Zeugolis, D. I., Khew, S. T., Yew, E. S., Ekaputra, A. K., Tong, Y. W., et al. (2008). Electro-spinning of pure collagen nano-fibres - just an expensive way to make gelatin? *Biomaterials, 29*(15), 2293–2305.

Zimina, E. P., Bruckner-Tuderman, L., & Franzke, C. W. (2005). Shedding of collagen XVII ectodomain depends on plasma membrane microenvironment. *Journal of Biological Chemistry, 280*(40), 34019–34024.

Zero Waste Processing and Sustainability of Leather

J. Raghava Rao, B. Chandrasekaran, and T. Ramasami

1 Introduction

Industrial processing generally involves the conversion of raw materials into useable and useful products for defined, economically remunerative applications. Industrial wastes are those substances or materials which discarded after a primary use without any or much economic recovery. Waste could be also considered as a disused resource. Zero waste processing by objective defines an industrial methodology that does not discard any material as worthless at any stage of conversion of a raw material into products of defined use or generate materials which are discarded as useless (Verma & Sharma, 2022).

Sustainability of any human activity demands that it delivers values to multiple stakeholders namely society, economy, and environment. Any sustainable activity needs to be economically viable, socially equitable, and environmentally bearable, all at the same time (Li et al., 2015). The key criterion of sustainability of an industrial activity is much more than the applications of zero waste processing methodologies. It needs to embrace also social equity and environmental aspects.

Sustainability of an industry engaged in conversion of a putrescible natural product like animal skin derived as by-product of meat and allied sector depends upon also the forcing of external circumstances. Availability of natural resource-based materials being limited, they often suffer inelastic supply chain related problems. This makes the production of low value and high volume products from such resources economically unviable. Whereas man-made materials lend themselves to mass production

J. R. Rao · B. Chandrasekaran (✉)
CSIR-Central Leather Research Institute, Adyar, Chennai 600020, India
e-mail: bchandru@clri.res.in

J. R. Rao
e-mail: jrrao@clri.res.in

T. Ramasami
Department of Science & Technology, Government of India, New Delhi, India

S. K. Janardhanan and L. A. Zugno (eds.), *Emerging Trends in Leather Science and Technology*, https://doi.org/10.1007/978-981-99-9754-1_2

and enjoy the benefits of scale of production and resource intensive research and development, products like leather made from skins or hides recovered from dead or slaughtered animals remain challenged at the market place.

Leather has been partially replaced or substituted by manmade and new materials for several industrial and low value applications on account of cost factors (Jones et al., 2021). Leather is becoming viability challenged for several applications due to external forcing factors like competition from synthetics, costs of compliance to environmental, and social governance and regulatory pressures. With economic recovery of value-added materials from wastes generated in leather processing remaining inadequate, viability challenge is becoming manifested.

2 Results and Discussion

2.1 *Principles of Sustainability Science*

Sustainability cannot be considered as a goal or destination. Principles of sustainability of an industrial sector are to be culturally internalised as a process or a journey; continuously changing for interconnecting to the external circumstances and contexts depending upon the developmental needs (Piwowar-Sulej, 2021). Sustainable development path of an industrial sector is more relevant than sustainability of the sector in a particular context of a country. In case of leather and allied products, the viability bottom line has been rendering the movement of manufacturing activity from context to context during the last 100 years with potentially larger negative carbon footprint as a sector (Laurenti et al., 2017). A deeper analysis of principles of sustainability issues is in order for the global leather sector.

2.2 *A Perspective on Sustainability of Global Leather Sector: An Analysis*

Raw hides and skins availability is larger in countries in Americas and Europe with high per capita meat consumption. Leather processing is a global activity; however, it is mostly in developing and emerging economies in Asia and Africa. Conversion of leather into consumer products is also in Asia and small industrial clusters involving a complex of Tiny, Micro, Small and Medium Enterprises. Major users of leather products are in High and Upper Middle Income countries. In other words, there is large movement of materials and products across the world in global leather trade. Large percentage of hides manufactured in India is from those recovered from bovine animals died of natural causes in their places of farming. Carcass recovery and utilization activities occur in villages and raw hides and skins are then cured and shipped to tannery clusters in far off locations. Raw material mobility and movement are integral aspects of leather processing activity globally (Joseph & Nithya, 2009).

2.3 Carbon Footprint of Material Movement in Global Leader Trade

Carbon footprint of movement of raw materials and final products from leather calls for closer scrutiny (Hashmi et al., 2022). Such footprints of movement of ton of materials per kilo meter based on mode of transport may be available. It might be appropriate to include also the carbon footprint of raw material movement in leather processing for each cluster. This analysis is needed from the bearable perspective of the planet.

2.4 Character of Leather Processing as a Co-product of Meat Industry

Leathers are a by-product of the meat industry; 99% of the leather comes from cow, sheep, goat, and pig livestock. Farmers do not raise the livestock for skins and hides. Global meat trade with a production volume of 345.17 million tons per year with an average return of 2.59 US$ per kg is estimated at US $ 897.5 billion in 2021(Parlasca & Qaim, 2022). Value recovered by the meat handling system from a hide or skin is about 1 to 3% of the total economic value realised from the sale of meat. However, the global leather trade is estimated at US $ 407.92 billion in 2021 with an average return of US $ 58.3 per kg of raw hides and skins. Conversion of raw hides and skins into leather and later into consumer products like footwear and accessories creates high volume employment, especially in low and lower middle income group countries. There is a tangible contribution to social equity in developmental processes of developing economies (Paul et al., 2022). In a country like India, contributions of leather and allied industry to social equity and women empowerment are substantial and documented. Social equity bottom line of the sustainability is significant in India falling under Lower-Middle Income group countries with an annual per capita income of US$ 6600 on PPP terms. This element may become questionable if and when the per capita incomes of nations cross US$ 15000 per annum.

2.5 Leather in New Material World

Global leather industry is seriously challenged by synthetics on account of cost issues, especially in high volume and low priced sector (Patel et al., 2022). Given that total human population of the world is at 7.5 billion and is increasing, it might be easy to estimate the total surface area required for covering the feet in the form of footwear. It is estimated that about 35 billon sq. ft of materials are required annually by the footwear industry alone. With a total annual production of about 20 billion sq. ft of leather, the role of synthetics as a replacements competition from non-leather, especially in addressing the viability aspect of sustainability (Muralidharan et al., 2022a, 2022b). It is becoming increasingly evident that leather is progressively being replaced by synthetics in many conventional applications on account of cost and viability challenges of the new material world.

The global leather sector may have to undertake an internal audit of its current wastage pattern for some of the leathers processed. This audit shall look into the loss of economic recovery potentials. It may be necessary to examine the viability bottom-line for certain price range of products and applications. A sectoral plan for the avenues where leather and allied products which could retain comparative advantages in a dynamic market competition from synthetics and other man-made materials needs to be developed. Sustainability of leather sector could only be achieved when the leather and allied products are able to sustain in a dynamic market competition passing through the competition shake out (Rinaldi et al., 2022).

2.6 Technology Paradigms in Leather Processing: Need for a Revisit on Account of Sustainability Challenge

Traditionally leather processing activity had remained water intensive. Several environmental concerns have been expressed by the regulatory authorities about leather sector. Leather processing activity had remained as part of cultural practices rather than planned industrial and technology enabled socio economic activity. Such cultural practices related activity have started to give way even in the context of developing economies with increased level of industrial activity. Environmental regulatory framework in most nation states has included tanneries among the red category of industries. Environmental burden for leather processing activity in the form of solid and liquid wastes and gaseous emissions could be worked out (Sivakumar, 2022). In countries like India, especially in a water starved state like Tamil Nadu, the tannery sector has been forced to comply with "Zero Liquid Discharge" norms since 2005. Sustainability of the sector on account of viability challenge from costs of compliance is raised by the industry. The practice of Zero Liquid Discharge for industrial waste waters is attended by the need for use of currently available energy intensive technology practices (Ricky et al., 2022). Energy viability of the ZLD and its carbon foot-print need to be examined as well. While practice of ZLD backed by energy intensive multiple effect evaporation of 'Rejects of Reverse Osmosis' may have avoided certain negative consequences of their discharge, energy intensive evaporation may suffer from high carbon footprint and hence may not be sustainable at the current level of energy intensity (Thi & Tóth, 2023). Alternative technology paradigms may be needed.

2.7 Zero Waste Processing of Skin Based Industries: An Aspect of Bio Economy of Skin

It is widely known that production of tonne of leather from, say cow hides, generates significant quantities of both liquid and solid wastes (Moktadir & Rahman, 2022). Typical data on the wastes generated from leather processing activity are presented in Table 1. Since any wasted substance could be viewed as disused resource for some other higher value applications, it is pragmatic to critically analyse the bio economy of animal skin and map avenues of zero waste processing of hides as a road to sustainability of leather.

2.8 Resource and Waste Aspects of Raw Hides and Skins Used in Tannery Sector

Raw hides and skins form the starting materials for leather making in a tannery. They arrive at a tannery in the forms of salted, dried, and refrigerated stock. Refrigerated stock of hides and skin carry the original moisture content of 60–65% by weight but not salt. Dried hides carry with them less water and no salt.

2.9 Salt as a Waste

Salted hides carry nearly equal proportions of water, salt, and hide matter. Refrigerated hides and skins suffer from limited pot life and carbon foot-print of material mobility. Hence the distance between the origin and location of the tannery would be generally short. Salted stock carries nearly one third of its weight as sodium chloride. The salt used for curing is almost totally wasted currently. The presence of salt in cured hides and skins add significantly to the cost of waste water treatment and practice of Zero Liquid Discharge. Bearability bottom-line of sustainability with the use of salt for curing is under challenge (Samidurai et al., 2022).

Table 1 Sources and types of pollutants generated in leather processing

Process step	Liquid wastes	Solid wastes	Gaseous emissions
Soaking	BOD, COD, SS, TDS, Organic Nitrogen	Hair, trimmings ad green fleshings	Ammonia
Unhairing, fibre opening and fleshing	BOD, COD, SS, TDS, alkalinity, sulphide, organic and ammoniacal nitrogen	Hair, lime, lime fleshing, lime split and trimmings, lime and organic matter containing sludge	Sulphide
Deliming, bating and degreasing	BOD, COD, SS, TDS, Organic and ammoniacal Nitrogen		Ammonia
Pickling and tanning	BOD, COD, SS, TDS, acidity, neutral salts, veg and chrome tans, syntans		
Mechanical operations such as sammying, splitting and shaving	BOD, COD, SS, TDS, acidity, neutral salts, veg and chrome tans, syntans	Splits and shavings from tanned leathers	
Retanning, dyeing and fatliquoring	BOD, COD, TDS, neutral salts, veg and chrome tans, syntans, dyes and fatliquors		
Mechanical operations such as buffing and trimming		Tanned leather trimmings and dust	
Finishing	Reminder finishing agents	Finished trimmings	Solvents and formaldehyde

Source UNIDO. 2019 (Buljan & Král, 2019)

2.10 Trimmings from Hides

It is customary in leather sector to trim the ends and defective regions at the periphery. It is estimated that trimmings recovered from raw hide or skin per tonne could be 40 kg. Economic recovery by developing technologies for product diversification for increasing viability are crucial. They may offer potentials for emerging as a co-product of leather (Ocak, 2022).

2.11 Hair and Flesh

Removal of hair and flesh from raw hide or skin is assisted by the deployment of depilating chemical agents and mechanical shearing forces. More than 95% of the

Table 2 Constituents of salted hides (per kg of salted hides)

Constituent	g/kg of salted hides
Water	450
Proteins	330
Collagen	240
Albumin	45
Keratin	45
Lipids	40
Carbohydrates	5
Total Dry organic matter	375

Source UNIDO (Buljan & Král, 2019)

chemicals used are wasted. Hair and flesh recovered from raw hides and skins are contaminated with chemicals and lower their economic value significantly. They are used for low end applications in the case of goat and sheep (Ramya et al., 2022; Venkatachalam et al., 2022). In the case of bovine leathers, the hair pulping technologies are used. Hence the entire amount of hair contained in the raw hides and skins is wasted and pose problems of pollution later. The wasted pulped hair adds to the cost of waste water treatment. Solid wastes obtained from hair and flesh per ton of raw hides and skins are 100 and 300 kg, respectively for bovine and ovine (Puhazhselvan et al., 2022).

2.12 *Minor Matrix Materials Contained in Raw Hides and Skins*

Minor matrix materials contained in skin or hide could contribute by weight less than 1.5%. They could be broadly categorized as proteoglycans, natural fat and lipids, carbohydrates, pigments, glycoproteins, cementing materials, autolytic enzymes including animal collagenases. A tentative list of such macro materials is presented in Table 2. Almost the entire amount of minor matrix materials is wasted currently.

2.13 *Hide Matter as Resource for Leather*

It is estimated that about 30% of the materials contained in raw hides and skins constitute hide matter that goes into leather making after the completion of beam house processes.

2.14 Waste Burden in a Typical Bovine Leather Processing and Conversion into Products

More than 70% of leathers produced in the world by way of area is from bovine hides. Based on global standards of leather processing, waste burden of a tannery could be estimated. It is presented in Table 3. During the leather processing, several chemical reactions are promoted. One of the serious steps in leather making is alternation of pH from neutral to alkaline and acidic to near neutral pH conditions. Large alternations in pH while processing leads to the formation of significant amounts of neutral salts (Zhang et al., 2022). Several mineral tanning and post tanning chemicals used also carry neutral salts. Significant share of costs of waste water treatment adopting Zero Liquid Discharge is contributed by the load of neutral salts contained in waste waters. Thickness adjustment of tanned leather for specific applications generate solid waste in the form of chrome shaving, buffing dust, trimmings, etc. (Noyon et al., 2022).

The waste burden adds to costs of production. Treatment and waste management costs could vary depending upon the location and technology. Some tentative costs of waste management in tanneries are listed in Table 4.

While finished leather manufacture from raw hides and skins generate certain amount of solid and liquid wastes as shown in Tables above, not all the finished leather made is worthy of being used for production of shoes. Depending upon the grade, the cutting value of finished leather differs markedly. The cutting value aspects of finished leathers for shoe production are reflected in Table 5.

With lower cutting value and economic recovery from only 65–70% of the surface area, leather cannot remain competitive against synthetics, where value returns are

Table 3 Typical waste burden from conventional leather processing

Parameter	Typical waste burden, conventional process kg/tonne *of wet salted hide*	A tannery with daily input of 10 tonnes of w.s. hides, every day discharges approx. (Kg)
Volume of wastewater	30 m^3/tonne	300 m^3
COD	180	1800
BOD_5	90	900
Suspended solids (SS)	120	1200
Cr^{3+}	5	50
S^{2-}	5	50
Total nitrogen (TKN)	14	140
Cl^-	180	1800
$(SO_4)^{2-}$	80	800
Oil and grease	10	100
TDS	400	4000
pH	6–9	

Source UNIDO (Buljan & Král, 2019)

Table 4 Typical structure of cost of tannery effluent treatment (US$/m^3)

Component	India	Italy	France	Brazil
Energy	0.28	1.21	0.06	0.5
Chemicals	0.2	1.10	0.21	0.14
Salary and labour	0.13	1.43	0.27	0.08
Repair and maintenance	0.05	0.66	0.07	0.08
Sludge and dewatering and disposal	0.05	4.07	0.36	0.03
Depreciation	0.14	1.10	–	–
Miscellaneous	0.54	1.43	0.06	–
Total	1.39	11	1	0.85

Source UNIDO (Buljan & Král, 2005)

Table 5 Cutting value of finished leathers based on the grades

Grade	% Cutting value	% waste
A	100–97	Up to 3
B	96–92	8
C	91–87	13
D	86–82	18
E	81–77	23

Source UNIDO (Buljan & Král, 2005)

nearly 100%. Hence for enhancing viability bottom line, leather, and allied products need to move into Zero waste processing as a principle rather than as a technology upgradation objective (Ramesh et al., 2022).

2.15 *Origin of Negative Public Perceptions of Waste Burden Form Leather Processing*

Environmentally benign leather processing is an essential aspect of the sustainability of the sector. Negative perceptions suffered by the industry are related to this water intensive processing methodology, high burden of Total Dissolved Solids, Chromium bearing sludge, offensive odour, fear of eco constrained dyes and other sensitive compounds, solid and liquid waste burden per tonne of leather processed (Mandal et al., 2023). Public perception of an industrial sector plays a critical component of regulatory processes and demands of a nation with a strong civil society. Transition of leather industry in favour of the best environmental, social governance practices by design and effort is necessary condition for addressing the challenge of public perception and migration from red category to environmentally benign grouping.

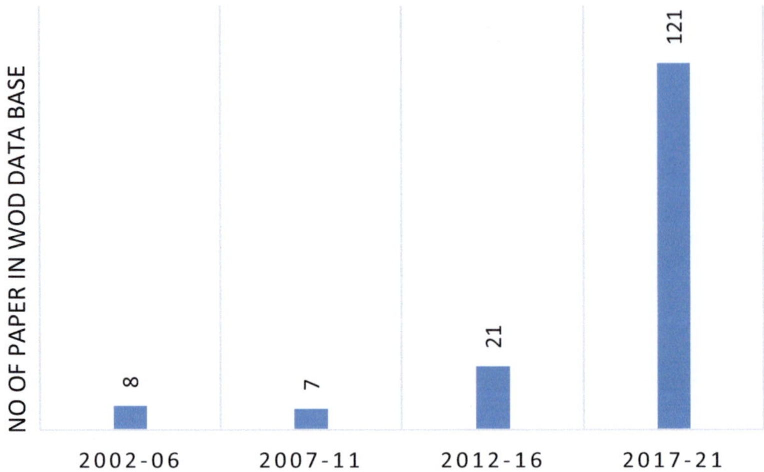

Fig. 1 Increasing focus on sustainability science of leather

This might have emerged as sustainability need of the leather sector in developmental paradigm (Ebabu et al., 2022).

2.16 *State of the Art Studies on Sustainability Science of Leather*

Study of Web of Science data base reveals that sustainability of leather is emerging as a major research theme in leather world. Trends in research activities leading scholarly publications are reflected in Fig. 1. Major themes of sustainability research seem to include cleaner production (Ortolano et al., 2014), exploring high value applications for skin constituents (Appala et al., 2022), energy viability of processing operations as well as Zero Liquid Discharge challenges, etc.

2.17 *Some Possible Strategies for Zero Waste Processing of Leather*

Twelve first principle based strategic approaches with net zero carbon footprint

1. Developing a bio economy principle for skin based industrial sector
2. Development of carbon spread sheet of all matrix materials of skin
3. Working out atom and energy economy of all steps in leather processing and waste treatment and management

4. Developing a production strategy for promoting auxiliary industries from the currently wasted matrix materials
5. Elimination of the use of eco constrained substances at all stages of processing
6. Switch over to salt free or less salt curing methodology for reducing TDS burden
7. Revisit beam house operations for waste elimination from skin/hide
8. Narrow pH processing strategy to minimise the TDS burden
9. Do Ecology options of leather processing based on waste minimization through in plant control measures, secure end of pipe treatment technologies and adopt zero liquid discharge, minimize sold waste generation and recovery of significant economic values from what is wasted as solids and generated from pollution abatement measures
10. Undertake A Total Life Cycle Analysis of leather and competing alternative materials
11. Developing a secure management of End of life Cycle of leather products
12. Developing a net zero carbon footprint strategy for hide and skin derived product portfolio.

2.18 Current Technology Options for Approaching Zero Waste Principle in Leather Processing

Cleaner Production in Leather Processing is the continuous application of an integrated preventive environmental strategy to processes and products so as to reduce risks to humans and the environment.

For processes, this means:

- Conserving raw materials and energy
- Eliminating the use of toxic raw materials
- Reducing the quantity and minimize all emissions and wastes before they leave the production plant.

For products, this means:

- Reducing impacts along the entire life cycle, from raw material extraction to disposal.

Cleaner production reduces risks to;

- The workers
- The community
- Consumers of products
- Future generations.

Cleaner production lowers the cost of:

- Production
- End of Pipe treatment

- Health care
- Cleaning up the environment.

Cleaner production improves:

- Process efficiency
- Product quality.

Zero waste processing methodology:

- Eliminates waste in the entire production chain
- Finds economically viable use of all useable resources
- Eco threat from process wastes is minimized, if not eliminated
- Tends to maximize the economic returns of solid waste.

2.19 Processing Alternatives

Conventional leather manufacturing is mainly based on do-undo process logic like curing (dehydration)—soaking (rehydration), liming (swelling)—Deliming (deswelling), pickling (acidification)—depickling (basification), neutralization—fixing. Further, the hides/skins are subjected to wide variations in pH 2.8–12.5 (Fig. 3). The change in pH demands the use of acids and alkali, which leads to the generation of a large amount of neutral salts. This results in a net increase in total dissolved solids (TDS) comprising chlorides, sulfates, and other minerals in tannery wastewater. In addition, toxic gases like hydrogen sulfide, and ammonia are also evolved during the process. The following Table 6 provides the best available technologies indicating the pros and cons of each and every technology.

2.20 In Process Control Measures for Achieving Zero Waste Leather Manufacturing

Leather making involves various physico-chemical treatment and deploys vast amount of water/chemicals. It is estimated that the conventional leather manufacturing consumes 30–50 m^3 of water and around 1000 kg of chemicals (Preservation to Finishing) for the processing of every tons of hides/skins. Further, the bulk chemicals contribute 60–70% of the total amounts of chemicals used and the major portion is discharged as liquid/solid waste. Hence, the adoption of various in-process control strategies like recycling and reuse, etc., may reduce the pollution load in wastewater. Besides, the in-process control strategies would pave way for resource efficient leather making and also reduce the chemical input. The various in-process control options for achieving zero waste leather manufacturing are discussed in detail from Table 7.

Table 6 Best available technologies for cleaner leather processing

Process alternative	Need For technology	Available technologies	Positive and negative aspects
Salt-free curing (Wu et al., 2017)	• To avoid storage of contaminated salt • To reduce the salinity/total dissolved solids in the wastewater • Water minimization	Cooled air	✓ Automation in handling ✓ No increase in weight ✓ Adapted to sorting ☒ High investment ☒ Need in trucks equipment ☒ Constant feeding with hides
		Ice addition	✓ Low investment cost ✓ Low running cost ✓ Adapted to low quantities ☒ Increase in weight ☒ Limited protection in time
		Dry ice	✓ Moderate investment cost ✓ Reduced increase in weight ☒ Risks with carbon dioxide ☒ Limited protection in time
		Salt + Alkali	✓ Reduced salinity ✓ Ease of application ☒ Limited protection in time
		Silica gel/polyol as an alternative dehydrant	✓ Reduced salinity ✓ Ease of application ☒ Relatively costlier than conventional salting process ☒ Limited protection in time

(continued)

Table 6 (continued)

Process alternative	Need For technology	Available technologies	Positive and negative aspects
		Application of various plant extracts	✓ Reduced salinity ✓ Ease of application ☒ Limited protection in time ☒ Availability of the material
		Organic formulation as preservation-cum-unhairing (PCU) agent	✓ Reduced salinity ✓ Ease of application ✓ Additional advantage of hair removal without using sodium sulphide ☒ Costlier than conventional salting process
Soaking		Processing of green hides	✓ No chemical contamination ✓ Good quality fleshing ✓ Hygienically safe handling ✓ Good organisation of transportation ✓ Well organised processing ☒ Salt need to be added in soaking to remove non collagenous proteins
		Counter current soaking	✓ Volume of soak liquor discharge is reduced by 60% ✓ Land needed for Solar Evaporation Pan is 60% less ☒ Increased TDS in waste liquor

(continued)

Table 6 (continued)

Process alternative	Need For technology	Available technologies	Positive and negative aspects
Lime and sulfide free unhairing (Thanikaivelan et al., 2002, 2003, 2004)	• To eliminate lime sulfide • Avoidance of dissolved hair in wastewater and its volume To prevent H_2S gas liberation	Enzymatic and enzyme assisted	✓ Avoidance/minimization of use of sodium sulphide ✓ After liming pelt is clean and smooth ✓ Pelts are free from wrinkles ✓ High area yield ☒ Proper process control is required ☒ Not applicable for all types of raw material
		Amine/oxidative unhairing agent	✓ Avoidance/minimization of use of sodium sulphide ☒ Costlier than conventional salting process
		Hair saving unhairing/online hair filtration	✓ Minimization of organic TDS and COD in the wastewater ✓ Possibility of reuse of hair as a fertilizer ☒ Capital intensive
		Reuse of lime liquor	✓ Reduction in BOD and COD ✓ Lower suspended solids ✓ Reduction in sulfide and lime input

(continued)

Table 6 (continued)

Process alternative	Need For technology	Available technologies	Positive and negative aspects
Deliming	• To reduce TDS • To reduce liberation of ammonia gas	Carbon-dioxide deliming (Deng et al., 2015)	✓ Reduction in discharge of nitrogen compounds in the effluent ✓ Reduced ammonia odour in the plant ✓ Achievements of BOD reductions of up to 50% ✓ Easy adaptability to automation ✓ Improved quality in terms of refined and cleaner grain and improvement in the degreasing action ✓ Reduced risk of acid shock and achievement in the gradual reduction of pH ☒ High investment cost ☒ Proper process control ☒ Not applicable for thicker hides
		Use of dry ice (Sathish et al., 2013)	✓ Reduction of discharge of nitrogen compounds in the effluent ✓ Reduced ammonia odour in the plant ✓ Achievements of BOD reductions of up to 50% ✓ Easier handling for the operator ☒ Not applicable for thicker hides

(continued)

Table 6 (continued)

Process alternative	Need For technology	Available technologies	Positive and negative aspects
Pickling/chrome tanning	• To reduce TDS/Cr load in spent liquor • Reduction in wastewater volume • Reduction in chromium containing leather solid waste	Low salt/salt free pickling (Non-swelling acids) (Marsal Monge et al., 2004, 2005)	✓ Reduction in TDS ✓ Improved uptake of chromium ✓ Decreased process time and improves productivity ☒ More process control ☒ Increased process time for thicker hides
		High exhaust chrome tanning (Materials and methods)	✓ Reduction in TDS ✓ Improved uptake of chromium ☒ Costlier than conventional BCS
		Direct chrome liquor recycling/reuse (Raghava Rao et al., 2003)	✓ Reduction in water consumption ✓ Reduction in TDS ✓ Improved uptake of chromium ☒ More process control
		Zero water (Floatless) chrome tanning technologies (Materials and methods)	✓ Reduction in water consumption ✓ Reduction in TDS ✓ Improved uptake of chromium ✓ Decreased process time and improves productivity ✓ No chromium-containing wastewater ☒ Increased process time for thicker hides

(continued)

Table 6 (continued)

Process alternative	Need For technology	Available technologies	Positive and negative aspects
		Inverse chrome tanning (Cai et al., 2015)	✓ Reduction in TDS ✓ Improved uptake of chromium ✓ Avoidance of chromium-containing unusable splits and shaving dust ☒ Not applicable for all types of raw materials and end products
		Chrome recovery and reuse (Kanagaraj et al., 2008)	✓ No chromium containing wastewater ✓ Easy to adapt ✓ Recovered chromium can be reused in tanning process
		Chrome free alternative tanning agents	✓ No chromium containing effluent ✓ No chromium containing solid waste ☒ Not suitable for all types of end products ☒ Reduced shrinkage temperature
Vegetable tanning	• Improvement of exhaustion	Low float tannage in drums	✓ COD is considerably reduced ✓ Shorter process time
		Recycling of spent liquor/floats	✓ COD is considerably reduced ☒ Proper fungicide to be used

(continued)

Table 6 (continued)

Process alternative	Need For technology	Available technologies	Positive and negative aspects
Post tanning	• Improved exhaustion of chemicals • Reducing TDS and COD	Early trimming of hides and skins	✓ Saving on process chemicals and water (processes based on weight) ✓ Improved solid waste management ✓ Increased production capacity ✓ Easier handling of the hides or skins thus reducing labour cost ✓ Increase in materials quality/selection
		Strict process control and monitoring	✓ Reduction in water and energy consumption ✓ Reduced TDS and salinity in wastewater
		Process design and chemical selection	✓ Simpler the process, easier to control
Finishing	• Reduction of VOC	Solvent free finishing	✓ Minimization of VOC ☒ Reduced fastness to rubbing ☒ Use of cross linkers
		Use of roller coater/airless spray guns and HVLP spray guns	✓ Minimization of VOC ✓ Improved transfer efficiency ✓ Less bounce back or wastage of chemicals ☒ Higher investment cost

Table 7 In-process control options for achieving zero waste leather manufacturing

In-plant control measures	Need	Available options	Advantages
Recycling (Nugraha et al., 2019)	Reduce water and chemical consumption Reduce pollution load in wastewater	Recycling of sectional Process liquor such as Reuse of main soak liquor in dirt soak Reuse of second wash lime liquor/deliming liquor in soaking (or) liming Reuse of pickling and tanning floats in successive cycles Reuse of tanning float in the subsequent pickling cycle	✓ Reduction in water and chemical consumption ✓ Reduction in TDS load ✓ Increases the chromium utilization from 70 to 90% ✓ Further reduces the sulfate load from 30–55 to 10–22 kg/ton ☒ Accumulation of bacterial load and higher preservative consumption in soak liquor recycling ☒ Recovery of digested hair is pre-requisite for lime liquor recycling ☒ Pickle/chrome liquor recycling needs to be closely monitored ☒ Leather quality may be affected after certain recycling processes
Renovation and recycling	To obviates the drawbacks associated with recycling process such as interference of bacterial load, soluble organic etc.	Advanced oxidation process such as Electrochemical oxidation followed by recycling of sectional wastewater	✓ Economically viable ✓ Complete removal of organic in wastewater ✓ The operating cost is comparatively lower than the conventional wastewater treatment system ☒ Cr(VI) formation may occur if Cr(III) is not removed completely

(continued)

Table 7 (continued)

In-plant control measures	Need	Available options	Advantages
Process rationalization (Saravanabhavan et al., 2006)	Reduce water, chemical consumption and also to reduce chromium containing solid waste generation	Reverse leather processing—Post tanning followed by tanning Inverse chromium tanning—Organic pre-tanning followed by mechanical operations and chrome tanning	✓ Reverse leather processing avoids "Do-Undo" process logic and results in water, chemical and TDS load reduction ✓ Inverse chrome tanning completely avoids chromium containing unusable splits, shavings etc. ☒ Assortment and selection of materials in tanning stage is not possible ☒ Chemical cost will be increased in tanning segment due to the use of organic tanning agent

(continued)

Table 7 (continued)

In-plant control measures	Need	Available options	Advantages
Chromium management	To avoid discharge of chromium composite wastewater and sludge	Chromium recovery and reuse	✓ It provides convincing solution to the problem of chrome discharge in waste waters and in avoiding chromium in the sludge produced in composite tannery wastewater treatment ✓ Improvement in biomethanation efficiency due to the segregation of sulfate containing spent chrome liquor ☒ Need to setup a separate system for chromium recovery ☒ Some capital col involved ☒ Increased discharge of neutral salts
Control on post-tanning discharge (Buljan & Krá, 2019)	To reduce pollution load To improve biodegradability of wastewater	Less salt syntan/dye stuff Bio-based restricted substances free post-tanning auxiliaries Formulation of liquid-based post-tanning auxiliaries Multifunctional post-tanning auxiliaries	✓ Reduction in TDS load ✓ Reduction in chemical input ☒ Constraints on storage and transportation of liquid auxiliaries ☒ Inconsistency in product stability and its performance

(continued)

Table 7 (continued)

In-plant control measures	Need	Available options	Advantages
Emission control in tanneries	To reduce the emission of harmful/toxic gases and VOC reduction	Adoption of various cleaner technologies to reduce gaseous emission Improvement in coating system such as (a) roller coating (b) high-volume low pressure system (c) foam and airless finishing Use of water-based finishing system Advanced extraction ventilation and odour abatement system	✓ Improvement in coating efficiency ✓ Reduction in chemical wastage (30–40%) ✓ Avoidance of solvent containing exhaust vapour into the atmosphere ✓ Recovery of solvent in adsorption plant ☒ Capital and maintenance cost involved
Noise control in tanneries	• To reduce noise pollution caused by machineries	• Proper maintenance and lubrication • Replace metal with hard plastic wherever its possible • Use of low noise spray guns • Silencing of exhaust outlets	✓ Reduction in noise pollution ☒ Involvement of additional cost and monitoring

2.21 *End of Pipe Control Measures for Achieving Zero Waste Leather Manufacturing*

In leather processing, the basic strategy is to clean skin of the unwanted interfibrillary material through a set of pretanning operations, preserve permanently by means of tanning, and add aesthetic properties during post tanning stages. The relative contributions of pretanning, tanning, and post tanning steps to the pollution loads in tanning sector are compared in the following Table 8. It is evident that a major source of pollution is the waste discharged into wastewater during the pretanning operations (Saxena et al., 2017).

The tanning industry in India is primarily in the small and medium enterprise (SME) sector. The financial and technical capacity to select, install, commission, operate and maintain individual and unit specific effluent treatment plant is limited to a maximum of about 15% of the total number of tanning units in the country. Therefore, the need for a collective approach to the end-of-pipe treatment of tannery wastewater through common effluent treatment plant (CETP) has been recognized (Kavitha & Ganapathy, 1970). 16 common effluent treatment plants have been commissioned in the country in different locations to serve about 70% of functional tanning units, with 12 of them being established in Tamil Nadu. Currently, there is no functional tannery in Tamil Nadu without being connected to pollution Control devices for end-of-pipe treatment. Of the 154 individual effluent treatment plants (ETP) and 16 CETPs in Indian leather sector, 140 ETPs and 12 CETPs are in Tamil Nadu. Efforts are now for reducing the energy inputs needed for waste water treatment making it to meet bearability and viability bottom lines.

Conventional tannery wastewater treatment consists of preliminary treatment including screening, grit chamber, and equalization. The primary wastewater treatment separates settleable solids in the primary settling tank (primary clarifier) through

Table 8 Pollution load (kg) for processing one tonne of raw hide/skin (Bajza & Vrcek, 2001)

Pollution parameters	Soaking	Beamhouse	Tan yard process	Post- tanning and wet finishing	Total pollution load, kg
Biochemical oxygen demand (BOD5 @ 20 °C)	12	44	3	12	71
Chemical oxygen demand (COD)	23	94	10	27	154
Sulfides (as S^{2-})	–	4	–	–	4
Suspended solids	28	65	9	6	108
Total chromium (as Cr)	–	–	5	1.5	6.5
Total dissolved solids	160	85	130	44	419
Chlorides (as Cl^-)	107	25	54	6	192
Sulfate as (SO_4^{2-})	8	9	45	22	84

coagulation/flocculation and the suspended organic solids settle at the bottom of the sedimentation tank as primary chemical sludge (Zhao et al., 2022). The secondary treatment process stabilizes most of the organic matter in wastewater as CO_2, water, and biomass. It can be achieved either by attached growth processes or by suspended growth processes. Various commonly used treatment technologies (Table 9) including activated sludge process (ASP) and up-flow anaerobic sludge blanket (UASB) process are discussed.

2.22 *Unburdening Solid Waste Problem of Leather Processing: Some Current Research Directions*

It is estimated that for every tonne of leather processed, around 700 kg of solid waste is generated from the manufacturing process. Leather is made from the skin making protein called as "collagen" and thus the need for removing the other unwanted materials present in the skin matrix becomes inevitable. Apart from these, the skin/leather matrix also undergoes various operations such as trimming, fleshing, shaving, buffing, etc. before being converted into final product and thus leads to the generation of huge amount of solid wastes. The unit operations involved in leather manufacturing process along with the proteinous solid waste generated at each operation along with the possible value added products produced is detailed below in Fig. 2.

Some of the protein rich solid wastes generated form leather process includes hair, trimmings from raw hides, tanned leather shavings and trimmings and buffing dust. Dumping of such solid waste can easily affect the ecosystem and is also not a suitable economical option. Instead, the waste can be converted into different products by doing necessary chemical, biological, or physical treatment promoting the concept of circular economy. This creates a need for the development of technologies that can lead to efficient reuse of such organic materials for various product development. Some such technologies that are developed to overcome the solid waste management issue have been briefly discussed in Table 10.

2.23 *Research Directions for Next Generation Technology for Maximizing Bio Economy of Skin Based Product Portfolio: Step Towards Sustainability of Leather*

2.23.1 Sustainability Principle Beyond Environmental Regulatory Compliance

Sustainability is defined as the development that meets the needs of the present without compromising the ability of the future generation in meeting their own needs (Keeble, 1988). Unless the economic growth addresses the needs of the present in

Table 9 Various treatment technologies for zero waste leather processing

Treatment	Need for technology	Available alternate technologies	Positive and negative aspects
Primary treatment	To transfer the floc to the surface of water through the attachment of air bubbles to the floc	Dissolved air flotation (DAF)-injecting compressed air in which oxygen is dissolved in the wastewater under pressure to achieve clarification and then air is released at atmospheric pressure in the floatation tank or basin (Gerba & Pepper, 2015)	✓ It is more efficient than settling in removing low-density particles, even at the higher hydraulic loadings ☒ Expensive as it requires pressurized air
	To avoid the formation of primary chemical sludge due to addition of coagulants and flocculants	Sequential oxic anoxic bioreactor (SOABR)—It comprises of three chambers and each chamber consists of anoxic and oxic zones. The bottom of the reactor was sloped at 45° with the base to increase the efficiency of flow dynamics of suspended particles present in the wastewater. The oxic zone removes the suspended particles by interception method and anoxic zone digest the sludge and suspended solids scavenged by the oxic chamber (Mannacharaju et al., 2020)	✓ Low electrical energy consumption ✓ Low operating cost ✓ No primary chemical sludge and also sludge generation is very minimum (0.01 kg/m^3) ☒ Design is little complicated, construction difficulties in field applications

(continued)

Table 9 (continued)

Treatment	Need for technology	Available alternate technologies	Positive and negative aspects
Secondary treatment	To stabilizes most of the organic matter in wastewater as CO_2, water and biomass	Moving Bed Biofilm Reactor (MBBR)—It is a combination of both suspended growth (ASP) and attached growth (bio filter processes). The concept of the MBBR process is to use the whole tank volume for biomass growth. It uses simple floating media carriers for attached growth of biofilms (Barwal & Chaudhary, 2014)	✓ Carrier media in MBBR remain in suspension within the reactors due to perforations or screen arrangement at the discharge ✓ The carrier matrix is designed in such a way to have a high specific surface area to accommodate biofilm growth ✓ The quantity of sludge production in MBBR is lower than the typical activated sludge process ☒ Difficulties in maintenance due to difficulties in determining MLSS

(continued)

Table 9 (continued)

Treatment	Need for technology	Available alternate technologies	Positive and negative aspects
		Sequencing batch reactor (SBR)-It consists of a set of tanks which is operated on a fill draw activated sludge system. It experiences five cycles in sequence namely fill, react, settle, draw and idle	✓ Operating flexibility ✓ Minimum requirement of footprint ✓ Less capital cost ☒ Requirement of high level of maintenance ☒ Sophistication and potential plugging of aeration devices
		Fluidized bed reactor (FBR). It works on the principle of fluidization of carrier matrices (of bacteria immobilized or bacteria adhered) by the injected air and wastewater (Nelson et al., 2017; Tisa, 2014)	✓ Excellent contact between organics in wastewater and carrier matrices embedded with bacteria, ✓ high mass transfer rate ✓ Low sludge generation

(continued)

Table 9 (continued)

Treatment	Need for technology	Available alternate technologies	Positive and negative aspects
		Membrane bioreactor (MBR)—It is a combination of biological treatment with membrane separation such as microfiltration or ultrafiltration	✓ Less carbon footprint than the conventional activated sludge process ✓ Effectively displaces conventional treatment plant process steps such as primary treatment, secondary activated sludge process disinfection ✓ Less generation of sludge ☒ High operating costs ☒ Frequent membrane fouling

(continued)

Table 9 (continued)

Treatment	Need for technology	Available alternate technologies	Positive and negative aspects
		Fluidized immobilized carbon catalytic reactor (FICCO)—It is a microorganism immobilized nanoporous carbon based fluidized reactor (Mannacharaju et al., 2018; Saranya et al., 2019)	✓ Performance of the reactor is not affected by suspended solids presence in the wastewater ✓ Low Foot print requirement compared to conventional treatment technologies ✓ No limitation on influent organic strength of wastewater ☒ pH of wastewater must be neutral ☒ Performance is limited with refractory organics
		Chemoautotrophic activated carbon oxidation reactor (CAACO)—It is a microorganism immobilized nanoporous carbon based packed bed reactor (Arceivala & Asolekar, 2007; Sekaran et al., 2013)	✓ Low foot print requirement compared to conventional treatment technologies ✓ Less residence time ✓ Very minimum sludge production ☒ Requires frequent backwash

(continued)

Table 9 (continued)

Treatment	Need for technology	Available alternate technologies	Positive and negative aspects
Tertiary treatment/ zero liquid discharge	To eliminate any liquid waste for recovery of water for reuse	Reverse osmosis (RO) (Nayar et al., 2019)	✓ Energy efficiency and technical maturity ☒ Not effective standalone technology ☒ Intensive pre-treatment requirement ☒ High membrane fouling
		Forward osmosis (FO) (Ahmed et al., 2019; Kolliopoulos et al., 2018)	✓ More cost-effective and can work at even higher feed salinities (up to 200 g/L TDS) ☒ Low water flux at very high feed salinities ☒ Reverse solute flux ☒ Membrane fouling issues ☒ Requirement of intensive pre-treatment

(continued)

Table 9 (continued)

Treatment	Need for technology	Available alternate technologies	Positive and negative aspects
		High-pressure reverse osmosis (HPRO) (Davenport et al., 2018)	✓ RO can concentrate the brine 1.7 times more ☒ Intensive pre-treatment required ☒ Membrane fouling issues
		Osmotically assisted reverse osmosis (OARO) (Peters & Hankins, 2019)	✓ Highest recovery rate while compared with RO/HPRO at salinity of 140 g/L TDS ☒ Economically high cost because of its multiple RO/FO stages ☒ Intensive pre-treatment required ☒ Membrane fouling issues ☒ Lack of a generic draw solution

(continued)

Table 9 (continued)

Treatment	Need for technology	Available alternate technologies	Positive and negative aspects
		Electrodialysis (ED)—It applies an electric potential as the driving force to remove dissolved ions through ion exchange membranes. In contrast to RO membranes that reject all ions, ion exchange membranes selectively permit the transport of counter ions but prevent the passage of co-ions (Qureshi & Zubair, 2016; Tong & Elimelech, 2016)	☒ Energy cost increases with increasing feed Salinity ☒ Intensive pre-treatment required
		Electrodialysis reversal (EDR)—Electrodialysis in which the polarity of the electrodes is reversed frequently for minimizing fouling and scaling (Zhao et al., 2019)	☒ Energy cost increases with increasing feed Salinity ☒ Intensive pre-treatment required
		Brine concentrator	✓ Exceptional technology for water recovery up to 99% ☒ Capital cost and operational cost both are significantly high (Panagopoulos & Haralambous, 2020)
		Brine crystallizer	

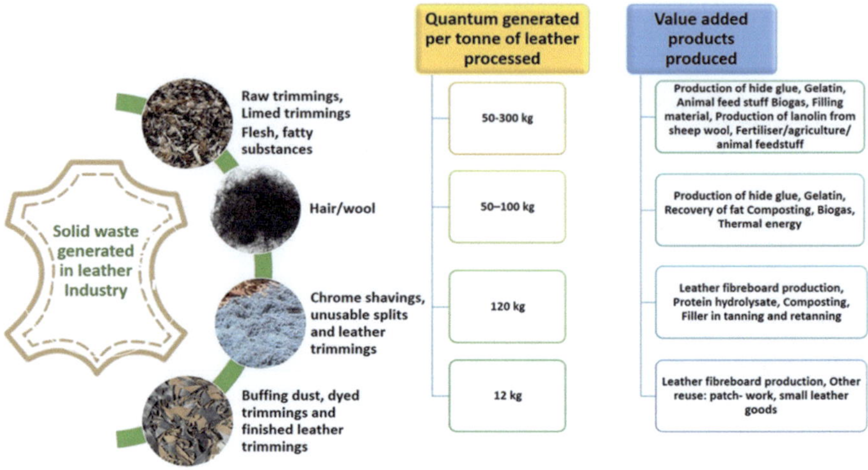

Fig. 2 Solid waste generation in different processes of leather making for 1 ton of RM along with value added products produced (Buljan & Král, 2019)

an intelligent and equitable manner, the natural resources that are necessary for the future needs of the society cannot be met. Sustainability includes much more than compliance to environmental regulatory pressures alone and emphasises practice of environmental and social governance also.

The comprehensive sustainability is built upon three pillars namely society, environment, and economy (Fig. 3). It encompasses the environmental bearability, economic viability, and social equity (Pomponi & Moncaster, 2017). Environmental bottom line of sustainability is the ability of maintaining the factors and practices that contribute to quality of environment on a long-term basis. Economic viability bottom line is about the ability state of utilization of the resources in a responsible and efficient manner so as to reap consistently the operational profit (Taghvaee et al., 2022). Social equity bottom line is about the ability of a social system to consistently achieve a good social well-being (Åhman, 2013). Attainment of sustainability is the process of moving to the right point of intersection of the three aspects viz. society, environment, and economy (Gupta & Racherla, 2018). Leather industry utilizes the co-product of the meat industry, the hides, and skins. Zero waste processing could help in all three elements of sustainability bottom line (bearable, viable, and equitable) which includes social perception of the sector being socially contributing to the equity in the socio economic development (Yorgancioglu et al., 2020).

Table 10 Technologies available to combat the solid waste generation issues

Solid waste generated	Need for technology	Available technologies	Positive and negative aspects
Raw trimmings and limed trimmings	• Avoid highly proteinous raw material end up in landfills • Rich in collagen that can be utilized for other applications	Pharmaceutical grade gelatine Sustainable packaging materials as a replacement for plastics (Masilamani et al., 2017; Muralidharan et al., 2022a, 2022b) Microbiological media as a carbon source (Ramesh et al., 2018; Renitha et al., 2015) Activated carbon (Yuvaraj et al., 2018) Multifunctional polymeric retanning agents (Sathish et al., 2019) Bioplastics (Muralidharan et al., 2020) Biofuel production (Devaraj et al., 2018)	✓ Retanning agents devoid of formaldehyde ✓ Better organoleptic properties and dye uptake compared to conventional retanning agents ☒ Need for infrastructure and sophisticated machineries
Hair	• Burning of accumulated hair generates toxic gases • Rich source of keratin protein that can be utilized for various purpose	Retanning agent based on keratin grafted acrylic polymer (Ramya et al., 2022) Protein filler for leather processing (Yu et al., 2015) Compost for agricultural purpose (Thankaswamy et al., 2018) Keratin hydrolysate for biomaterial field and as additives in cosmetic products Extraction of amino acids	✓ Complete internalisation of solid waste ✓ Better replacement for conventional urea-based fertilizers ☒ Storage, collection and transport of material

(continued)

Table 10 (continued)

Solid waste generated	Need for technology	Available technologies	Positive and negative aspects
Chrome shavings and tanned leather trimmings	• Problems associated with disposal of chromium containing leather pieces • Leaching of chromium into ground and contamination of areas around landfill sites	Chrome tanning agent produced using shavings as reducing agent (Rao et al., 2002) Protein syntan for leather processing Adsorbents for removal for toxic chemicals (Saravanabhavan et al., 2007) Composite sheets (Ashokkumar et al., 2011) Protein hydrolysate for animal feed (Pati et al., 2013) Chromium(III) pigments by controlled incineration (Sreeram et al., 2011) Scaling inhibitors (Qiang et al., 2013) Wound dressing material (Ramnath et al., 2012)	✓ Elimination of chrome containing materials in landfills ✓ Avoiding separation of chromium from proteinous waste ☒ Lack of up scaled technologies ☒ Need for infrastructure such as reactors for the product synthesis at in-house level
Finished leather trimmings and buffing dust	• Disposal of highly non-degradable materials	Leather boards (Senthil et al., 2015) Energy generation (biofuel) Leather composite sheets Regenerated leather along with plant fibres (Teklay et al., 2017) Fabrication of super capacitor (Konikkara et al., 2016)	✓ Development of small scale industries leading to employment generation

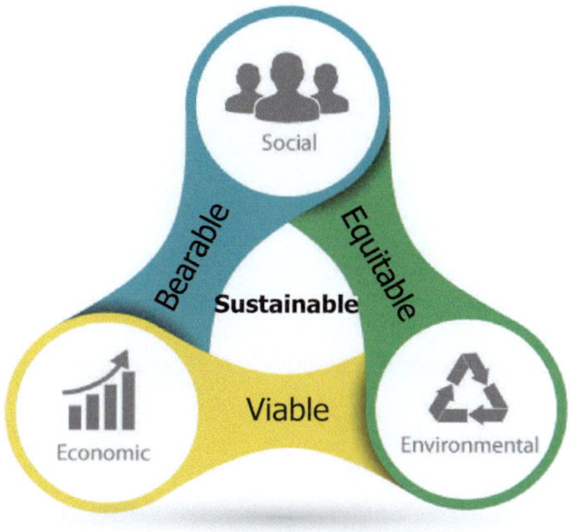

Fig. 3 Three pillars of sustainability (www.nicepng.com)

2.24 *Sustainable Leather Manufacturing*

In the view of sustainable leather manufacturing some of the recent developments are sequenced and the same is shown in Fig. 4 as Best Available Technological options for Sustainable Leather Manufacturing.

This process line could completely eliminate the salt, sulfide, and lime from leather manufacturing and also reduce the water consumption up to 60–70%. The number of unit process and process duration involved in leather manufacturing can also be reduced, which in turn accelerates the leather production. Therefore, implementation of these processes could help moving towards sustainable leather manufacturing.

2.25 *Sustainable Development Goals: Towards Eventual Destination*

At the Sustainable Development Summit on 25 September, 2015, UN, the Member States adopted the 2030 Agenda for Sustainable Development, including a set of 17 Sustainable Development Goals (SDGs), otherwise known as the Global Goals. The SDGs are a new, universal set of goals, targets, and indicators that UN Member States are expected to use to frame their agendas and political policies over the next 15 years.

The new development framework that aims to transform our world would guide all global, regional, and national development endeavours for the next 15 years towards the aimed transformation through set 17 goals (Fig. 5).

It is recommended that through Goal 9, the Member States of the United Nations call upon the international community to 'build resilient infrastructure, promote inclusive and sustainable industrialization and foster innovation'. The inclusive and sustainable industrial development (ISID) can therefore serve as a primary engine not

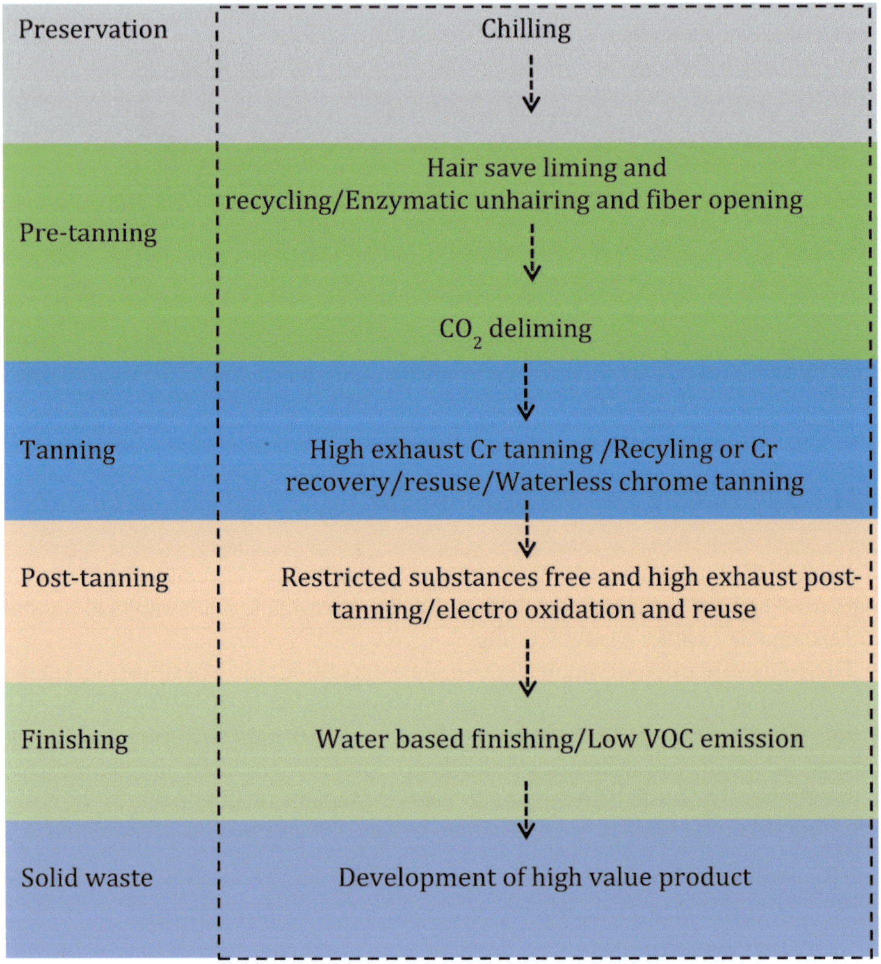

Fig. 4 Best available technological options for sustainable leather manufacturing

only for creating jobs and promote economic growth but also of technology transfer, investment flows and skills development, as also acknowledged in the Addis Ababa Action Agenda of the Third International Conference on Financing for Development held in July 2015.

In addition to Goal 9, all other SDGs incorporate some industry-related aspects and targets. As a core driver of the global development agenda to eradicate poverty and advance sustainable development, ISID therefore makes a critical contribution towards addressing the economic, social, and environmental dimensions of development in a systemic and holistic manner. The concerted move of the nations in this direction will ensure the reach of the eventual destination.

Fig. 5 Sustainable development goals (SDGs)—UNDP

2.26 Carbon Footprint (CF), Life Cycle Analysis (LCA) for Leather

Carbon footprint (CF) is the total amount of CO_2 and other greenhouse gases, emitted over the full life cycle of a process or a product (e.g. leather) (Rondoni & Grasso, 2021). It is expressed in grams of CO_2 equivalents. The CF for leather includes material and operations from the raw material starting from the slaughterhouse to the end of leather product life cycle (Chen et al., 2014). The issue of boundaries is crucial. In principle, cleaner and more efficient technologies are very important for the reduction of CO_2 emissions; however, there are also other factors which significantly influence the total leather CF, among them being mainly the following:

- Transport
- Water consumption
- Efficiency in use
- Energy footprint
- Nature of raw material
- Biodegradability
- Use/recyclability.

At the moment there is still no globally agreed methodology for a CF calculation and it is far beyond the scope of this paper to discuss the difference of various methods proposed. Until recently two methods of Life Cycle Analysis (LCA) computation prevailed.

The carbon footprint of conventional leather processing involves use of transportation, use of energy, use of chemicals, wastewater generated, and others. The carbon footprint calculations reported including cattle breeding is 110 kg CO_2e/m^2.

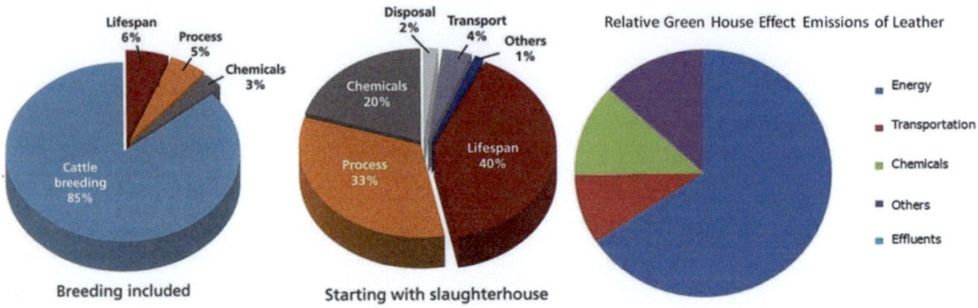

Fig. 6 Contribution of various factors towards carbon footprint (Buljan & Král, 2019)

After slaughtering, it is estimated at 17 kg CO_2e/m^2 of leather. The details of all the factors are given in Fig. 6.

The annual Carbon Footprint of hide sector in India is estimated about 1.813 mil ton CO_2e/year. The various factors that need to be considered for improving the CF in Leather are as follows:

- Sourcing and manufacturing
- Transportation (e.g. by surface or sea instead of air)
- Improve energy efficiency and use of renewable (solar energy) and non-traditional sources (heat pumps)
- Higher the complexity of the chemicals, the higher the CO_2 emission
- Powder products are higher in CO_2 emissions than liquid.

2.27 *Promoting Circular Economy*

There are instances that the tannery processing various types of leathers from Raw to finishing are able to convert their non-leather forming proteinous wastes into retanning agents for their own consumption (Moktadir et al., 2020). While attempts are being made to minimise waste generation at the source incorporating atom efficient process technologies, utilising/converting waste generated from one process into a resourceful input for any other purpose are encouraged as part of the circular economy (Karuppiah et al., 2021).

2.28 *Industry 4.0 and Its Potential Impact for Sustainable Leather Manufacture*

Most of the manufacturing sectors have embraced Industry 4.0 with the aim to reduce wastages and enhance productivity and contributing to sustainable manufacturing with no or low effect on environment. Thus Industry 4.0 envisages the overall manufacturing activity in totality to be practiced as embracing a cultural phenomena than focussing into any particular process (Beltrami et al., 2021). Industry 4.0 aims potentially to improve the following:

- Productivity and competitiveness
- Increase energy and resource efficiency effectiveness
- Contribute to Environment
- Enable the transition to a circular economy
- Enable industrial economy in which end of life products are reused, remanufactured and recycled
- Contribute Sustainable production-consumption patterns
- Provide opportunities for developed and developing countries to achieve economic growth and sustainable development.

This calls for integration of all the available tools that have been employed not only in leather manufacture but also adopting such methods that are relevant to any other manufacturing sector.

2.29 *Modern Tools for Industry 4.0*

Automation and AI based technologies for effective control of various production parameters uniformly even at remote places simultaneously are a reality today. This calls for a paradigm shift in leather processing industry, focusing mainly on decentralised production concepts in lieu of mass production centres. Energy and waste optimisation technologies with principles of circular economy and AI assisted web-controlled automation ensure safety and reliability. The current Industry 4.0 based manufacturing Technology employs.

- Cyber-Physical Systems (CPS)
- Internet Of Things (IOT)
- Internet Of Services (IOS)
- Internet of Data (IOD).

This would emphasize digitisation of various material, process, and production parameters and analyse them for providing solutions for sustainable manufacturing. Details are given in Table 11.

2.30 *Automation for Resource Optimization*

The leather industry over the years have evolved in incorporating machines for material handling and processing including chemical and water addition, pneumatically controlled processing drums, and other machinery. Many of these machines have offered scope for embedding AI based devices to collect information and process for

Table 11 Industry 4.0—potential application in leather sector

	Customer	Design	Procurement	Production	Quality	Maintenance
Inter-operability		Concurrent engineering		Production data form sub contractors and to customers		
Virtualization		Concurrent engineering	Concurrent engineering	Simulation modeling		
Decentralization	Restructure based on customer focused end-to end operation		Order from modules	Modular production cells		
Real-time capability	Sustainability reporting			Modular production cells	Big data drive quality management	Predictive maintenance
Service orientation	Post sales service					
Modularity				Modular production cells		

Source LERIG (2019)

optimised productivity (Horváthová et al., 2019). Use of AI Based data analytics that could offer problem solving/solution providing has emerged as an important option.

2.31 Industry 4.0—Impact on Leather Sector

The leather sector in India has consistently adapted to the customers' product and system needs as well. Customers embarking on Industry 4.0 will have a priority of digitizing the supply chain with the following requirements and more:

- Collaborative Design and Concurrent Engineering (Azevedo et al., 2010)
- Predictive Maintenance
- After Sales Service
- Data Enhanced Products
- Globally Visible Production
- Environmental and Compliance status.

2.32 Need for Adapting Industry 4.0

While most of the industry practice waste minimization (Marques et al., 2017), resource optimisation, Automation (wherever possible) and other regulatory mechanisms related to environment, compliances for ethical and social needs, integrating all these aspects and presenting the manufacturing activity as a model for low carbon emission could well be achieved with the active implementation of Industry 4.0. This also facilitates various certifying agencies for grading the concerned manufacturing industry with respect to overall efficiency, viz. gold, silver, bronze or green, yellow, amber, red, etc. Already many leather chemical manufacturers are coming out with developing chemicals and integrating their usage with low carbon emission processes. Thus implementation of Industry 4.0 necessitates the need for advanced training of existing manpower in most of the developing nations (Singh & Gupta, 2013).

2.33 Technology Paradigms Needed for Sustainability of Leather Sector

Needed technology paradigms call for (a) zero waste processing (Chojnacka et al., 2021), (b) near total utilization of animal hides and skins with optimal returns, (c) near total utilisation of leather, and (d) Compliance to sustain environment social responsibility (Omoloso et al., 2020). New technological breakthroughs like sustainable bio, instead of chemical, processing may be necessary. Leather sector cannot afford to miss the positive inputs of Industrial revolution 4.0.

3 Conclusion

3.1 Way Forward for Sustainability of Leather in New Material World

An excellent framework for moving towards sustainability of leather sector has been provided as part of the initiative of the UNIDO (Buljan & Král, 2019). It has flagged the issue and provides some useful directions for moving the sector along the path of sustainability contributing to all three bottom lines viz. bearability, viability, and social equity. There is now a compelling need for the various components of research and development to carry out a holistic analysis of the carbon footprint of global leather industry with sustainable movement of materials, large wastage burden and energy viability of environmental regulatory compliance. An objective audit of technologies in practice for their sustainability coefficient is necessary. Appropriate ranking and conscious movement of the sector from its viability and internal competitiveness focus to a transparent migration to responsible environmental social governance practices may be necessary. Sustainability needs of the leather sector in future cannot rely on the technology tools of yesterday and last century. It would be appropriate that every leather cluster has an Innovation park that could promote conversion of innovative ideas for generating new technology based entrepreneurship for achieving zero waste production. The sector may need to embrace tools of industrial revolution 4.0 and beyond proactively and earn the trust of informed civil societies as a responsible industrial sector that contributes significantly to the social equity through employment and empowerment and to global good and wellness.

Acknowledgements The authors thank Dr. R. Aravindhan, Dr. M. Sathish, and Dr. R. Renganath Rao of CSIR-CLRI for their support in compiling the information in preparing the document.

References

Åhman, H. (2013). Social sustainability—Society at the intersection of development and maintenance. *Local Environment, 18*(10), 1153–1166.

Ahmed, M., Kumar, R., Garudachari, B., & Thomas, J. P. (2019). Performance evaluation of a thermoresponsive polyelectrolyte draw solution in a pilot scale forward osmosis seawater desalination system. *Desalination, 452*, 132–140.

Appala, V. N. S. G., Pandhare, N. N., & Bajpai, S. (2022). Biorefining of leather solid waste to harness energy and materials—A review. *Biomass Conversion and Biorefinery*. https://doi.org/10.1007/s13399-022-02455-8

Arceivala, S. J., & Asolekar, D. S. R. (2007). *Wastewater treatment for pollution control and reuse*. AccessEngineering | McGraw-Hill Education—Access Engineering. McGraw-Hill Education.

Ashokkumar, M., Thanikaivelan, P., Krishnaraj, K., & Chandrasekaran, B. (2011). Transforming chromium containing collagen wastes into flexible composite sheets using cellulose derivatives: Structural, thermal, and mechanical investigations. *Polymer Composites, 32*(6), 1009–1017.

Azevedo, A., Francisco, R. P., Bastos, J., & Almeida, A. (2010). Virtual factory framework: An innovative approach to support the planning and optimization of the next generation factories. *IFAC Proceedings Volumes, 43*(17), 320–325.

Bajza, Z., & Vrcek, I. V. (2001). Water quality analysis of mixtures obtained from tannery waste effluents. *Ecotoxicology and Environmental Safety, 50*(1), 15–18.

Barwal, A., & Chaudhary, R. (2014). To study the performance of biocarriers in moving bed biofilm reactor (MBBR) technology and kinetics of biofilm for retrofitting the existing aerobic treatment systems: A review. *Reviews in Environmental Science and Bio/technology, 13*(3), 285–299.

Beltrami, M., Orzes, G., Sarkis, J., & Sartor, M. (2021). Industry 4.0 and sustainability: Towards conceptualization and theory. *Journal of Cleaner Production, 312*, 127733.

Buljan & Král. (2005). Costs of tannery waste treatment, UNIDO.

Buljan, J., & Král, I. (2019). The framework for sustainable leather manufacture. UNIDO.

Cai, S., Zeng, Y., Zhang, W., Wang, Y., & Shi, B. (2015). Inverse chrome tanning technology based on wet white tanned by Al–Zr complex tanning agent. *Journal of the American Leather Chemists Association, 110*(4), 114–120.

Chen, K.-W., Lin, L.-C., & Lee, W.-S. (2014). Analyzing the carbon footprint of the finished bovine leather: A case study of aniline leather. *Energy Procedia, 61*, 1063–1066.

Chojnacka, K., Skrzypczak, D., Mikula, K., Witek-Krowiak, A., Izydorczyk, G., et al. (2021). Progress in sustainable technologies of leather wastes valorization as solutions for the circular economy. *Journal of Cleaner Production, 313*, 127902.

Davenport, D. M., Deshmukh, A., Werber, J. R., & Elimelech, M. (2018). High-pressure reverse osmosis for energy-efficient hypersaline brine desalination: Current status, design considerations, and research needs. *Environmental Science & Technology Letters, 5*(8), 467–475.

Deng, W., Chen, D., Huang, M., Hu, J., & Chen, L. (2015). Carbon dioxide deliming in leather production: A literature review. *Journal of Cleaner Production, 87*, 26–38.

Devaraj, K., Aathika, S., Mani, Y., Thanarasu, A., Periyasamy, K., et al. (2018). Experimental investigation on cleaner process of enhanced fat-oil extraction from alkaline leather fleshing waste. *Journal of Cleaner Production, 175*, 1–7.

Ebabu, W., Israil Hossain, Md., El-Naggar, M. E., Kechi, A., Hailemariam, S. S., et al. (2022). Exploration of functional polymers for cleaner leather industry. *Journal of Inorganic and Organometallic Polymers and Materials, 32*(1), 1–14.

Gerba, C. P., & Pepper, I. L. (2015). Chapter 25—Municipal wastewater treatment. In *Environmental microbiology* (3rd ed., pp. 583–606). Academic Press.

Gupta, S. K., & Racherla, U. S. (2018). Interdependence among dimensions of sustainability: Evidence from the Indian leather industry. *Management of Environmental Quality: An International Journal, 29*(3), 406–415.

Hashmi, N., Aqib Jalil, S., & Javaid, S. (2022). A multi-objective model for closed-loop supply chain network based on carbon tax with two fold uncertainty: An application to leather industry. *Computers & Industrial Engineering, 173*, 108724.

Horváthová, M., Lacko, R., & Hajduová, Z. (2019). Using industry 4.0 concept—Digital twin—To improve the efficiency of leather cutting in automotive industry. *Quality Innovation Prosperity, 23*(2), 01.

Jones, M., Gandia, A., John, S., & Bismarck, A. (2021). Leather-like material biofabrication using fungi. *Nature Sustainability, 4*(1), 9–16.

Joseph, K., & Nithya, N. (2009). Material flows in the life cycle of leather. *Journal of Cleaner Production, 17*(7), 676–682.

Kanagaraj, J., Chandra Babu, N. K., & Mandal, A. B. (2008). Recovery and reuse of chromium from chrome tanning waste water aiming towards zero discharge of pollution. *Journal of Cleaner Production, 16*(16), 1807–1813.

Karuppiah, K., Sankaranarayanan, B., Ali, S. M., Jabbour, C. J. C., & Bhalaji, R. K. A. (2021). Inhibitors to circular economy practices in the leather industry using an integrated approach: Implications for sustainable development goals in emerging economies. *Sustainable Production and Consumption, 27*, 1554–1568.

Kavitha, P. R., & Ganapathy, G. P. (1970). Removal of solids from the tannery effluent by a suitable technology: A case study Vellore District, Tamil Nadu, India. *Journal of the Industrial Control Pollution, 31*(1).

Keeble, B. R. (1988). The Brundtland report: 'Our common future.' *Medicine and War, 4*(1), 17–25.

Kolliopoulos, G., Martin, J. T., & Papangelakis, V. G. (2018). Energy requirements in the separation-regeneration step in forward osmosis using TMA–CO2–H2O as the draw solution. *Chemical Engineering Research and Design, 140*, 166–174.

Konikkara, N., Kennedy, L. J., & Vijaya, J. J. (2016). Preparation and characterization of hierarchical porous carbons derived from solid leather waste for supercapacitor applications. *Journal of Hazardous Materials, 318*, 173–185.

Laurenti, R., Redwood, M., Puig, R., & Frostell, B. (2017). Measuring the environmental footprint of leather processing technologies. *Journal of Industrial Ecology, 21*(5), 1180–1187.

Li, J., Pan, S.-Y., Kim, H., Linn, J. H., & Chiang, P.-C. (2015). Building green supply chains in eco-industrial parks towards a green economy: Barriers and strategies. *Journal of Environmental Management, 162*, 158–170.

Mandal, S., Jayanthi, D., & Muralidharan, C. (2023). Reduction in chemicals leaching in the leather industry effluent using hydrated layered double hydroxide as a fixative. *International Journal of Environmental Science and Technology, 20*(2), 1527–1538.

Mannacharaju, M., Natarajan, P., Villalan, A. K., Jothieswari, M., Somasundaram, S., et al. (2018). An innovative approach to minimize excess sludge production in sewage treatment using integrated bioreactors. *Journal of Environmental Sciences, 67*, 67–77.

Mannacharaju, M., Somasundaram, S., & Ganesan, S. (2020). Treatment of post tanning wastewater with minimum sludge generation using sequential anoxic/oxic bioreactor and its microbial community profile. *Journal of Water Process Engineering, 35*, 101244.

Marques, A., Guedes, G., & Ferreira, F. (2017). Leather wastes in the Portuguese footwear industry: New framework according design principles and circular economy. *Procedia Engineering, 200*, 303–308.

Marsal Monge, A., Ramón, P., Font Vallès, J., Cot Cosp, J., Frías, V., et al. (2004). Auxiliary agents with non-swelling capacity used in pickling/tanning processes. Part 3. *Journal of the Society of Leather Technologists and Chemists, 88*, 191–196.

Marsal Monge, A., Rius Carrasco, A., Cot Cosp, J., Lalueza Baro, J., Ramón, P., et al. (2005). Salinity reduction in the production of nappa skins by using agents with non-swelling capacity in pickling/tanning. *Journal of the Society of Leather Technologists and Chemists, 89*(6), 232–236.

Masilamani, D., Srinivasan, V., Ramachandran, R. K., Gopinath, A., Madhan, B., et al. (2017). Sustainable packaging materials from tannery trimming solid waste: A new paradigm in wealth from waste approaches. *Journal of Cleaner Production, 164*, 885–891.

Moktadir, Md. A., Ahmadi, H. B., Sultana, R., Zohra, F.-T., Liou, J. J. H., et al. (2020). Circular economy practices in the leather industry: A practical step towards sustainable development. *Journal of Cleaner Production, 251*, 119737.

Moktadir, Md. A., & Rahman, M. M. (2022). Energy production from leather solid wastes by anaerobic digestion: A critical review. *Renewable and Sustainable Energy Reviews, 161*, 112378.

Muralidharan, V., Arokianathan, M. S., Balaraman, M., & Palanivel, S. (2020). Tannery trimming waste based biodegradable bioplastic: Facile synthesis and characterization of properties. *Polymer Testing, 81*, 106250.

Muralidharan, V., Janardhanam, S., Palanivel, S., & Madhan, B. (2022a). Sustainable fabrication of bio-derived hybrid films using biomolecules extracted from animal skin. *Process Biochemistry.* https://doi.org/10.1016/j.procbio.2022.04.012

Muralidharan, V., Palanivel, S., & Balaraman, M. (2022b). Turning problem into possibility: A comprehensive review on leather solid waste intra-valorization attempts for leather processing. *Journal of Cleaner Production, 367*, 133021.

Nayar, K. G., Fernandes, J., McGovern, R. K., Al-Anzi, B. S., & Lienhard, J. H. (2019). Cost and energy needs of RO-ED-crystallizer systems for zero brine discharge seawater desalination. *Desalination, 457*, 115–132.

Nelson, M. J., Nakhla, G., & Zhu, J. (2017). Fluidized-bed bioreactor applications for biological wastewater treatment: a review of research and developments. *Engineering, 3*(3), 330–342.

Noyon, Md. A. R., Dey, T. K., Jamal, M., Rathanasamy, R., Chinnasamy, M., et al. (2022). Fabrication of LLDPE based biodegradable composite incorporated with leather shavings and buffing dust: An approach for waste management. *Journal of Applied Polymer Science, 139*(47), e53184.

Nugraha, A., Indrasti, N., & Suparno, O. (2019). Water reuse strategy in leather tannery process. *International Journal of Recent Technology and Engineering, 7*(6S5), 958–964.

Ocak, B. (2022). Collagen hydrolysate/chitosan/keratin-based ternary films from bio-wastes of the leather and poultry industries against plastic waste generation. *Biomass Conversion and Biorefinery*. https://doi.org/10.1007/s13399-022-03297-0

Omoloso, O., Wise, W. R., Mortimer, K., & Jraisat, L. (2020). Corporate sustainability disclosure: A leather industry perspective. *Emerging Science Journal, 4*(1), 44–51.

Ortolano, L., Sanchez-Triana, E., Afzal, J., Ali, C. L., & Rebellón, S. A. (2014). Cleaner production in Pakistan's leather and textile sectors. *Journal of Cleaner Production, 68*, 121–129.

Panagopoulos, A., & Haralambous, K.-J. (2020). Minimal liquid discharge (MLD) and zero liquid discharge (ZLD) strategies for wastewater management and resource recovery—Analysis, challenges and prospects. *Journal of Environmental Chemical Engineering, 8*(5), 104418.

Parlasca, M. C., & Qaim, M. (2022). Meat consumption and sustainability. *Annual Review of Resource Economics, 14*(1), 17–41.

Patel, K., Munir, D., & Santos, R. M. (2022). Beneficial use of animal hides for abattoir and tannery waste management: A review of unconventional, innovative, and sustainable approaches. *Environmental Science and Pollution Research, 29*(2), 1807–1823.

Pati, A., Chaudhary, R., & Subramani, S. (2013). Biochemical method for extraction and reuse of protein and chromium from chrome leather shavings: A waste to wealth approach. *Journal of the American Leather Chemists Association, 108*(10), 365–372.

Paul, B., Patnaik, U., Jaganth, G., & Murari, K. (2022). Employment, technology and value chain: A case of indian leather industry. *Science, Technology and Society, 27*(4), 563–586.

Peters, C. D., & Hankins, N. P. (2019). Osmotically assisted reverse osmosis (OARO): Five approaches to dewatering saline brines using pressure-driven membrane processes. *Desalination, 458*, 1–13.

Piwowar-Sulej, K. (2021). Human resources development as an element of sustainable HRM—With the focus on production engineers. *Journal of Cleaner Production, 278*, 124008.

Pomponi, F., & Moncaster, A. (2017). Circular economy for the built environment: A research framework. *Journal of Cleaner Production, 143*, 710–718.

Puhazhselvan, P., Pandi, A., Sujiritha, P. B., Antony, G. S., Jaisankar, S. N., et al. (2022). Recycling of tannery fleshing waste by a two step process for preparation of retanning agent. *Process Safety and Environmental Protection, 157*, 59–67.

Qiang, X., Sheng, Z., & Zhang, H. (2013). Study on scale inhibition performances and interaction mechanism of modified collagen. *Desalination, 309*, 237–242.

Qureshi, B. A., & Zubair, S. M. (2016). Exergy and sensitivity analysis of electrodialysis reversal desalination plants. *Desalination, 394*, 195–203.

Raghava Rao, J., Chandrababu, N. K., Muralidharan, C., Nair, B. U., Rao, P. G., et al. (2003). Recouping the wastewater: A way forward for cleaner leather processing. *Journal of Cleaner Production, 11*(5), 591–599.

Raja Chidambaram, M. (2019). Industry 4.0 - Impact on Leather Sector, In *Next-Gen technologies for leather sector: Approaches towards industry 4.0, Leather Research Industry Get-together (LERIG), CLRI, Chennai, India.*

Ramesh, R. R., Muralidharan, V., & Palanivel, S. (2018). Preparation and application of unhairing enzyme using solid wastes from the leather industry—An attempt toward internalization of solid wastes within the leather industry. *Environmental Science and Pollution Research, 25*(3), 2121–2136.

Ramesh, R. R., Ponnuvel, M., Ramalingam, S., & Rathinam, A. (2022). Compact glyoxal tanning system: A chrome-free sustainable and green approach towards tanning-cum-upgradation of

low-grade raw materials in leather processing. *Environmental Science and Pollution Research, 29*(23), 35382–35395.
Ramnath, V., Sekar, S., Sankar, S., Sankaranarayanan, C., & Sastry, T. P. (2012). Preparation and evaluation of biocomposites as wound dressing material. *Journal of Materials Science: Materials in Medicine, 23*(12), 3083–3095.
Ramya, K. R., Sathish, M., Madhan, B., Jaisankar, S. N., & Saravanan, P. (2022). Effective utilization of tannery hair waste to develop a high-performing re-tanning agent for cleaner leather manufacturing. *Journal of Environmental Management, 302*, 114029.
Rao, J. R., Thanikaivelan, P., Sreeram, K. J., & Nair, B. U. (2002). Green route for the utilization of chrome shavings (chromium-containing solid waste) in tanning industry. *Environmental Science & Technology, 36*(6), 1372–1376.
Renitha, T. S., Sridevi, J., Gowthaman, M. K., Kamini, N. R., Ramanaiah, B., et al. (2015). Wealth from waste—Beef extract for microbiological media from tannery solid waste. *RSC Advances, 5*(13), 9891–9897.
Ricky, R., Shanthakumar, S., Ganapathy, G. P., & Chiampo, F. (2022). Zero liquid discharge system for the tannery industry—An overview of sustainable approaches. *Recycling, 7*(3), 31.
Rinaldi, F. R., Di Bernardino, C., Cram-Martos, V., & Pisani, M. T. (2022). Traceability and transparency: Enhancing sustainability and circularity in garment and footwear. *Sustainability: Science, Practice and Policy, 18*(1), 132–141.
Rondoni, A., & Grasso, S. (2021). Consumers behaviour towards carbon footprint labels on food: A review of the literature and discussion of industry implications. *Journal of Cleaner Production, 301*, 127031.
Samidurai, S., Khambhaty, Y., & Alagamuthu, T. S. (2022). Bio-preservation of raw hides/skins: A review on greener substitute to conventional salt curing. *Environmental Science and Pollution Research, 29*(43), 64513–64535.
Saranya, P., Selvi, P. K., & Sekaran, G. (2019). Integrated thermophilic enzyme-immobilized reactor and high-rate biological reactors for treatment of palm oil-containing wastewater without sludge production. *Bioprocess and Biosystems Engineering, 42*(6), 1053–1064.
Saravanabhavan, S., Sreeram, K. J., Raghava Rao, J., & Unni Nair, B. (2007). The use of toxic solid waste for the adsorption of dyes from waste streams. *Journal of Chemical Technology & Biotechnology, 82*(4), 407–413.
Saravanabhavan, S., Thanikaivelan, P., Rao, J. R., Nair, B. U., & Ramasami, T. (2006). Reversing the conventional leather processing sequence for cleaner leather production. *Environmental Science & Technology, 40*(3), 1069–1075.
Sathish, M., Madhan, B., & Raghava Rao, J. (2019). Leather solid waste: An eco-benign raw material for leather chemical preparation—A circular economy example. *Waste Management, 87*, 357–367.
Sathish, M., Madhan, B., Saravanan, P., Raghava Rao, J., & Nair, B. U. (2013). Dry ice—An eco-friendly alternative for ammonium reduction in leather manufacturing. *Journal of Cleaner Production, 54*, 289–295.
Saxena, G., Chandra, R., & Bharagava, R. N. (2017). Environmental pollution, toxicity profile and treatment approaches for tannery wastewater and its chemical pollutants. In P. de Voogt (Ed.), *Reviews of Environmental Contamination and Toxicology* (Vol. 240, pp. 31–69).
Sekaran, G., Karthikeyan, S., Nagalakshmi, C., & Mandal, A. B. (2013). Integrated Bacillus sp. Immobilized cell reactor and Synechocystis sp. Algal reactor for the treatment of tannery wastewater. *Environmental Science and Pollution Research, 20*(1), 281–291.
Senthil, R., Hemalatha, T., Manikandan, R., Das, B. N., & Sastry, T. P. (2015). Leather boards from buffing dust: A novel perspective. *Clean Technologies and Environmental Policy, 17*(2), 571–576.
Singh, S. C., & Gupta, D. (2013). Sustainability: A challenge for Indian Leather Industry*, 2*(4).
Sivakumar, V. (2022). Towards environmental protection and process safety in leather processing—A comprehensive analysis and review. *Process Safety and Environmental Protection, 163*, 703–726.

Sreeram, K., Rao, J., & Nair, B. (2011). Chromium (III) pigments: Use of leather wastes as alternative starting material. *Journal of the American Leather Chemists Association, 106*(07), 219–225.
Taghvaee, V. M., Nodehi, M., Saber, R. M., & Mohebi, M. (2022). Sustainable development goals and transportation modes: Analyzing sustainability pillars of environment, health, and economy. *World Development Sustainability, 1*, 100018.
Teklay, A., Gebeyehu, G., Getachew, T., Yaynshet, T., & Sastry, T. P. (2017). Conversion of finished leather waste incorporated with plant fibers into value added consumer products—An effort to minimize solid waste in Ethiopia. *Waste Management, 68*, 45–55.
Thanikaivelan, P., Rao, J. R., Nair, B. U., & Ramasami, T. (2002). Zero discharge tanning: A shift from chemical to biocatalytic leather processing. *American Chemical Society* (ACS Publications). https://doi.org/10.1021/es025618i.
Thanikaivelan, P., Rao, J. R., Nair, B. U., & Ramasami, T. (2003). Biointervention makes leather processing greener: An integrated cleansing and tanning system. *Environmental Science & Technology, 37*(11), 2609–2617.
Thanikaivelan, P., Rao, J. R., Nair, B. U., & Ramasami, T. (2004). Progress and recent trends in biotechnological methods for leather processing. *Trends in Biotechnology, 22*(4), 181–188.
Thankaswamy, S. R., Sundaramoorthy, S., Palanivel, S., & Ramudu, K. N. (2018). Improved microbial degradation of animal hair waste from leather industry using Brevibacterium luteolum (MTCC 5982). *Journal of Cleaner Production, 189*, 701–708.
Thi, H. T. D., & Tóth, A. J. (2023). Investigation of carbon footprints of three desalination technologies: Reverse osmosis (RO), multi-stage flash distillation (MSF) and multi-effect distillation (MED). *Periodica Polytechnica Chemical Engineering, 67*(1), 41–48.
Tisa, F. (2014). Basic design of a fluidized bed reactor for wastewater treatment using fenton oxidation. *International Journal of Innovation, Management and Technology, 5*(2), 93–98.
Tong, T., & Elimelech, M. (2016). The global rise of zero liquid discharge for wastewater management: Drivers, technologies, and future directions. *Environmental Science & Technology, 50*(13), 6846–6855.
Venkatachalam, M., Rathinam, A., Rao, J. R., & Krishnan, C. (2022). Bioconversion of animal hair waste using salt- and sulphide-tolerant Bacillus sp. KLP1 and depilation using keratinase. *International Journal of Environmental Science and Technology, 19*(7), 6389–6398.
Verma, S. K., & Sharma, P. C. (2022). Current trends in solid tannery waste management. *Critical Reviews in Biotechnology, 0*(0), 1–18.
Wu, J., Zhao, L., Liu, X., Chen, W., & Gu, H. (2017). Recent progress in cleaner preservation of hides and skins. *Journal of Cleaner Production, 148*, 158–173.
Yorgancioglu, A., Başaran, B., Sancakli, A., Yorgancioglu, A., Başaran, B. et al. (2020). Value addition to leather industry wastes and by-products: hydrolyzed collagen and collagen peptides. In *Waste in textile and leather sectors*. IntechOpen. https://doi.org/10.5772/intechopen.92699
Yu, Y., Zeng, W., Wang, Y., & Shi, B. (2015). Inhibitory effect of protein filling agent from bovine hair on the oxidation of Cr(III). *Journal of the American Leather Chemists Association, 110*(12), 385–391.
Yuvaraj, P., Rao, J. R., Fathima, N. N., Natchimuthu, N., & Mohan, R. (2018). Complete replacement of carbon black filler in rubber sole with CaO embedded activated carbon derived from tannery solid waste. *Journal of Cleaner Production, 170*, 446–450.
Zhang, X., Chattha, S. A., Song, J., Zhang, C., & Peng, B. (2022). An integrated pickling-bating technology for reducing ammonia-nitrogen and chloride pollution in leather manufacturing. *Journal of Cleaner Production, 375*, 134070.
Zhao, D., Lee, L. Y., Ong, S. L., Chowdhury, P., Siah, K. B., et al. (2019). Electrodialysis reversal for industrial reverse osmosis brine treatment. *Separation and Purification Technology, 213*, 339–347.
Zhao, J., Wu, Q., Tang, Y., Zhou, J., & Guo, H. (2022). Tannery wastewater treatment: Conventional and promising processes, an updated 20-year review. *Journal of Leather Science and Engineering, 4*(1), 10.

CLRI Achievements in Utilization of Solid Waste from the Meat and Leather Industry

T. P. Sastry, Teklay Asgedom Teferi, and Senthil Rethinam

1 Introduction

It is important to utilize every single part of slaughtered animal or carcass more efficiently without wastage. Normally, apart from skin/hide 30–40% of the animal is utilized for food which includes meat, edible offals, processed foods, etc. Remaining 60–70% body parts are not properly utilized or underutilized. In the case of carcass, the wastage is more because after removing skin/hide from the carcass, the de-flayed animal is discarded in open places which create environmental pollution. If the animal/carcass byproducts are not properly utilized, a valuable source of revenue is not only lost but it will also add to the cost of disposal of the solid wastes from the slaughter houses/meat industries.

Similarly, considerable amount of tannery waste is available in India, which include raw hide trimmings, hair, fleshings, splits, chrome shavings, trimmings of crust and finished leather and buffing of leather. The wastes correspond to 45–50% of dry weight of hides and skins (Steadman et al., 1977). These wastes are basically proteinaceous in nature existing in tanned and un-tanned conditions. In many countries, these solid wastes are underutilized and major portions of these wastes are dumped as a landfill.

T. P. Sastry (✉)
Central Leather Research Institute, Adyar, Chennai 600020, India
e-mail: sastrytp56@gmail.com

T. A. Teferi
Leather Industry Development Institute, Addis Abba, Ethiopia

S. Rethinam
Saveetha Dental College, Saveetha University, Chennai 600077, Tamilnadu, India
e-mail: senthilrethinam.sdc@saveetha.com

S. K. Janardhanan and L. A. Zugno (eds.), *Emerging Trends in Leather Science and Technology*, https://doi.org/10.1007/978-981-99-9754-1_3

1.1 Animal Byproducts

Apart from meat, there are several valuable byproducts available in large quantities from slaughterhouses. Table 1 illustrates approximate yield of meat and byproducts based on live weight of animal.

We can observe many valuable byproducts in Table 1, some of them are edible and others are non-edible (Mann, 1962). Edible products are liver, heart, tongue, tail, kidney, brain, glands, stomach and testicles. On the other hand, we have non-edible products like rumen contents, excreta in large intestines, horns and hooves. In some of the developing countries, intestines are used for edible purposes instead of using them for value-added products such as surgical catguts, collagen sheets, sports guts, etc. Blood is forbidden in some parts of the world while it is being used for making sausages and blood puddings in other countries. Out of the byproducts available from an animal 40% are edible and 20% are none edible.

Table 1 Approximate yield of meat and byproducts based on live weight of cattle

Serial number	Name of the product/byproduct	Percentage yield on live weight
1	Meat	35.00
2	Bones (green)	25.00
3	Hide	6.50
4	Blood	4.0
5	Horns and hooves	0.60
6	Small and large intestines	2.0
7	Stomach and rumen	13.0
8	Liver	1.20
9	Heart	0.35
10	Kidneys	0.20
11	Lungs	1.20
12	Brain	0.17
13	Spleen	0.15
14	Pancreas	0.07
15	Bile	0.06
16	Tongue	0.30
17	Free fat	1.50
18	Hair, teeth and other losses	8.70
	Total	100

2 Resources of Raw Byproducts

Mainly, the raw byproducts are available from 1. Slaughter houses and 2. Meat processing units. Apart from this, huge amount of raw materials can be collected from animal carcasses available in most of the developing countries.

As reported by the Ministry of Food Processing, a total of 3,616 recognized slaughter-houses exist in India as on 2014 (https://www.financialexpress.com/opinion/how-unlicenced-slaughter-houses-pose-grave-problem-in-india/3/).

Huge revenue loss is sustained by many developing countries due to non-utilization and under-utilization of various animal byproducts emanating slaughter houses.

2.1 Collection, Processing and Utilization of Slaughter House Byproducts

The following are some of important byproducts from slaughter houses:

(i) Hides and skins: They are utilized for making leather and leather products
(ii) Rumen contents
(iii) Horns and hooves
(iv) Blood
(v) Fat
(vi) Soft tissues such as intestines, lungs, heart, kidney, liver and other condemned offals.

2.2 Ruminal Contents

Ruminal contents (RC) consist of partially digested food eaten by the animals before slaughter, weighs approximately 23 kg and measures about 25 L. RC contains digestive juices and millions of bacteria. It is either wasted by allowing it to go to sewage system or underutilized as manure for plants in India. This situation may be same in other developing countries. RC is rich in essential amino acids, B-complex vitamins and minerals. Approximately 4000 tons of crude protein could be recovered if RC of bovines and ovines are properly collected and processed for animal feeds.

Useful feed components were prepared from RC at Central Leather Research Institute (CLRI), Chennai, using ensiling techniques (Rao et al., 1987). Good growth rates were observed when RC-with maize meal was incorporated in poultry rations (Rao et al., 1992).

Vermicomposting of RC

Vermicomposting is a useful technique for recycling organic and farm wastes to increase humus nutrient content of soil. This process was demonstrated in Carcass Recovery Centers, established by CLRI all over India. This process can be easily carried out using local earthworms, cow dung, partially digested leaves, farm residues and dried RC. 100 kg of organic waste yields about 30 kg rich dry organic compost. In the studies conducted at CLRI the RC was used with semi-digested leaves in 5:1 proportion using local earthworms. The compost obtained when applied to ornamental plants has shown excellent growth. This process if applied at slaughterhouses will be boon to rural masses as a sustainable technology.

2.3 *Bones*

Bones constitute about 16–25% on the live weight of animal. A number of factors such as age, sex, breed and nutritional status of the animal affect the yield of bones. Animal bones are largely obtained from fallen carcasses rather than from slaughter houses. Small quantity of bones is also obtained from meat processing plants during deboning operations. The dry bones are collected extensively through middle men working in rural areas.

2.3.1 Use of Bones

Bone constitutes both inorganic and organic components, i.e., hydroxyapatite (70%) and collagen (30%), respectively. Bone marrow contains fats and other nutrients. Many types of industries based on bone are established such as (i) handicrafts, (ii) fat extraction units, (iii) rendering industry, (iv) bone crushing units, (v) mineral-based industries like bone meal and dicalcium phosphate, (vi) bone ash and bone char industries, (vii) protein-based products (i.e., ossein and gelatin) industries.

Handicrafts

Traditionally, bones have been used in handicrafts for the manufacture of inlay works, jewelry, fancy and decorative articles. But due to advent of plastics, the use of bones in these articles has declined drastically.

Fat extraction units

Fat or tallow extraction is done using green bones; however, the raw material is not available in sufficient quantities in India. The situation may be same in other developing countries. Green bone marrow contains 96% fats and to extract this, the ends of the bones should be chopped. The precut bones along with sufficient water are processed in a cooker at appropriate temperature, later the contents are allowed to cool and settle. The fat layer separates at top is removed. The fat obtained from the fresh bones is white or light yellow in color and is suitable for edible purposes. The treated bones are later taken out from the cooker and washed well with water and dried on a platform in the open sun.

i. **Rendering industry**

Inedible fat is extracted from bones during rendering process of meat cum bones. This fat is used in animal feed, candle and soap industries.

ii. **Bone crushing industry**

Bones are available in nature having different shapes and sizes and cannot be used as such in the bone processing units like bone crushing, ossein and gelatin. The crushing of bone is done using bone crushing machine, which is essentially a hammer mill. Crushed bone pass through a rotating screen or vibrating screen with various mesh sizes to separate crushed bone into various sizes. Commercially the popular sizes in which bones are crushed are ¾ inch, 5/8 inch and 3/8 inch. These sizes are used both for manufacture of ossein as well as gelatin/glue. The material passing through 3/16 inch mesh but resting 3/32 inch sieve is called bone grist, which is an ideal material for the manufacture of bone char. The remaining material that passes 3/32 inch mesh is called bone meal. Sinews are tough fibrous collagen layer adhering or covering the bone. Sinews are also used as starting material for the preparation of gelatin.

Yield

Approximate yield of various products obtained after crushing of dry bones is as follows (Table 2).

iii. **Mineral-Based Products**

The following mineral-based products can also be made from bones:

(a) Sterilized bone meal
(b) Bone ash
(c) Bone char
(d) Dicalcium phosphate.

Table 2 Approximate yield of various products

Serial number	Product	% yield
1	Crushed bones	50–55
2	Grist	10–12
3	Bone meal	20–25
4	Sinews	5–10
5	Loss	5–10%

a. **Sterilized bone meal**

This is manufactured by treating crushed bones in bone digester and subjected to steam pressure of 2.81 kg/cm^2 for a period of 2 h. Thus, the bones get sterilized and become soft. Later, the stock is taken out from bone digester dried and milled to powder.

b. **Bone ash**

The ashing of the bones can be done in rotary kilns like ones used in cement industry. Conventional kilns used for fixing limestone and pottery are also used. The temperature required for complete ashing of bone is in the range of 800–900 °C for 4–5 h. Bone ash contains exclusively inorganic part of bone resembling hydroxyapatite. Bone ash is pure white in color having about 30% calcium and 15% phosphorus.

Bone ash finds use in agriculture and livestock feed. It is also used for making bone China in ceramic industry. Bone ash is also used as starting material in the manufacture of dicalcium phosphate.

c. **Bone char (Active carbon)**

Bone char is prepared by carbonizing granular bones at 600–800 °C in a furnace in the absence of air. Low quality bones, cooked bones and other granular bones that are discarded are used in the production of bone char. Active carbon is mainly used for de-colorization and adsorption purpose in sugar and pharmaceutical industries. During the manufacture of bone char, large quantities of gases are evolved which on scrubbing result in the production of bone tar, bone oil, etc.

d. **Dicalcium phosphate (DCP)**

Manufacture of DCP is described along with manufacture of ossein.

vi. **Protein-based products**

a. **Preparation of ossein**

Ossein is a proteinous (collagen) part of bone and is used as a raw material for the production of gelatin. Ossein is prepared by demineralization of bones. The raw material used is crushed bones passing through 3/8 inch mesh and resting on 3/16 inch mesh. Bones are de-mineralized using vats made of stainless steel vats or fiber reinforced plastic. Hydrochloric acid or sulfuric acid is used for de-mineralization having concentrations 6–16% using counter current methods. The acid after treatment is saturated with inorganic portion of bone which is drained into precipitation tanks and recovered by neutralization. The completion of de-mineralization is checked by ash content of ossein. The ash content should be below 2%, otherwise the di-mineralization should continue. The acid used for the extraction should be pre-chilled and the temperature during extraction should not be allowed to exceed 40 °C as it would result in hydrolysis of ossein extracted. The resultant ossein is washed with water to remove excess of acid and dried using warm air below 50 °C. Ossein can be stored at dry conditions and room temperature for about 2 years.

b. **Dicalcium phosphate**

The acid solution emerging out after demineralization contains mineral or inorganic portion of bone. Calcium hydroxide is added to this solution with stirring till the solution attains a pH of 5.4–6.3 and stirring continued for one hour. Dicalcium phosphate settles at the bottom of the vats as a thick white precipitate. The supernatant is pumped out and the DCP is removed and dried at 100 °C to remove moisture and stored at room temperature.

c. **Yield of Products**

Bone = One Ton

Ossein = 220–250 kg

DCP = 600–650 kg

Acid required = 0.750 Ton (Commercial HCl with Specific gravity 1–1.8)

Lime = 200 kg.

Dicalcium phosphate is also prepared using bone ash as starting material. The bone ash is dissolved in HCl and the solution is neutralized with calcium hydroxide to precipitate DCP in the very fashion as explained above.

3 Value-Added Biomedical Materials from Slaughter House and Animal Byproducts

Central Leather Research Institute (CLRI) has developed many value-added biomedical materials using slaughter house animal byproducts. The byproducts like intestinal offals, blood and bones are being underutilized or wasted in most of the slaughter houses in our country. The utilization of these products will not only reduce the environmental pollution but value-added products can be prepared, which helps in revenue and new employment generation. Some of the important products prepared using animal byproducts at CLRI are described in this article.

(A) **Collagen sheet as biological burn/wound dressing material**

The biological dressing over the raw areas of the skin protects viable epithelial elements from destruction and may even provide scaffolding for the epithelial ingrowth. The biological dressing would also guard against exogenous infection and fluid loss from exudation. At present major medical problem is clinical treatment of skin loss due to sever and maximum burns and wounds. It is very important to cover the open wound surface before the onset of infection. Management of larger wounds poses major challenge in the medical profession. Coverage of uninfected wound with auto graft has been found to be the best alternative for preventing fluid loss from burn areas and controlling air borne infection, however, the availability of auto graft when area of skin loss is large is rather limited or at times is not possible. Like autograft, homo graft also does not provide any satisfactory solution if the wound is already infected.

Various other biological dressings such as cadaver skin, porcine skin, amniotic membrane and synthetic skin have been used to cover the granulating surfaces during the past many years. However, due to many statutory problems, the availability of the material in our country is rather limited.

Collagen sheets (Fig. 1) prepared from cattle intestine were used as biological dressing in the management of burn wounds as earlier as 1978. Extensive studies were done using this material with excellent results. These collagen sheets were prepared from the collagenous part of bovine intestine after successful usage of collagen sheet as burn/wound dressing material experimentally and clinically (Fig. 2), it is now commercially available in the market.

(B) Fibrogel as dressing material for the management of chronic wounds

The Fibrogel is prepared from the fibrin part of bovine blood. In most of the slaughter houses in our country, blood is being wasted and causing bad smell and environmental pollution. Approximately 200 g of fibrin can be isolated from one liter of blood. Improved wound healing in a moist environment as compared to dry environment was first described by Winter (1962). A dry wound environment will cause tissue desiccation (drying) and the collagen substance (matrix) become dehydrated and the cells die. A hard crust (scab) forms. Keratinocytes can only migrate across a moist, vascular surface so they must burrow beneath the crust to reach viable tissue.

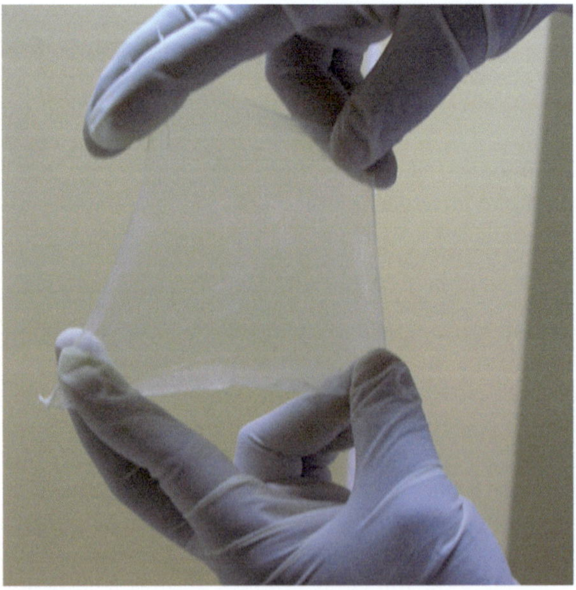

Fig. 1 Collagen sheet

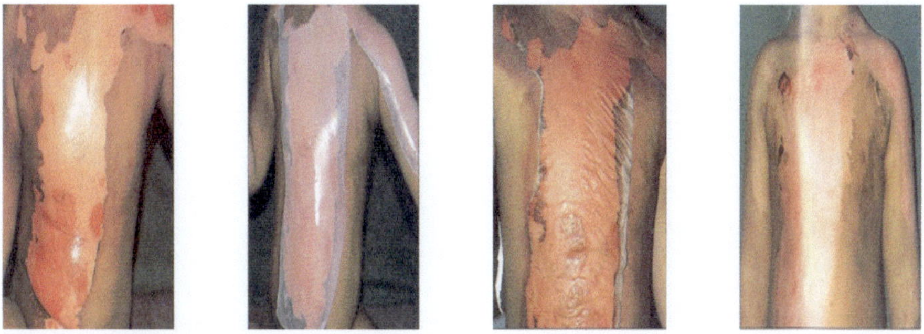

Fig. 2 Patients treated with collagen sheet

This consumes energy and time and delays wound healing. Realizing the merits of maintaining a moist wound healing environment, moisture retentive dressings were developed that would better support healing, as opposed to conventional gauze dressings. It is important to emphasize that a moist wound environment is not a wet wound environment. Excess exudates will predispose the patient to hyper granulation tissue formation in the wound bed and macerate periwound skin. Therefore, an important objective in providing topical wound care is to select dressing that will manage exudates, while still maintaining a moist wound surface (Harris & Roslstad, 1994).

Central Leather Research Institute developed a process for the preparation of wound dressing using physiologically clotted fibrin (PF) and gelatin (G) as raw materials. Thus prepared wound dressing is a composite of physiologically clotted fibrin and gelatin-PFG. This material as such exhibits good water absorption capacity that may be attributed to the water imbibing capacity of fibrin, which contains hydrophilic groups like –OH, –COOH, $–NH_2$ etc. and its porous structure.

After successful application of Fibrogel as wound dressing material on animal models, the same was applied on the wounds/chronic ulcers of humans with encouraging results (Fig. 3). This product is commercialized.

(C) **Collagen sheets from fish scales as chronic wound dressing**

Fish scales are major solid wastes in the aqua food industries. Fish scales are composed of connective tissue protein, collagen, covered with calcium salts. The amount of protein ranges from 41 to 84% and the remaining is calcium phosphate and calcium carbonate. Normally collagen is extracted from bovine skins and bones and is used in many medical formulations. Due to bovine spongiform encephalopathy (BSE) disease, the collagen prepared from bovine sources may be objectionable in future for its use in medicine. The export of these collagens from bovine sources to other countries has been curtailed due to the same problem. Collagen sheet prepared from fish scales will overcome these problems and as the raw material is abundantly available in marine food industry we can not only prepare value-added products but also can reduce the pollution problem (Sankar et al., 2008).

The products from the fish scales are safe as is the edible gelatin from bovine bones. As pharmaceutical gelatin is prepared from bovine bones, we can produce the

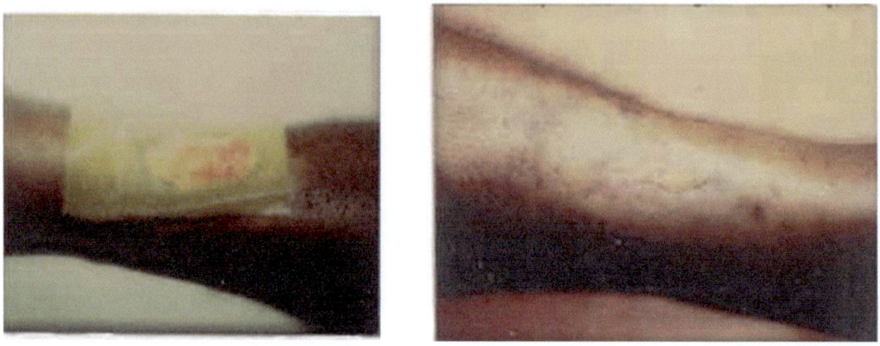

Fig. 3 Healing pattern using Fibrogel, Fibrogel applied on chronic wound (left), Healed wound (right)

collagen sheet from fish scales and sterilize the final product. Utilizing this collagen, CLRI has prepared collagen sheet which is white and transparent membrane. This is an excellent chronic wound dressing material, which is used both in animals and humans (Fig. 4) experimentally.

(D) Bone graft using animal bones

CLRI has developed an osteoinductive material (induces new bone formation) using animal bones. Demineralized bone matrix (DBM) is an osteoconductive and osteoinductive commercial biomaterial and approved medical device used in bone defects with a long track record of clinical use in diverse forms. True to its name and as an acid-extracted organic matrix from bone sources, DBM retains much of the proteinaceous components native to bone, with small amounts of calcium-based solids, inorganic phosphates and some trace cell debris. Many of DBM's proteinaceous components (e.g., growth factors) are known to be potent osteogenic agents. Commercially sourced as putty, paste, sheets and flexible pieces, DBM provides a degradable matrix

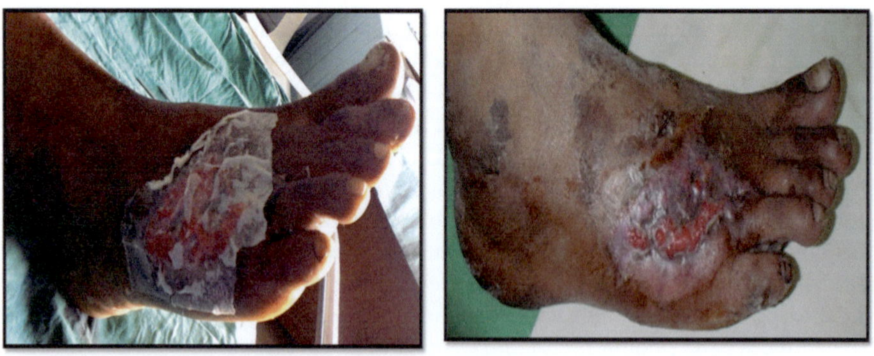

Fig. 4 Healing pattern of fish collagen sheet on a chronic foot ulcer

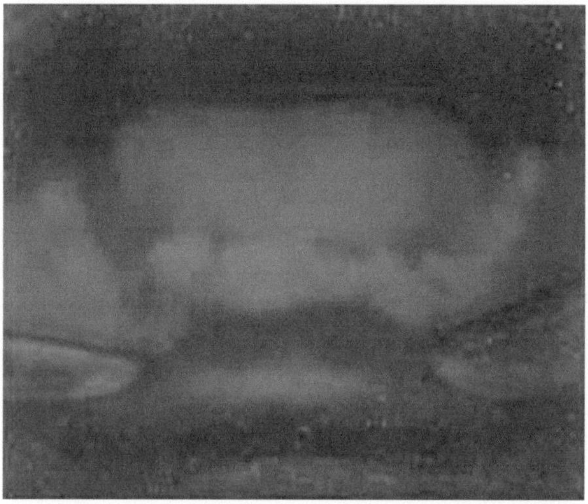

Fig. 5 Bone graft applied in oral surgery

facilitating endogenous release of these compounds to the bone wound sites where it is surgically placed to fill bone defects, inducing new bone formation and accelerating healing. Given DBM's long clinical track record and commercial accessibility in standard forms and sources, opportunities to further develop and validate DBM as a versatile bone biomaterial in orthopedic repair and regenerative medicine contexts are attractive. At CLRI demineralized bone matrix (DBM) is isolated from the bones and with the addition of calcium apatite bone graft is prepared. This product was used clinically both in animals and humans. This bone graft is used in oral surgery for augmentation of new bone in maxillofacial surgery (Fig. 5). DBM is also an osteoinductive material (Urist, 1965) and is commercialized and available in market.

4 Products from Tannery Wastes

a. **Cattle tail hair**

In many developing countries, cattle hairy tail end is cut and thrown away as a waste. In some other countries, hairy part of cattle hair is allowed to rot for 2–3 days, where after the hair is pulled out. Some countries export the tail hair while in other countries like India it is used for manufacture of cheap paint brushes. The cattle ear lobe hair is used for making the finest artist brushes. The body hair available in tannery is washed free of lime and used in the manufacture of coarse carpets, druggets and carpet under felts. Some countries like India export tannery hair (Mahendrakumar, 1994).

Bristle is the term used for the hair growing along the back and neck of pigs, hogs and wild boars. The bristles are hand pulled out twice in a year from live animal. This practice is done in China and northern parts of India. Paint brush industry uses these bristles extensively in both countries.

b. **Hide and skin trimmings**

Untanned hides and trimmings are very valuable materials and constitute about 4–8% soaked weight of hides and skins. These materials are used a starting material for manufacture of glue and gelatin in many developing countries (Srinivasan et al., 2001). In some countries, this raw material is used for making collagen hydrolysate, which is used as an ingredient in cosmetic applications (Lipsett, 1980). The hide trimmings are used for making dog chews and treats. The hides/skins are limed, dehaired, delimed, bleached and the bleached pelts are sized and shaped according to the requirement, dried and sold as dog chews (Fig. 6).

c. **Chrome shavings and vegetable shavings (EI shavings)**

These materials are major byproducts from tanneries and leather goods industry. Chrome shavings and vegetable shavings are used for manufacture of leather boards in many countries. In certain others, these materials are exposed to high pressure or cooked in open vessels and resultant mass is dried, powdered and used as a leather meal. This leather meal is used as an ingredient in poultry feed rations and fish feeds.[1]

d. **Value-added products from tannery and finished leather waste**

i. **Regenerated Leather (RGL)**

According to the estimates, 20–30% leather is discarded as a waste during footwear production. Also, enormous amount of leather products viz., chappal, shoes and bags etc., and end products of leather industry constitute major solid wastes (Kanagaraj et al., 2006). Central Leather Research Institute has developed a process for the preparation of regenerated leather (RGL) from used leather products (Senthil et al., 2015a, 2015b). Since RGL alone did not exhibit required mechanical properties, leather composites were prepared by incorporating various plant fibers such as coconut (Cocos nucifera), sugarcane (Saccharum officinarum), banana (Musaparadisiaca) enset (Enseteventricosum), Hibiscus (Hibiscus cannabinus), jute (Corchorus trilocularis), palm (Phoenix dactylifera) and sisal (Agave sisal) in various proportions into the leather fibers (LF). Among the composites prepared, leather fiber and coconut fiber (50:40 ratio) exhibited significant values in terms of tensile strength, elongation at break, tearing strength, water desorption and flexibility. In another set of experiments, Teklay et al. (2017) prepared leather composites using two binders, i.e., resin binder (RB) and natural rubber latex (NRL) along with various plant fibers in different proportions. Among the different composite sheets prepared, LF: jute fibers (70:30) using NRL; LF: Hibiscus fibers (90:10), LF: palm fiber (80:20) and LF: sisal fiber (60:40) prepared using RB exhibited better mechanical properties than their respective controls. LF-Hibiscus fibers (70:30) prepared using RB and all the composites prepared using NRL exhibited better elongation properties. The machinery and process involved in preparing RGL are shown schematically in Fig. 7. The products made using RGL and its composites are shown in Figs. 8, 9, 10 and 11.

[1] The use of chrome and vegetable shavings for animal feed is highly regulated and can be restricted/prohibited in many countries.

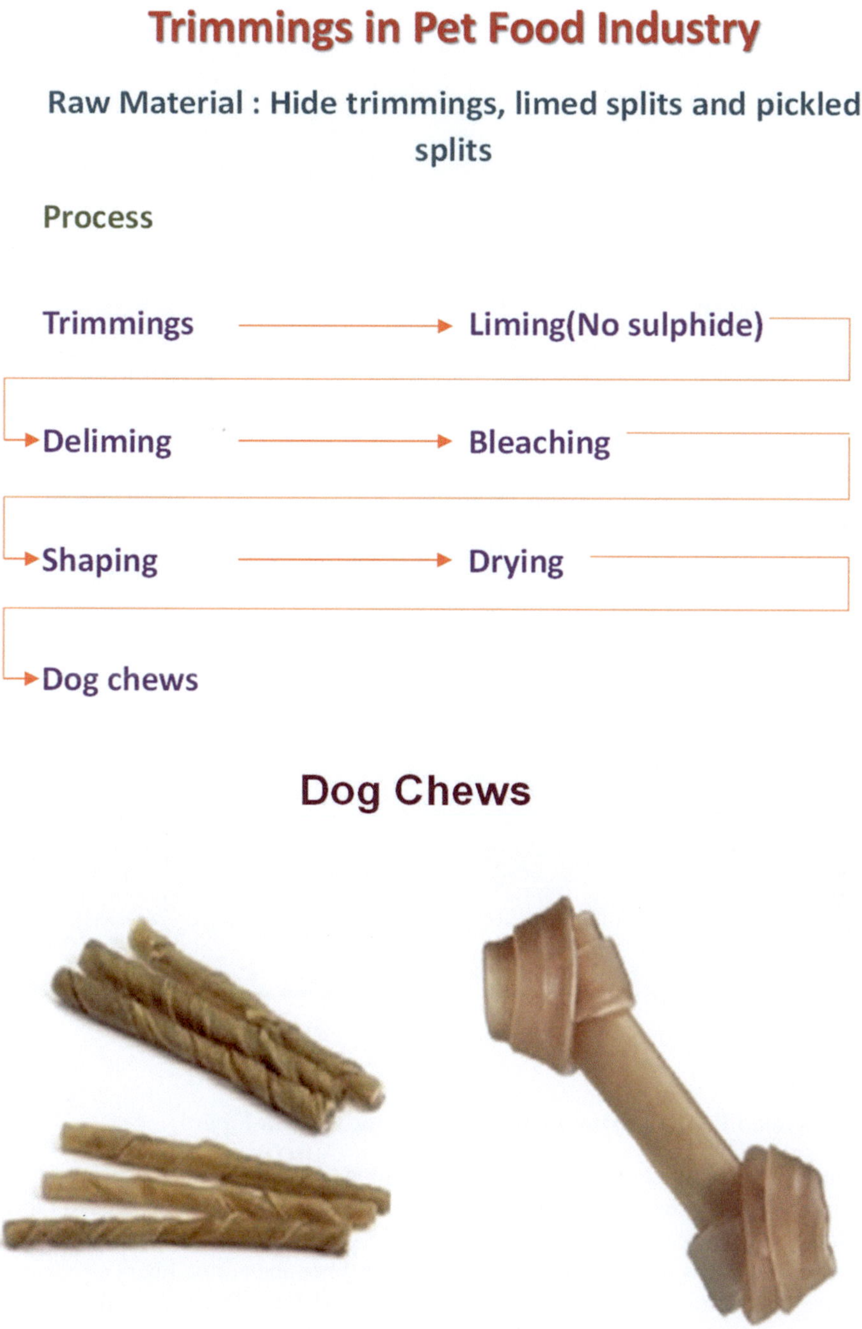

Fig. 6 Dog chews

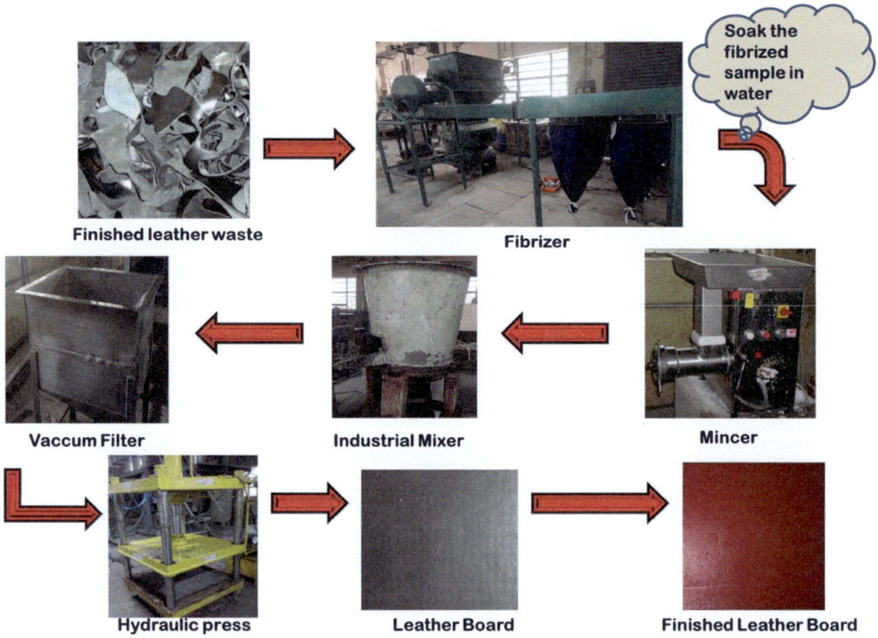

Fig. 7 Schematic diagram of process involved in preparing RCL

Fig. 8 Regenerated leather from leather waste

ii. Blended Fabrics Using Leather Fibers and Cotton Fibers

Central Leather Research Institute has developed a process for preparing leather blended fabrics using leather fibers and cotton/polyester fibers. The blended fabrics prepared were characterized for their phycico-chemical characteristics. Leather blended Yarn was prepared by mixing of leather fiber and cotton fibers or polyester fibers PF. PFs, among which a combination of 50:50 (LF:CF/PF) displayed significant physico-chemical properties. Also, 50:50 (LF:CF/PF) ratio of LBFs possessed the

Fig. 9 Ladies handbag made using RCL

Fig. 10 Key chain holder made using RGL

Fig. 11 **a** Raw composite sheet, **b** finished composite sheets and **c** final product

Fig. 12 Flowchart for the preparation of leather cotton blended fabrics

required mechanical properties for the manufacture of consumer materials (Senthil et al., 2015a, 2015b). Hence, a versatile technology has been developed, which could offer double benefits for waste management and income generation. A flowchart for preparation of leather cotton blended fabrics is given in Fig. 12.

iii. **Preparation of parchment like leather material from chrome shavings**

The history of parchment is hidden in the mists of antiquity. A parchment is essentially a piece of un-tanned hide or skin, preserved usually by drying. It has been used by used by the mankind from the time immemorial. The usual process of preparation of parchment leather is as follows. The fresh skins are kept in lime liquor for a few days and then unhaired, adhering flesh is removed, split to uniform thickness, shaped followed by sun drying to get transparent parchment. The major limitation to prepare parchment leather is availability of raw material because the raw material is mainly used to produce leather, which is more valuable material than the parchment. Moreover, the sun drying, involved in this process, may also result in the interior putrefaction of the proteinous matter. Utilizing chrome shavings for preparation of parchment leather can alleviate these limitations.

At the end of chrome tanning, the leathers are leveled with splitting and shaving methods. The chrome shavings have the composition of moisture—45 to 60%, and chromium 2.5–3.0%. These materials on storage dry rather quickly and once the chrome-tanned materials are dry, the pieces are hard to wet back. The most popular utilization mode for chrome shavings in our country is the manufacture of leather boards. In some tanneries, the chrome shavings are used as landfills, this practice is hazardous as Cr (III) waste will be oxidized to Cr (VI), which is a toxic and contaminates groundwater. To alleviate these problems, CLRI (Sastry et al., 2005) has

Fig. 13 Products made using parchment like leather material from chrome shavings, Lamp shade (Left), Puppet (Right)

developed a process for the preparation of parchment leather from chrome shavings, which is a value-added product compared to leather boards. Moreover, the parchment leather made out of chrome shavings is more transparent compared to traditional parchment leather. The material prepared by this novel process is envisaged to have potential application in puppet industry and other interior decorative objects like lamp shades, etc. (Fig. 13). Small-scale industries can come up in feature for producing the parchment leather from chrome shavings. Another important aspect of this process is that the chrome liquor, which comes out as by product, can be used in tanning operations and this results in total utilization of chrome shavings.

5 Conclusion

Slaughterhouses, meat processing and leather industries generate byproducts such as blood, bones, and trimmings, often discarded or under-utilized in many developing countries. This chapter provides an overview of converting these byproducts into valuable products such as biomedical devices and regenerated leathers, using processes developed by the Central Leather Research Institute. Developing suitable technologies is crucial to maximize the economic benefits of these byproducts.

References

Harris, A., & Roslstad, B. S. (1994). Hypergranulation tissue: A nontraumatic method of management. *Ostomy Wound Management, 40*(5), 20–22.

Kanagaraj, J., Velappan, K. C., Chandrababu, N. K., & Sadulla, S. (2006). Solid waste generation in the leather industries and its utilization for cleaner environment—A review. *Journal of Scientific & Industrial Research, 65*, 541–548.

Lipsett, V. A. (1980). New uses of hide collagen utilization. *Journal of the American Chemists Leather Association, 75*, 439–455.

Mahendrakumar. (1994). International workshop on animal byproducts utilization held at CLRI (Vol. 194, pp. 27–45).

Mann, I. (1962). Processing and utilization of animal byproducts, (FAO Agricultural development paper No.75). FAO.

Rao, N. M., & Fontenot, J. P. (1987). Ensiling of rumen contents and blood with wheat straw. *Animal Feed Science and Technology, 18*, 67–73.

Rao, N. M., Gupta, P., & Krishna, N. (1992). Ensiling and nutritional evaluation of bovine rumen contents fortified with different nitrogen sources. *Indian Journal of Animal Sciences, 62*(9), 876–879.

Sankar, S., Sekar, S., Mohan, R., Rani, S., Sundaraseelan, J., & Sastry, T. P. (2008). Preparation and partial characterization of collagen sheet from fish (Lates calcarifer) scales. *International Journal of Biological Macromolecules, 42*, 6–9.

Sastry, T. P., Sehgal, P. K., & Ramasami, T. (2005). Value added eco-friendly products from tannery solid wastes. *Journal of Environmental Science and Engineering, 47*(4), 250–255.

Senthil, R., Hemalatha, R., Uma, T. S., Das, B. N., & Sastry, T. P. (2015a). Recycling of finished leather wastes: A novel approach. *Clean Technologies and Environmental Policy, 17*, 187–197.

Senthil, R., Inbasekaran, S., Gobi, N., Das, B. N., & Sastry, T. P. (2015). Utilization of finished leather wastes for the production of blended fabrics. *Clean Technologies and Environmental Policy, 17*, 1535–1546.

Srinivasan, T. S., Rao, D. V. S. P., Rao, N. M. (2001). Processing of tannery byproducts. In: *Technologies for value realization of carcass byproducts in developing countries—A Handbook* (pp. 352–369).

Steadman, T. R., Hillman, M. E. D., Pickett, G. E., Scantland, D. A., Jacomet, J. A., & McClure, T. A. (1977). Final report on potential opportunities for increasing the utilization of tannery offal.

Teklay, A., Gebeyehu, G., Getachew, T., Yaynshet, T., & Sastry, T. P. (2017). Conversion of finished leather waste incorporated with plant fibers into value added consumer products—An effort to minimize solid waste in Ethiopia. *Waste Management, 68*, 45–55.

Urist, M. R. (1965) Bone: Formation by autoinduction. *Science,* 893–899.

Winter, G. D. (1962). Effect of air drying and dressings on the surface of a wound. *Nature, 193*, 293–294.

An Overview of Leather Processing

Anthony D. Covington and William R. Wise

1 Introduction

This brief review of leather processing provides a 'snapshot' of the current state of leather science and the consequent leather technology which creates the processes. Any overview of the global leather industry with regard to its methodology of processing must consider the following elements: the current understanding of the science underpinning the technology, the drivers for change, and the shortcomings in the required science (Covington & Wise, 2020a, 2020b). Around the world, a general principle dominating the industry is sustainability, defined by the United Nations Brundtland Commission as 'meeting the needs of the present without compromising the ability of future generations to meet their own needs'. This goes further than the previous preoccupation of the industry, the ecological impact of operations. In other words, the imperative for modern leather making is that it should be environmentally sound, both in general production and in the creation of new, innovative collagen-based biomaterials: implicit in this view is the elimination of errors in the thinking upon which processing is founded.

The key to achieving these aspirations is an understanding of the scientific principles underpinning the technology of leather making. Not taking advantage of the latest leather science only leaves a 'trial and error' approach to development, based on the false premise that experience is enough: getting the science wrong leads to the same outcome. Either way, progress is either static or inefficient and costly. Recent developments in thinking (Covington & Wise, 2020a, 2020b) have effectively

A. D. Covington (✉) · W. R. Wise
Institute for Creative Leather Technologies, University of Northampton, University Drive, Northampton NN1 5PH, UK
e-mail: tonycovington@btinternet.com

W. R. Wise
e-mail: will.wise@northampton.ac.uk

S. K. Janardhanan and L. A. Zugno (eds.), *Emerging Trends in Leather Science and Technology*, https://doi.org/10.1007/978-981-99-9754-1_4

created a paradigm shift, allowing a quantum step forward in the potential for modifying processing and prediction of the outcome of fixation reactions on collagen. This is a new logical mindset rather than a set of rules and recipes.

Sustainability should not be considered in terms of setting out a list of preferred technologies, BATNEEC (best available technology not entailing excessive cost) is not absolute, much as 'clean' technologies have been in the recent past, when the favoured processes depended on the fashion of the time, rather than clear thinking about principles. The key to sustainable development and a successful future is sound leather science.

One of the features of leather science in recent decades has been the correcting of misapprehensions and errors in the technology of leather making. Some of these are addressed in this brief overview of the present status of leather processing: the importance here is to avoid taking research and development down non-productive 'blind alleys'.

2 Beamhouse

2.1 Collagen Structure

An understanding of the fundamental nature of the substrate at the heart of leather making is critical to formulating strategies for consolidating and improving processing. Unfortunately, there are significant gaps in that knowledge. Whilst the hierarchy of collagen structure is broadly defined—amino acid sequences in the alpha chains, the triple helix unit and quarter stagger formation, leading to fibril formation and thence to fibre structure—some elements of structure remain unclear. Examples are presented in Figs. 1 and 2, illustrating that fibrils, which appear solid under the transmission electron microscope, exhibit different appearances when acid swollen or extracted from bone (Covington & Wise, 2020a).

Fig. 1 Acid swollen fibrils

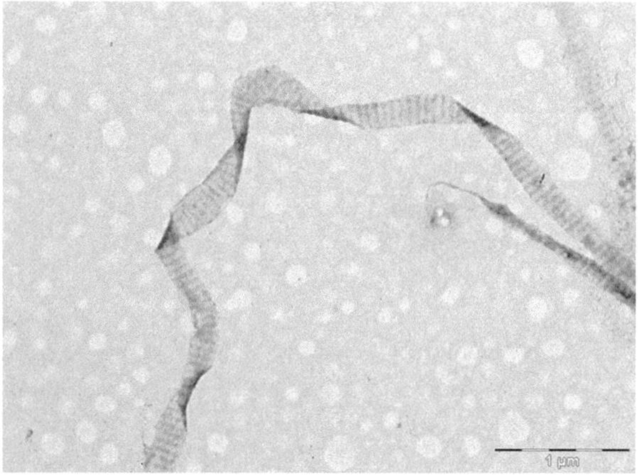

Fig. 2 Fibril extracted from bone. *Courtesy* HEC Koon

If the fibrils are not hollow in its native state (as might be assumed), the nature of the chemical species acting as the filler within fibrils is unknown: it is therefore not accounted for in any stage of processing and hence might or might not contribute to the desirable properties and performance of the resulting leather. In addition, the full and complete role of the water bound to collagen is not well understood and, as a consequence, it is not known if the in silico modelling of collagen in a range of situations is accurate. The issue of collagen hydration is exemplified by interrogating the Protein Data Bank, where multiple conflicting results are encountered.

2.2 Skin Structure

Skin is not a homogeneous, uniform material, although until the 20th century it was effectively treated as such by tanners. Now tanners can have a clearer view of the components of skin, structural and non-structural, major and minor, with knowledge of the roles they play in the properties and performance of skin and the potential roles they might play in leather, both wanted and unwanted. One example is elastin, which influences softness, stretch, and area, depending on the impact of processing on it (Addy & Covington, 2003). Other minor components such as proteoglycans and glycosaminoglycans are currently not addressed for their effects on leather handle and hydration: typically they are labelled as part of the 'ground substance' and consequently removed to some degree in traditional opening up at high pH.

2.3 Curing/Preservation

Holding rawstock in good condition is a problem facing the industry in all parts of the world, both developed and developing economies. It is recognised that the current greatest associated ecological problem is the discharge of neutral electrolyte, of which the biggest contributor is sodium chloride used for pelt preservation. Here, we can distinguish between long term preservation and short term preservation. The industry standard is salting with sodium chloride: it has the advantage that curing carries a low risk, because the technology is simple and effective when done properly, viable for many months. But the associated waste stream from subsequent processing is highly damaging to the environment, not least to plant growth.

Alternative long term preservation technologies are limited to alternative electolytes, drying to low moisture content, freezing or extremes of pH, which all have processing implications which have to be considered (Covington & Wise, 2020a).

(a) Other osmolytes

Inorganic salts as alternative candidates are limited: potassium chloride has been shown to be practical, albeit with some associated problems, including lower solubility (Bailey, 1995). Sugar-based reagents can control bacteria, as in the use of honey as a traditional wound dressing; the cost, availability, and tendency to allow fungal growth are problematic (Covington & Wise, 2020b).

(b) Drying

Reducing the moisture content limits the ability of bacteria to thrive; this is one of the features of the salting effect. Drying is familiar in the traditional shade drying technology of areas such as East Africa. The apparent simplicity of the technique is deceptive: the temperature cannot be too high and the air movement about the pelts must be free, otherwise bacterial growth is faster than the water removal. Alternative dehydrating methods are available, such as superabsorbent polymers used in diapers, but cost and limitation to short term preservation are shortcomings (Covington & Wise, 2020a).

(c) Reduced temperature

Bacterial growth is dependent upon the environmental temperature, since the damaging organisms grow optimally at around 35 °C: chilling pelts will preserve them, but only in the short term. This is practical and well known, but only in the developed economies. Cooling below freezing point will extend the period, as is well known in the domestic experience. Handling and defrosting are major operational difficulties, not to mention the damaging effect of ice crystal growth, causing disruption and loosening of the fibre structure.

(d) pH

Bacteria are sensitive to the pH of their environment, growing best at physiological conditions, around pH 8: growth is increasingly reduced the further away they are from that condition. High pH is not feasible because of the rate of damaging the pelt. Storage pickling is well known for sheepskins and is effective and practical: the modern version relies more on saturated brine than on the acid, which brings us back to the electrolyte problem and might make this technology counterproductive for all rawstocks.

All other modes of preservation can be described as short or medium term, depending on the degree of treatment, such as the amount of salt applied or the extent of drying. Since not all applications of preservation require continuing effectiveness for a year or more, a few weeks or months might suffice, less extreme treatment can be perfectly acceptable; this opens up a wider range of approaches.

(a) Reduced quantity of salt, with or without additives

Preservation period and effectiveness is proportional to the salt offer on pelt weight. Substitution of all or some of the usual salt offer by the use of inorganic or organic additives is already well known in the trade, but limits the effect, whilst having a minor improvement in the ecological impact, considering also the added impact of those additives. Also, it is a general truism that the reduction in use of a pollutant is not a solution to the problems posed by the pollutant. This is a blind alley of development.

(b) Other preservatives

Whilst there are practical preservation options available to the developed economies, they do not necessarily apply to the developing economies. One approach which seems to offer potential is the use of powdered locally indigenous plant material: this could be in the form of salt-accumulating plants grown in brackish waters or land based. The presence of salt may be regarded as a disadvantage, but the effect of other plant components, such as terpenes, can be useful in creating short term preservation by a biologically degradable reagent (Covington & Wise, 2020b). Availability in sufficient quantity for local industry might be a limiting factor.

It may be that preservation solutions, particularly local options, will be constituted by combinations of elements of the approaches set out above: practical technologies can only be created by understanding the principles and outcomes from interacting conditions.

2.4 Opening Up

'Opening up' is an umbrella term defining the removal of (assumed) unwanted components of the skin, including splitting of the fibre structure, although it does not carry an indication of which components should be targeted or to what extent the changes should be made. Treatment of the pelt with alkali—often calcium hydroxide—is a violent and indiscriminate chemical process, catalysing multiple and non-specific hydrolytic reactions, including weakening the pelt, but there can still be a consistent outcome in terms of the overall softening outcome. The process step depends on consistency in all parameters: pH, float to goods ratio, chemical offer, temperature, degree of mechanical action, time, in no particular order of precedence.

A priority for leather science is to codify the targets for opening up, with limits for component removal: this would allow prediction of appropriate enzymes for specific elements of the total desired outcome.

2.5 Sulfide Unhairing

The common technology of keratin degradation by lime + sulfide, 'hair burning' unhairing, has been in place for over a century, with an associated mechanism available in the literature and accepted for over half a century. However, the mechanism is in error, simply because the previously accepted value for pKa_2 the second dissociation constant of sulfide was in error, underestimated up to the beginning of the 21st century: the consequence is that sulfide ion, S^2, actually does not exist in aqueous solution. This means that the kinetics have to be revised in the following way (Covington & Wise, 2020b).

Rate $(S^{2-}) = k_s[S^{2-}]^a$ would be the form of the rate equation for the aqueous sulfide reaction.

Rate $(HS^-) = k_{HS}[HS^-]^a[OH^-]^b$ is the form of a modified rate equation.

The inevitable conclusion is that the revised mechanism of hair burning must be a function of the hydrosulfide ion, HS^-, with the involvement of hydroxyl ion, an extra level of complexity. Whilst this revelation may not immediately impact on the technology of unhairing, it begs the question as to how the mechanism and its implications should be reassessed, including other hair burning chemistries and even some hair saving techniques.

2.6 Hair Saving

Conventional hair burning unhairing processes have high ecological impact and cannot be sustained, so they are gradually being replaced by hair saving techniques. These chemistries (with some biochemical assistance) are well known in the art, but because they are necessarily more complicated, they do carry additional risks with regard to the efficacy of hair removal and possible adverse effects on the pelt itself. They too should be reviewed in the light of the role of the sulfide species, if that is part of the technology.

2.7 *Liming*

The principles of the other beamhouse process steps, involving the removal of skin components, are generally assumed to be understood: the opening up of the collagenic fibre structure, the coarse or fine splitting of the units of structure—the nature of the fibrils notwithstanding—is well known from microscopic observations. However, (the importance of) the role of swelling is also assumed from observation but not from a chemical mechanism: Heidemann postulated that swelling is not necessarily a *sine qua non* of ideal opening up. It is well known that osmotic liming is a high risk process step, but less risky options have rarely been investigated, for example Rabinovich's ideas of lyotropic interactions to modify collagen properties, which might be more targeted and therefore safer (Rabinovich, 2011).

2.8 *Bating*

The role of bating seems straightforward: proteolytic enzymes (from a variety of sources) are applied to pelt under the appropriate conditions, ideally around pH 9 and 35 °C, so that the non structural proteins can be degraded, thereby eliminating to some degree the adhesions between elements of the collagen structure. However, it has always been recognised as a risky step, because enzymes are highly efficient and hence unwanted side effects can occur with unwanted rapidity. Consequently, it is common to observe tanners applying bating enzymes to bovine hide under conditions that are markedly far from optimum in order to mitigate the risks.

For raw stocks that do not strictly require bating through the cross section, the desired bating reaction is actually confined to a cleaning effect on the grain. It would be less risky to the quality of the product to address the actual requirements of the step i.e. just to remove residual epidermis and hair debris: one solution would be use a keratinase product (ensuring the reagent is free of protease) so the rest of the hide structure is unaffected, especially the valuable grain enamel (Covington, 2021).

2.9 *Pickling and Swelling*

Pickling is an apparently necessary step in preparation for chrome tanning: it is a process step that conventionally needs a high offer of neutral electrolyte to avoid the damaging effect of swelling under acid conditions. There are approaches to mitigate the impact, such as chaser pickle or the use of non swelling acids or pretanning, all of which are associated with consequences which might be counterproductive: chaser pickle tanning has a risk of uneven chrome fixation, non swelling acids are a form of low reactivity tanning reagent, causing pretanning, which alters the properties of the leather (Heidemann, 1993). Equally, all of these approaches can be useful and successful in processing, but only as long as the underlying principles and technological outcomes are clearly understood by the tanner.

2.10 *Tanning*

The process of tanning results in changes to the pelt, some chemical and some physical, but perhaps most notable is the rise in hydrothermal stability. However, this outcome is highly variable, as indicated in Table 1, so it does not define the process step. The definition of tanning must take into account those processes known as 'leathering', in which collagen is modified into a leather-like product but the shrinkage temperature remains unchanged, for example in oil tanning or alum tawing (Covington, 2018).

Shrinkage temperature, a measure of the hydrothermal stability of the modified collagen, is a useful parameter in process control. Tanning causes a range of changes to collagen, all of which exhibit variations depending on the chemistry of the process step, so they cannot define the process. Tanning is strictly defined as the conversion of a putrescible material into one which can resist biochemical attack, being rendered imputrescible: this is a consistent outcome independent of tanning chemistry.

Any theoretical understanding of the mechanism of collagen stabilisation must be overarching, with no exceptions. A major feature of discussions on tanning principles to date has been the assumption of 'crosslinking' as the underpinning rationale. This

Table 1 Variations in tanning outcome measured by hydrothermal stability

Tanning reagent(s)	Indicative denaturation temperature (°C)
None	65
Metal salts: e.g., Al(III), Ti(IV), Zr(IV) etc	70–85
Plant polyphenol: gallotannin or ellagitannin	75–80
Plant polyphenol: flavonoid	80–85
Synthetic tanning agent: polymerised, derivatised phenols	75–85
Aldehyde: formaldehyde or glutaraldehyde	80–85
Aldehydic: phosphonium salt or oxazolidine	80–85
Basic chromium(III) sulfate	105–115
Semi metal: hydrolysable polyphenol then e.g., Al(III)	110–120
Vegetable retan: metal e.g., Al(III) then hydrolysable polyphenol	80–85
Synthetic polymer then aldehydic reagent e.g., melamine-formaldehyde then oxazolidine	105–115
Aldehydic reagent then synthetic polymer e.g., oxazolidine then melamine-formaldehyde	80–85

way of thinking assumes there is a general mechanism whereby there is direct chemical linking between triple helices, in a manner analogous to sewing or scaffolding. This is a critical aspect of leather science which is addressed in detail below.

2.11 Chromium(III) Tanning

The industry global standard is chromium(III) tanning; it took over the leather industry around the turn of the 19th century because of the advantages over vegetable tanning, notably the speed of processing and the versatility of properties it offered. In the latter half of the 20th century it was shown that the reaction centred on reaction between the chromium(III) molecular ions and the carboxylate groups on collagen (Sykes, 1956), but the nature of the molecular ion was not addressed.

The technology of 'masking' has been widespread, the creation of modified tanning properties by complexing the chromium(III) species with organic ligands: the underlying assumption was that the complexing reaction, often using formate, could control the tanning reaction by reducing the reactivity of the chromium species. It was not until late in the 20th century that it was shown that under typical industrial masking conditions, tanners were actually counterproductively making the chromium complex more reactive (Covington, 2010). In addition to the affinity of the chromium species for collagen carboxyls, it is necessary to understand the relative relationships between the chrome with the solvent and the chrome with the substrate—the bigger picture. Understanding the wider implications of chromium complexation leads to a rationale of the technologies involving other added ligands and opens up the potential for accurately controlling the changing chrome tanning reactivity during the process.

It was not until the turn of the 20th century that advanced X-ray analysis showed that the tanning species at the end of the process is unipoint fixed linear tetrachromium molecular ions and consequently precisely how the mechanism of conferring high hydrothermal stability works, based on the additional chemistry of the counterion (Covington et al., 2001). Note the term 'unipoint fixed'—not fixed at both ends of the complex—therefore not crosslinking in the typically understood sense of the word.

Chromium tanning as an industrial process is fairly 'foolproof' and capable of being applied successfully in relatively non technical factories. It is constantly under attack as an alleged polluting technology because it is based on transition metal chemistry. A detailed knowledge of the mechanism of the tanning process is critical to sustainability, by both maximising fixation efficiency and minimising the impact of discharges to the environment.

2.12 Link-Lock

If the data presented in Table 1 constitute a reasonable summary of tanning outcomes, it is apparent that they can be grouped into two populations: one at about 85 °C and the other at about 110 °C i.e. moderate and high hydrothermal stability. The lower shrinkage temperature group may be characterised as processes involving a single reagent. The higher shrinkage temperature group is less easy to generalise: two components appear to be necessary, but then the order of addition is critical and chromium tanning appears to be anomalous.

Chrome tanning with the chloride or perchlorate salts results in leather with a shrinkage temperature about 85 °C, addition of sodium sulfate raises the shrinkage temperature to 110 °C or higher, the same as tanning with (basic) chromium(III) sulfate. Other counterions can do the same thing, notably the effect of treating wet blue with pyromellitate (1,2,4,5-tetra carboxy benzene) which can raise the shrinkage temperature to about 130 °C in a reaction that is so fast it can only be electrostatic in nature and not covalent complexation. Therefore, chromium tanning is actually a two component system and its stabilising effect is dependent on the chemistry of the counterion. Furthermore, high shrinkage temperature is confirmed to be only achievable by a two component tanning reaction (Covington, 2010).

The high hydrothermal stability tannages can be described in the following terms: the first component links to the collagen either covalently or via multiple electrostatic interaction such as hydrogen bonding, then the second component reacts with the first, resulting in the locking of the molecules into a matrix around the triple helices. The effect is to make it difficult for the helical structure to collapse when the hydrogen bonding is broken hydrothermally: this is observed as high shrinkage temperature by the link-lock mechanism. Alternatively, the effect of any single tanning agent is merely to hinder such a collapse, so all such reactions result in similar moderate shrinkage temperature increase, practically independent of the chemistry.

If the components of a high shrinkage temperature process are offered in a different order the outcome may be different: the new first component reacts with the collagen, but the second component only reacts with the first component on collagen, no matrix is formed and the net effect is tanning by the first component alone, giving only moderate hydrothermal stability. The focus on high hydrothermal stability is important for many commercial applications of leather, where heat is a necessary part of the manufacturing process, such as heat lasting of shoe uppers or steam pressing of leather garments.

One consequence of the link-lock mechanism is how the notion of 'crosslinking' is viewed: this has been a commonly used term, which refers to the direct linking of triple helices, as envisaged by Gustavson (1953). That representation of collagen stabilisation has been discredited, if only because of the entropic penalty of such a defined and specific interaction. Link-lock replaces his mechanism with a more general and simpler model of reagent function (Covington et al., 2008). The concept of crosslinking is not helpful in explaining tanning outcome, indeed it is misleading and counterproductive, so the use of the term should be discontinued.

2.13 *Post Tanning or Multi-component Processing*

It is common in the industry to refer to pretanning, tanning, and retanning or post tanning as separate and specific process steps: this is not useful, the terms merely indicate the order of application of whatever reagents. It is more constructive to consider these stabilising reactions as a multicomponent system in which the substrate (raw collagen or chemically modified collagen) and the solvent are two additional components at any stage of the overall process.

The substrate has a role in the chemistry and the outcome of change because reactions may occur on the collagen in isolation or at reagents already fixed on the collagen creating a changing substrate. These reactions and interactions can be characterised in the following ways: independent, in which case the reagents react with different sites on the collagen nor do they interact with each other; antagonistic, when the reagents compete for reaction sites on the collagen and factors such as order of addition, offer and reactivity determine the winner; synergistic, when there is strong interaction between reagents that is more important than their reactions on collagen, when the requirements for the link-lock mechanism are fulfilled.

The aqueous solvent is typically overlooked as a component in the tanning reaction mechanism, indeed it is a powerful driving force leading to the kinetics of reactions and fixation outcome. Some conventional processes already rely on manipulating the solvent properties to achieve the desired outcome, well known in the art, as the following two examples demonstrate. Reactive dyeing depends on a high concentration of neutral electrolyte to provide charges in solution to repel the hydrophobic dye molecules onto the substrate; pickling with ethanolamine hydrochloride ($HOCH_2CH_2NH_3^+Cl^-$), a commercially available option, creates a hydrophobic aqueous medium which drives hydrophilic chromium(III) species onto the charged collagen environment.

The advent of low temperature eutectic liquids and ionic liquids constitute alternative solvents to water (or organic compounds): the latter is exemplified by the data on the effect of crystal lattice energy on the melting point of some ionic salts presented in Table 2, offering the potential to revolutionise tanning processes by matching the properties of the solvent and solute, allowing the use of the chemistries of reagents which currently cannot be used in water or modified aqueous media (Wise et al., 2015).

It must also be recalled that any fixation reaction on collagen is a contributor to the multicomponent system, since they all rely on exactly the same range of reaction mechanisms that define tanning agents: electrostatic, ionic, covalent, hydrogen bonding, and hydrophobic interactions. Hence dyeing, fatliquoring, and other specialised processes, such as water and fire resistance treatments, should be included in the analytical thinking.

(a) Retanning

Retanning can be defined as the conventional tanning process step which follows the primary tanning step, often this is chromium(III) but includes vegetable tannage

Table 2 Melting points of some ionic salts

Cation	Anion	m.pt. (°C)
Na^+	Cl^-	801
K^+	Cl^-	772
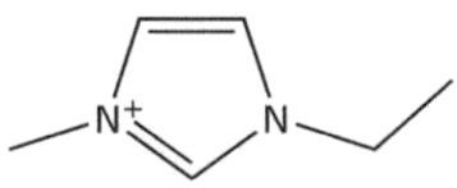	Cl^-	87
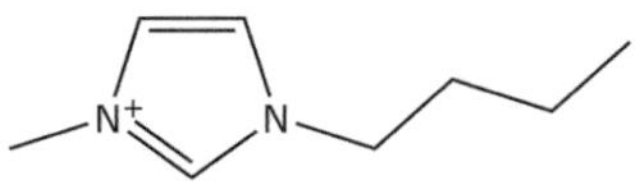	Cl^-	65
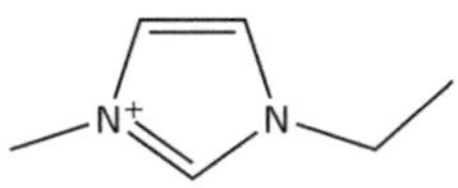	NO_3^-	38
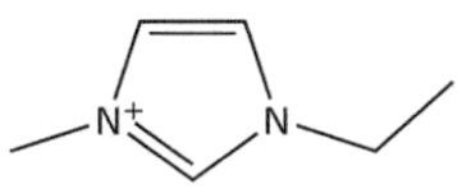	$AlCl_4^-$	7
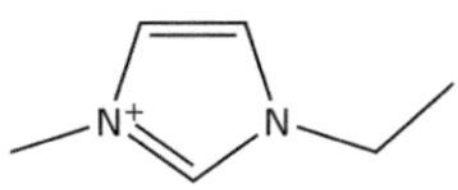	$CF_3CO_2^-$	−14

and increasingly also applies to processes referred to as 'wet white', differentiating them from the 'wet blue' product of chrome tanning. Wet white tanning typically comprises two components, a syntan and an aldehydic reagent, but usually they are not separated in the jargon to tan and retan; since the processing breaks at this point, the part processed wet white product is then retanned in the conventional sense. Note, the components of the wet white process typically act independently because they are not chosen for their affinity for each other, hence they do not fulfil the criteria for a link-lock reaction and the shrinkage temperature is limited to moderate values.

In terms of the chemistry, there is no difference between retanning and the conventional tanning step: retanning is conducted with a wider range of reagents than main tanning, but that is merely the requirements of the technology involved in making the required leather with its associated properties and performance. The only difference between tanning and retanning, apart from the timing, is the nature of the substrate and the necessity of recognising that the reaction kinetics and outcome are likely to need some modification to the reaction conditions, as indicated above.

Retanning includes specialist steps, conferring such properties as water and fire resistance, which have in the past proved to be difficult to achieve consistently. That is less problematic in the current industry, but the interest in exploiting more exotic, effective, and hydrophobic water-sensitive reagents points the way to technologies employing nonaqueous solvents.

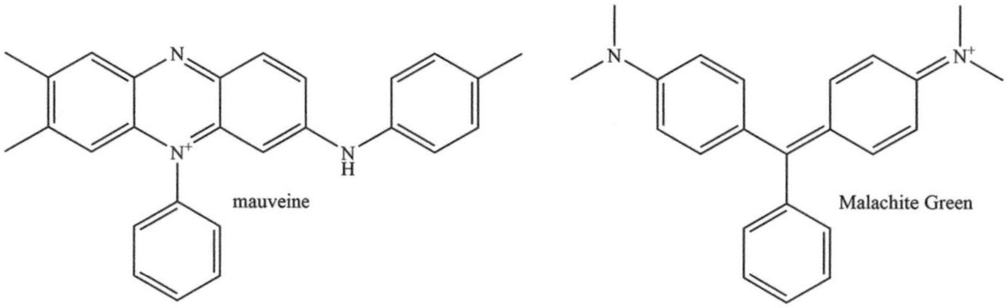

Fig. 3 Structures of Mauveine and Malachite green

(b) Dyeing

Colouring chemistry witnessed an explosion of new reagents following the advent of the synthetic dye mauveine towards the end of the 19th century. In the modern industry, tanners are familiar with mordant, acid, basic, direct, premetallised, reactive and sulfur dyes and their chemistries. However, there is a mounting problem concerning the environmental impact of some of these reagents, particularly with regard to aromatic amines in the dyes or in their decomposition products. Consequently, alternative colourants are being sought. Notably, aryl carbonium based dyes are contenders: interestingly, the original synthetic dye, mauveine, falls into this class of chemicals, as does Malachite Green shown in Fig. 3. Fungal sources of colourants offer novel chemistries, but have yet to be shown to be commercial options (Covington & Wise, 2020b).

(c) Lubricating

The conventional fatliquoring step, usually the last in post tanning, has not changed much over the decades. Its primary purpose is to prevent fibre sticking during drying, with a secondary purpose of conferring the required degree of softness to the product. Reagents may be of animal, vegetable or synthetic origin, made emulsifiable in aqueous solution by partially sulfating or sulfiting. With the exception of discontinuing the use of sperm whale oil in the latter part of the 20th century, the technology has changed little. There are alternative approaches known, each based on the principle of preventing fibre sticking, such as fixing hydrocarbon or derivative chains to collagen or physical separation with inert microspheres, but no technology has superseded the traditional options (Covington & Wise, 2020a).

2.14 Prediction of Processing Outcome or Tanning Requirements for Desired Outcome

Tanners can to a great extent predict the outcome of experimental processes, providing they understand the principles of multicomponent reactions, the chemistries of the reagents and the technological properties conferred by the reagents. This new view of processing, as a series of interacting fixation reactions, is useful because it means the traditional sequence of reagent additions to the system may not be the most efficient and effective—all process steps should be candidates for change.

Creating new materials, which includes hybrid biomaterials, is most efficiently and effectively achieved by applying a mechanistic approach to proposed processes, which allows accurate prediction of experimental outcomes. By reversing the analysis of predicting process goals, it is possible to start with an idea of desired properties and performance in a new product and to design an appropriate process to achieve those outcomes (Covington, 2011). This can extend to the development of new reagents and new chemistries, delivered in new solvent media. However, there are limits to what is possible, since the basis of all such developments is the organic substrate collagen.

2.15 Finishing

It is recognised in the leather industry that finishing is a separate technology to the rest of the process. The application of a coating to leather creates a two-component material, in which each contributes to the properties and performance: the finish provides protection to the vulnerable grain surface and can confer a wide range of aesthetic effects to enhance the commercial viability of the leather, while the leather itself defines the properties that make it a desirable consumer product.

The technology is fundamentally the same as all other coatings—basecoat + topcoat. In leather making, the crucial step is the adhesion between the basecoat and the leather substrate: the chemistries must be compatible to ensure they function as one, then the chemistries of the two components of the finish must be compatible for the durability of the finish layer. The difference between leather finishing and painting for example is the requirement for leather to move and flex in use, making greater demands on the properties of the finish.

Modern options for both layers are urethanes, acrylates, nitrocellulose and proteins: within each type there is a wide range of chemical modifications offered by the supply houses (Covington & Wise, 2020a). It is a feature of surface coatings that rapid product developments and availability lead to fast changes in leather finishing technology, so this is an area of leather production which cannot be easily defined at any one time.

The big change in finishing technology came towards the end of the 20th century, with the elimination of solvent based chemistry in favour of aqueous reagents. The leather industry quickly converted, although development was initially hampered by the total reliance on aqueous chemistry. However, the coming of non-aqueous solvents, ionic and deep eutectic solvents, opens up the possibility of using new and more reactive reagents in finishing for new leather properties.

3 Overview

The keys to success in the modern leather industry are getting the science right and providing practitioners with the understanding of how that science can lead to workable new technologies.

A feature of the technology currently defining the sustainability of the global leather industry is keeping ahead of the competition, by maintaining the economic viability and constantly rebuffing the attacks directed at the industry from many sides. In order to do so, the environmental impact of processing must be constantly addressed, properties and performance of leather products must be improved to conform to the increasingly discerning market and new collagenic materials for new markets must be devised and created. Understanding the principles of leather science will allow these developments to be undertaken efficiently and effectively.

Whilst it is not yet necessary for the leather industry to undergo a revolution, nevertheless it is useful to know that we have the paradigm tools for it to happen ...if and when required.

References

Addy, V. L., & Covington, A. D. (2003). NovoCor AX—So much to gain and nothing to lose. In *Leather Magazine* (p. 14).

Bailey, D. G. (1995). Preservation of cattle hides with potassium chloride. *Journal of the American Leather Chemists Association, 90*(1), 13.

Covington, A. D., Song, L., Suparno, O., Koon, H. E. C., & Collins, M. J. (2008). Link-lock: An explanation of the chemical stabilisation of collagen. *Journal of the Society of Leather Technologists and Chemists, 92*(1), 1.

Covington, A. D., Lampard, G. S., Menderes, O., Chadwick, A. V., Rafeletos, G., & O'Brien, P. (2001). Leather tanning studies using extended X-ray absorption fine structure. *Polyhedron, 20*, 461.

Covington, A. D. (2010). The mechanism of chrome tanning. *Global Journal of Inorganic Chemistry, 1*(2), 119.

Covington, A. D. (2011). Prediction in leather processing: A dark art or a clear possibility. *Journal of the Society of Leather Technologists and Chemists, 95*(6), 231.

Covington, A. D. (2018). Tanning or leathering. *Journal of the Society of Leather Technologists and Chemists, 102*(1), 46.

Covington, A. D., & Wise, W. R. (2020a). *Tanning chemistry. The science of leather* (2nd ed.). Royal Society of Chemistry.

Covington, A. D., & Wise, W. R. (2020b). Current trends in leather science. *Journal of Leather Science and Technology, 2*, 28.

Covington, A. D. (2021). The biochemistry of leather making. *Journal of the Society of Leather Technologists and Chemists, 105*(1), 53.

Gustavson, K. H. (1953). Uni- and multipart binding of chromium complexes by collagen and the problem of crosslinking. *Journal of the American Leather Chemists Association, 48*(9), 559.

Heidemann, E. (1993). Fundamentals of leather manufacturing. Eduard Roether KG.

Rabinovich, D. (2011). Everything you wanted to know about collagen models—But were too afraid to ask! *Journal of the American Leather, Chemists Association, 106*(8), 242.

Sykes, R. L. (1956). Modification of some reactive groups of collagen and the effect on the fixation of tervalent chromium salts. *Journal of the American Leather Chemists Association, 51*(5), 235.

Wise, W. R., Abbott, A. P., Alaysuy, O., Antunes, A. P. M., Douglas, A. C., & Guthrie-Strachan, J. (2015). Processing of leather using deep eutectic solvents. *ACS Sustainable Chemistry & Engineering, 3*, 1241.

Challenges in the Treatment of Tannery Effluents

Jakov Buljan and Sengoda Gounder Rajamani

1 Introduction

1.1 Mass Balance and Sources of Pollution in a Traditional Tannery

Leather manufacture is a water-intensive process and generates a significant volume of liquid and solid waste. The main reason is that the only true leather-building substance in raw hide is corium collagen, everything else (hair, epidermis, fat, some proteinaceous matter etc.) has to be removed. To achieve these goals, substantial quantities of both general and specialty chemicals are needed. Again, due to inherent characteristics of physical–chemical processes applied, the major part of them is not actually used up and ends as waste to be discharged (Aloy et al., 1976). Untreated liquid, solid and to some extent air emissions generated by the tanning industry can thus pose a serious threat to the environment, particularly to surface and groundwater.

The schematic tannery operation and sources of pollution are shown in Fig. 1.

1.2 Collagen

When evaluating the efficiency of leather manufacture, it is quite sobering to realize that only some 50% of the corium collagen, the only true leather-building substance,

J. Buljan
United Nations Industrial Development Organization, UNIDO, Austria and Mlinovi 62, 10000 Vienna, Zagreb, Croatia
e-mail: buljan.jakov@gmail.com

S. G. Rajamani (✉)
CSIR-Central Leather Research Institute (CLRI), Chennai, India
e-mail: dr.s.rajamani@gmail.com

S. K. Janardhanan and L. A. Zugno (eds.), *Emerging Trends in Leather Science and Technology*, https://doi.org/10.1007/978-981-99-9754-1_5

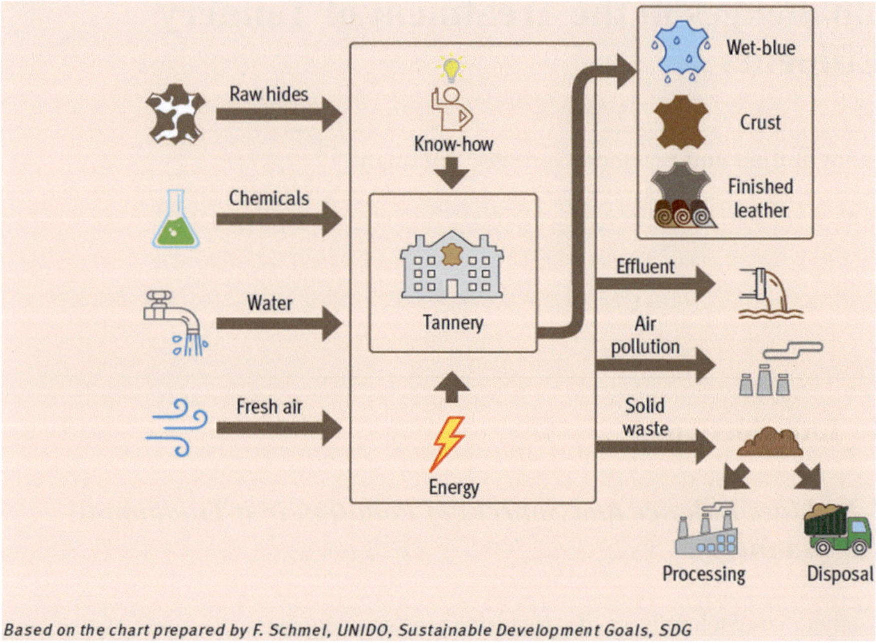

Fig. 1 A simplified schematic chart of the tanning operations

ends up as leather. Details of collagen distribution in finished leather and solid waste are given in the following Table 1.

1.3 Mass Balance of Chemicals in Leather Processing

General chemicals (acids, bases, salts) are required in leather process to create the medium for reactions but are not actually absorbed (Buljan et al., 1997). Finishing products are essential to lend the desired appearance. The mass balance of chemicals in leather process is given in the following Table 2.

1.4 Approximate Weight and Area Proportions, Light European Hides

In case of processing, one tonne of wet salted raw hides results in approximately 138 m^2 of finished (shoe upper) leather and some 60 m^2 of chrome tanned splits that can also be turned into leathers of acceptable quality (embossed grain, suede) as shown in the following diagrams.

Table 1 Collagen distribution: wet salted hide, finished leather and solid waste *(Starting material: 1,000 kg wet salted European raw hides, avg. 25.6 kg/piece, splitting in chrome)*

Component	Amount of collagen*		
	Kg	*% of corium collagen*	*% of total collagen*
Input			
Corium collagen (Leather building collagen)	280	100	92
Subcutis collagen	24	–	8
Total collagen input	**304**	**–**	**100**
Output			
Grain leather	113	40.0	37.2
Split leather	36	13.0	11.8
Total collagen in finished leather	**149**	**53.0**	**49.0**
Total collagen in solid waste (shavings, trimmings etc.)	**155**	**47.0**	**51.0**
Total collagen output	**304**	**100**	**100**

[*] *Note the distinction between the* **corium collagen** *and* **total collagen** *(coriumplus subcutis/flesh collagen)*

Table 2 An example of mass balance in leather processing

Kg/ton of wet salted hide			
Chemicals	Added in process	In/on leather & splits	Wasted
Chrome extract (Cr_2O_3)	101 (25)	19 (12)	82 (13)
Organic tannins	25	20	5
Fatliquors	22	17	5
Dyestuffs	5	4	1
Acids, bases, salts	191	–	191
Tensides	3	–	3
Enzymes	5	–	5
Finishing products	100	12	88
Total	452	72	380

Note EU (Best available Techniques, 2013) figures about chemical addition are even higher

1.5 The Scope for and Limitations of Cleaner Technologies

Pollution prevention, which ultimately leads to lower treatment costs, remains the supreme priority (Rajamani, 2019). By applying industrially proven low-waste advanced methods such as the use of salt-free preserved raw hides and skins, hair-save liming, low-ammonia or ammonia-free deliming and bating, advanced chrome management systems, etc., it is possible to significantly decrease the nature and amount of emissions (Money, 1999; Buljan, 2007; Rajamani, 2022) as shown in the following Fig. 2.

It means that despite all preventive measures, there is still a considerable amount of pollution load to be dealt with by the end-of-pipe methods (Buljan & Kràl, 2019).

Contrary to widespread misperception that vegetable tanning is environmentally harmless (in reality its effluent has very high, difficult to treat COD) and that is why we here deal only with chrome tanning because it is by far the most prevailing leather tanning method (Mosiewicz & Forestal, 1976).

Effluent treatment is a complex process and requires specialists for planning, designing and operating them. However, it is important for tanners to acquaint themselves with the main pollutants, their impact on environment and the basic principles and prevailing methods of tannery effluent treatment. That makes them better equipped for a dialogue with the factory's own environmental unit, wastewater treatment specialists as well as with environmental authorities and NGOs (Buljan, 2014).

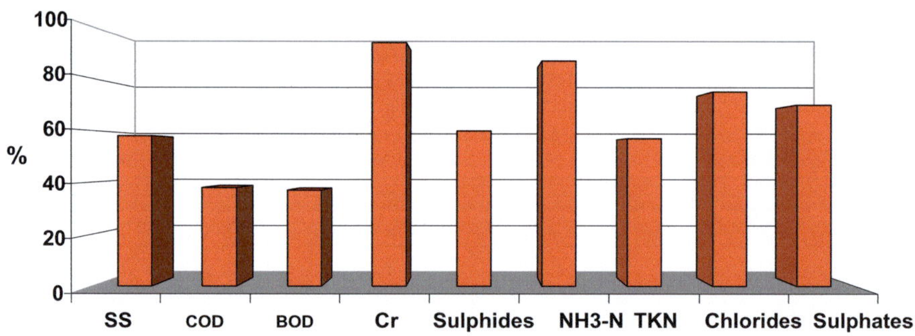

Fig. 2 An estimate of decrease of pollution loads in wastewater after introducing advanced technologies

2 Pollution Parameters—Description, Environmental Impact, Legislation

The effect of excessive pollutant levels commonly found in tannery effluents can be severe, their impact is here described for general guidance. Also, there is no single measure of pollution load; a few key parameters considered together are needed to offer a reasonably accurate picture (Table 3).

2.1 Sources of Pollution

To avoid possible confusion arising due to differences in water consumption, it is practical to indicate the amount of pollutants generated per tonne of raw hide input (Carre et al., 1983). While generally lower water consumption is very desirable (nowadays in well-managed tanneries significantly below 30 m^3/tonne), it obviously results in considerably higher concentration of pollutants (Fig. 3).

The characteristics of the wastewater discharge from conventional process (Bosnic et al., 2016) are given (Table 4).

2.2 Legislation

The average pollution load from raw to finishing process with water consumption of 45 m^3/tonne is given in the following Table 5.

2.3 Banned Substances in Tannery Effluents

In addition to the controlling of traditional parameters such as COD, BOD_5, TKN, chrome, sulphide etc. an important trend in European legislation is stricter regulation of the presence of biocides and pesticides in tannery effluents such as DDT, hexachlorohexane, aldrin, dieldrin, endrin, isodrin, naphthalene, etc.: some prohibited substances, notably organochlorine compounds, are stable enough to survive wastewater treatment process. Only registered and approved degradable biocides can be used and their choice is becoming more and more limited.

Virtually all pesticides found in the tannery wastewater are residues of pesticides used to treat live animals whereas biocides are regularly used in preservation and soaking, pickling, tanning and post-tanning processes.

Table 3 The main pollution parameters: description, environmental impact

Parameter	Description	Environmental impact
Biochemical oxygen demand, BOD_5	The quantity of oxygen required during the stabilization of decomposable organic and oxidizable inorganic matter by aerobic biological action under standard conditions over five days	The greater the BOD, the lower the amount of dissolved oxygen available for aquatic life
Chemical oxygen demand, COD	The quantity of oxygen consumed for the total oxidation of the oxidizable matter (organic and inorganic) with dichromate as the oxidising agent	Depletion of oxygen available for aquatic life, including chemical compounds not easily degraded by microorganisms normally present in the water recipient
Total suspended solids, TSS	Made up of solids of two distinct characteristics: (i) Solids with a rapid settling rate (settleable solids) and semi-colloidal solids	Substances present in the TSS significantly contribute to the BOD and COD levels. Responsible for the amount of sludge generated in the course of effluent treatment
Chromium, Cr^{3+} (III)	Emanates from incomplete uptake of chrome salts used for tanning—from technical, economic and environmental aspect the most superior method of leather making Discharged from processes in soluble form; however, precipitates and stabilized colloids are formed	Very fine, chrome-stabilized colloids are also formed; partially tanned protein solids are highly resistant to biological breakdown. Also, chromium hydroxide precipitates and persists in the ecosystems. Often confused with potentially more hazardous Cr^{6+} (VI) form
Sulphide, S^{2-}	Results from the use of sodium sulphide and sodium hydrosulphide and from the breakdown of hair in the unhairing process. At the pH below 9.5, hydrogen sulphide gas evolves. Strict segregation of alkaline and acidic streams is essential. Characterized by a smell of rotten eggs but, very unfortunately, only at lower concentrations	Flammable, corrosive, highly irritating gas. Mixtures with air may explode violently. Low concentration induces headaches and nausea. At higher levels, death sets literally within seconds. The most frequent cause of fatalities in the tanning industry. If discharged to surface water, even in low concentrations very toxic
Total Kjeldahl nitrogen, TKN	A widely accepted measure of presence of nitrogen in effluent: it includes both the organic nitrogen (i.e. nitrogen contained in amino acids/proteinaceous compounds) and nitrogen present in ammonium salts	Excessive growth of plants and algae, clogging of waterways, flows impaired. As the plants die their decomposition leads to depletion of oxygen and ultimately to anaerobic conditions and bad smell

(continued)

Table 3 (continued)

Parameter	Description	Environmental impact
Chlorides $(Cl)^-$	Originate from the large quantities of common salt used in hide and skin preservation and/or in the pickling process. Highly soluble and stable, unaffected by effluent treatment	Excessive concentrations inhibit the growth of plants, bacteria and fish in surface waters. Also, increased salt content in groundwater is now becoming a serious environmental hazard
Sulphates, $(SO_4)^{2-}$	Emanate from the chrome tanning salts, sulphuric acid, ammonium sulphate and some retanning agents; from sulphide removal by oxidation &use of alum sulphate in the coagulation process	Remain as sulphates or are broken down by anaerobic bacteria to produce the toxic, corrosive and malodorous hydrogen sulphide Contribute to TDS
Oil and grease	Released from within the skin structure whereas some fatliquors are added as lubricants for desired feel as well as some water repellence	Oil and floating grease agglomerate to form 'mats' causing blockage problems. Thin layers on water surface significantly reduce oxygen transfer from the atmosphere
Total dissolved solids (TDS)	Indicating the amount of all inorganic and organic substances dissolved in wastewater; essentially, everything present in water other than pure H_2O and suspended solids. In the tannery effluents, the main relevant components are **sulphates** and **chlorides**. Colloquially referred to as salinity	High TDS concentrations exert varying degrees of osmotic pressures and make otherwise fully treated wastewater unfit for irrigation and livestock watering; also, high TDS often becomes lethal to the biological inhabitants of an aquatic environment. Not affected by the usual effluent treatment
pH	Quantitative measure of the acidity or basicity of aqueous or other liquid solutions. The scale on which 7 is neutral, lower values are acidic, higher values alkaline	Both highly acidic and highly alkaline values negatively affect the flora and fauna of the receiving water bodies

3 Effluent Treatment, Typical Treatment Stages

Wastewater treatment is a multi-stage process to purify wastewater before it enters a body of water, is applied to the land or reused. The goal is to reduce or remove organic matter, solids, nutrients, chrome and other pollutants from wastewater, the receiving body of water can cope with as established by relevant authorities (Spellman, 1996). Before turning to treatment, it is important to emphasize:

- Design of an effluent treatment plant (ETP) is always tailored to requirements of a specific site; thus, there are no two identical ETPs.

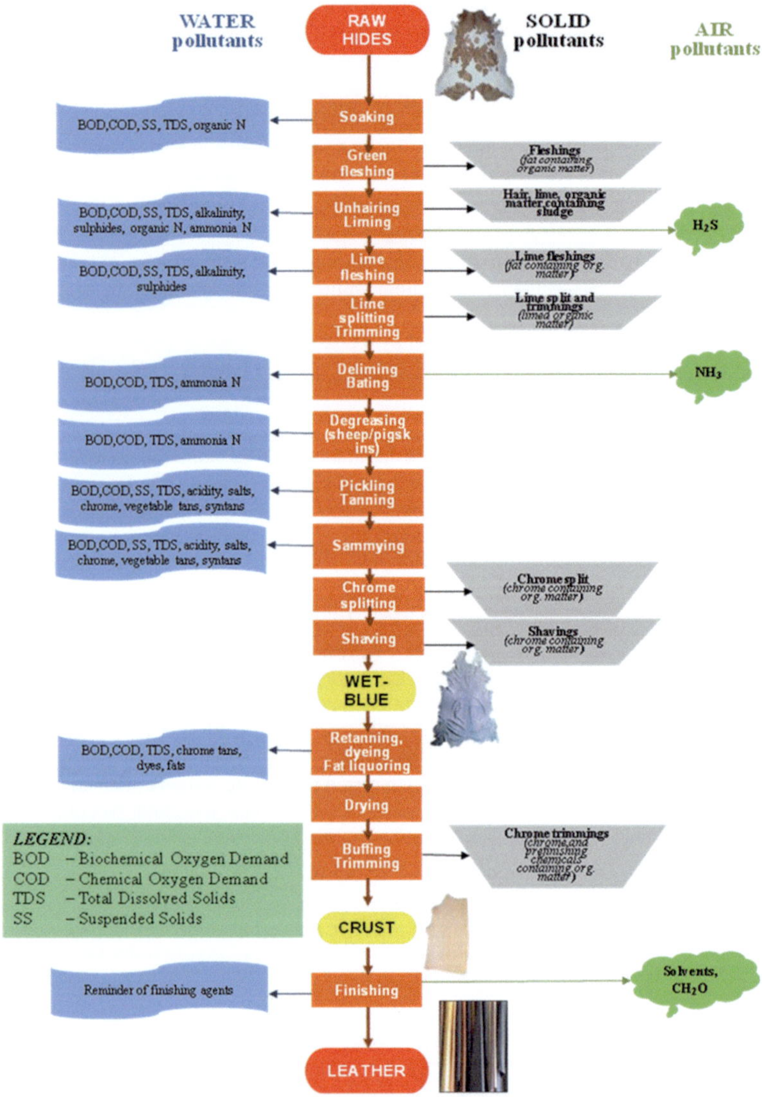

Fig. 3 The main points of waste discharge along the leather-making process

- Since matter can be neither created nor vanish, pollutants contained in effluent cannot disappear; they are only converted into environmentally more acceptable or easier to dispose form (sludge).
- Somewhat paradoxically, obvious is often overlooked: the same amount of pollutants at lower water consumption means lower hydraulic load (volume) but higher concentration, not always easy to treat.

Table 4 An example of pollution load, conventional process

Parameter	Typical pollution load, conventional process kg/tonne *of wet salted hide*	A tannery with daily input of 10 tonnes of w.s. hides, every day discharges approx. (Kg)
Volume of wastewater	30 m^3/tonne	300 m^3
COD	180	1800
BOD_5	90	900
Suspended solids (SS)	120	1200
Cr^{3+}	5	50
S^{2-}	5	50
Total nitrogen (TKN)	14	140
Cl^-	180	1800
$(SO_4)^{2-}$	80	800
Oil and grease	10	100
TDS *	400	4000
pH	6–9	

* The main constituents of the TDS are **chlorides** (Cl^-) **and sulphates** $\left(SO_4^{2-}\right)$

Table 5 An example of the average total pollution load—water consumption of 45 m^3/tonne*

Parameter	Unit	Average total pollution load	Typical limits, surface waters
BOD_5	mg O_2/l	2000	30–40
COD	mg O_2/l	4000	125–250
Suspended solids SS)	mg/l	2000	35–100
Cr^{3+}	mg Cr/l	150	1.5–2.0
S^{2-}	mg S/l	160	1.0–2.0
Total nitrogen (TKN)	mg N/l	280	100
Cl^-	mg Cl/l	5000	Locally specific
SO_4^{2-}	mg SO_4/l	1400	Locally specific
Oil and grease	mg/l	130	Locally specific
TDS	mg/l	10,000	Locally specific
pH		6–9	5.5–9.5

* Nowadays already nearly halved, typically 20–25 m^3/tonne (Knoedler, 2014)

- It is important for a tanner—who emanates wastewater, to understand the relation between the leather process technologies and wastewater treatment in order to reduce the overall cost of treatment.

The three main categories of tannery wastewater, each of very distinctive characteristics, are:

- Effluent emanating from the beamhouse—liming, deliming/bating, water from fleshing and splitting machines; they contain sulphide, their pH is high, but they are chrome-free.
- Effluent emanating from the tanyard—tanning and retanning, sammying; high Cr content, acidic.
- Soaking and other general effluent, mainly from the post-tanning operations—fatliquoring, dyeing; low Cr content.

3.1 Segregation of Tannery Streams

The segregation of sectional streams and separate in-house pre-treatment are necessary for an efficient effluent treatment system. A typical tannery layout with locations of segregation and treatment is shown in the following Fig. 4.

It is very important to segregate these streams and to pre-treat them separately respecting their characteristics to avoid possible safety risks (formation of deadly hydrogen sulphide) and to reduce cost of treatment and sludge disposal (to avoid contamination of sludge with Cr). In other words, mixing of liming and tanning streams gives rise not only to obnoxious smell typical for poorly managed tanneries: the resulting lethally poisonous gas hydrogen sulphide (H_2S) is still by far the most frequent killer in tannery accidents, mainly in inadequately ventilated spaces, especially in pits and channels. Accordingly, the practice of "delegating" sulphide oxidation to the CETP homogenization tank is questionable.

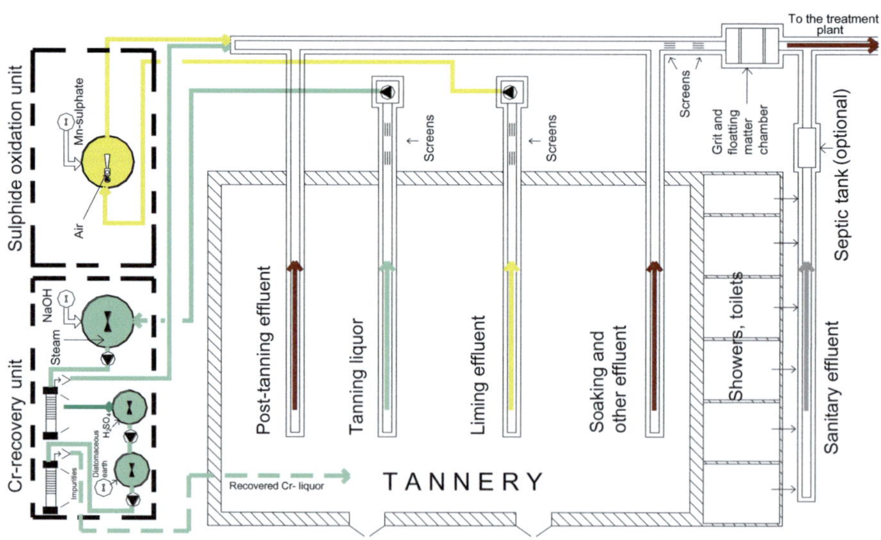

Fig. 4 Schematic layout of the in-house segregation of streams, including chrome recycling and oxidation of sulphide in liming effluent

The volume and pollution load of sanitary wastewater is comparatively insignificant.

Somewhat arbitrarily, it is common to distinguish the following main stages of effluent treatment:

- On-site treatment
- Physical-chemical treatment (primary)
- Biological treatment (secondary)
- Advanced (tertiary) treatment
- Sludge handling and disposal.

3.2 On-Site Treatment

Typically, in the case of CETPs servicing tannery clusters often found in developing countries, it is essential to have pre-treatment units installed in individual tanneries before the effluent is discharged into collection network. Their role is to remove large particles, sand/grit and grease but also to significantly reduce the content of chrome and sulphides.

The typical pre-treatment system for sulphide oxidation and chrome recovery system (Buljan & Rajamani, 1996) is shown in the following Fig. 5.

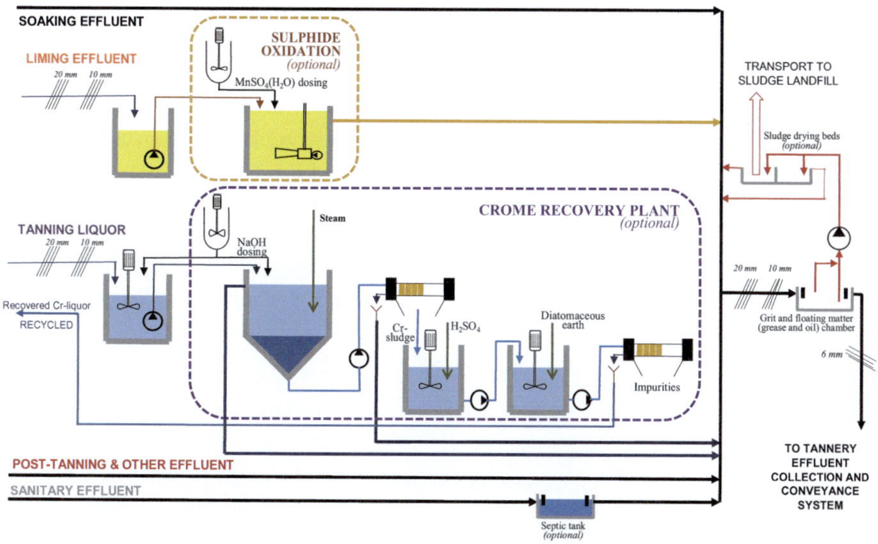

Fig. 5 Schematic flowchart of the in-house segregation of streams, chrome recycling, treatment of liming effluents and pre-treatment of mixed effluent in individual tanneries

4 Physical–Chemical Treatment (Primary)

Objectives:

- To eliminate the coarse material normally present in the raw wastewater that could clog/block pumps, pipes, and possibly, sewer lines.
- To well mix and balance different tannery streams and thus produce homogenized "raw material" that can be treated in a constant way.
- To adjust pH and eliminate toxic substances (sulphides) and/or shock loads that could negatively affect the rather sensitive biological treatment.
- To significantly decrease the BOD/COD load and thus reduce the cost of and simplify the biological treatment phase.

The basic steps of the primary treatment are:

- Screening (bar, self-cleaning screens)
- Pumping/lifting, fine screening
- Equalization and homogenization
- Chemical treatment (coagulation, flocculation)
- Solids separation, sludge dewatering.

Quite often grease and sand trap is also included, usually before fine screening and pumping.

4.1 *Screening*

Large ETPs require more sophisticated, aerated arrangements for removal of grit and floating matter as shown in the following Fig. 6, 7 and 8.

It is not possible to transfer effluents throughout the ETP by gravity only, at least one or more pumping/lifting stations are needed, the first typically located before the rotary screen. Depending on specific requirements—capacity/flow different types of pumps are used. Small and medium scale ETPs generally rely on different models of submersible pumps as shown Fig. 9.

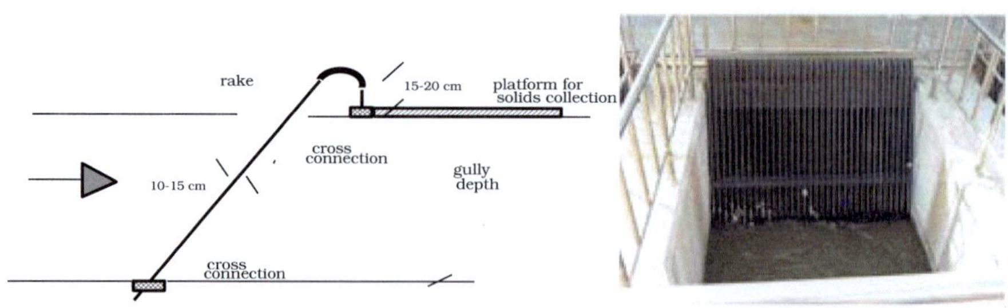

Fig. 6 Bar screen, side view

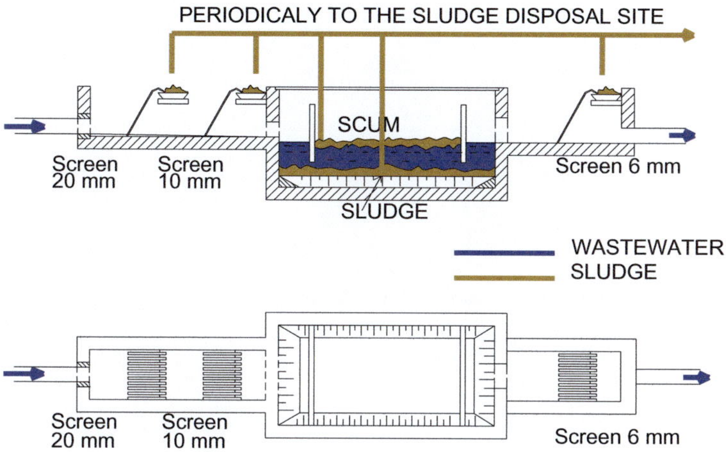

Fig. 7 Simple, non-aerated grit and floating matter removal chamber

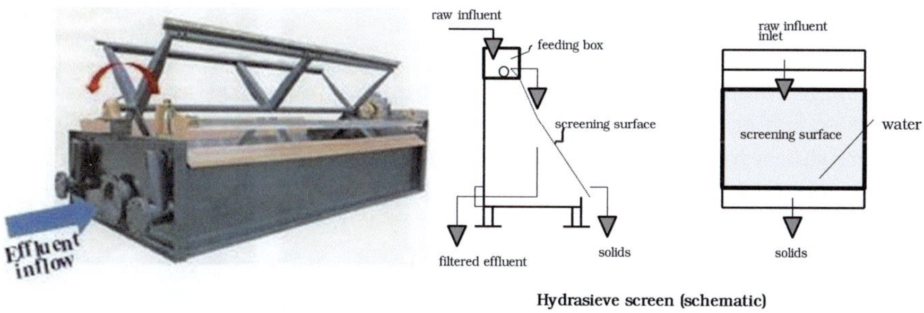

Fig. 8 Self-cleaning, rotary brushes screen, Parkwood type self-cleaning screen, Hydrosieve type

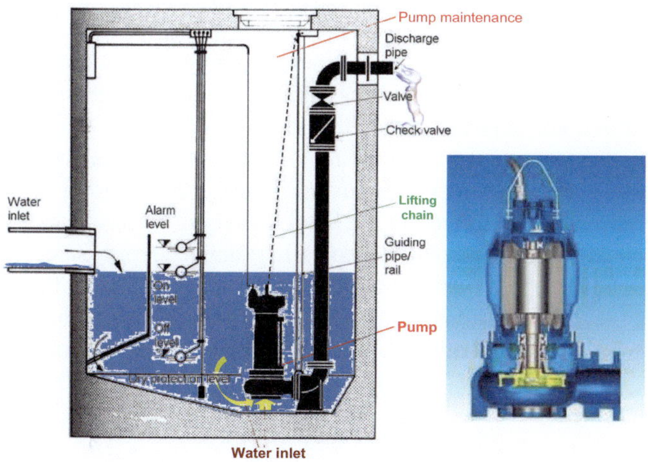

Fig. 9 Submersible pump

Fig. 10 Archimedes-type lifting pumps

For large volumes screw (Archimedes) pumps are preferred (Fig. 10).

Fine screening should drastically reduce the amount of fine suspended solids. Here are shown rotary drum-type screens with outer and inner flow as shown in the following Fig. 11.

4.2 Equalization—Homogenization

The main aim is the homogenization and subsequently a steady flow of the balanced effluent for further treatment. Accordingly, the volume of the equalization tank normally corresponds to the total daily effluent discharge; at the same time, volumes of raw and treated effluent are more or less equal. Also, normally there is no significant temperature difference between raw and treated effluent.

It is very important to keep all particulate matters in suspension, i.e. to avoiding settling of solids. This is achieved by using mixing cum aeration devices such as diffused air systems, Venturi ejectors and fixed or floating aerators (lately avoided due to lower efficiency and problem of aerosols). In practice, to play safe, the volume of the equalization tank approximately corresponds to the total daily effluent discharge plus minimum storage requirement in the equalization tank for mixing-cum aeration. A typical view of the equalization-cum homogenization tank is given Fig. 12.

The energy required to keep the solids in suspension is some 50 W/m^3, which is also sufficient for any residual sulphide oxidation.

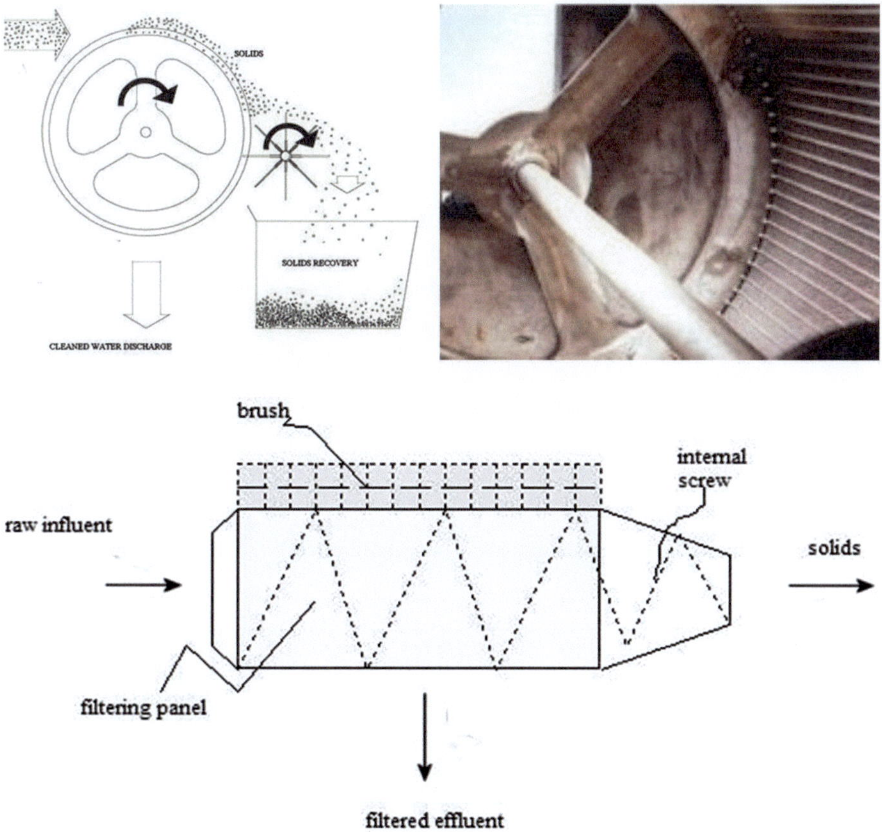

Fig. 11 Rotary drum fine screen, outer and inner flow

4.3 Chemical Treatment (Coagulation, Flocculation)

Chemicals are added in order to improve and accelerate the settling of suspended solids, especially of the fine and colloidal matter. Coagulation is the destabilization of colloids by neutralizing the forces that keep them apart; as a result, the particles collide to form larger particles (flocs). Industrial aluminium sulphate (alum) is most commonly used coagulation agent but iron sulphate, iron chloride and lime are also used. Obviously, there is a side effect: coagulation agents add to the effluent's TDS load.

Polymers added during the flocculation stage help forming the bridges among the flocs and bind the particles into large agglomerates or clumps, which are then removed from the liquid by sedimentation, filtration, straining or floatation as shown in the following Fig. 13.

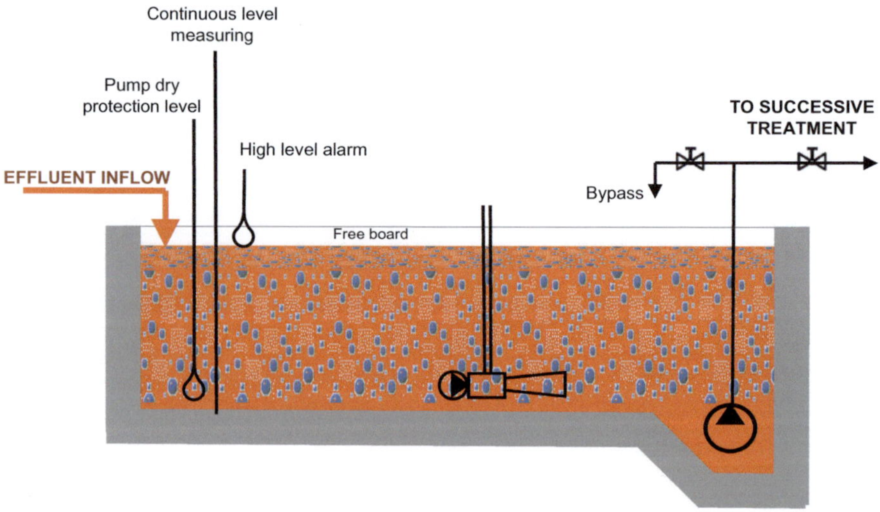

Fig. 12 Schematic view of the equalization—homogenization tank

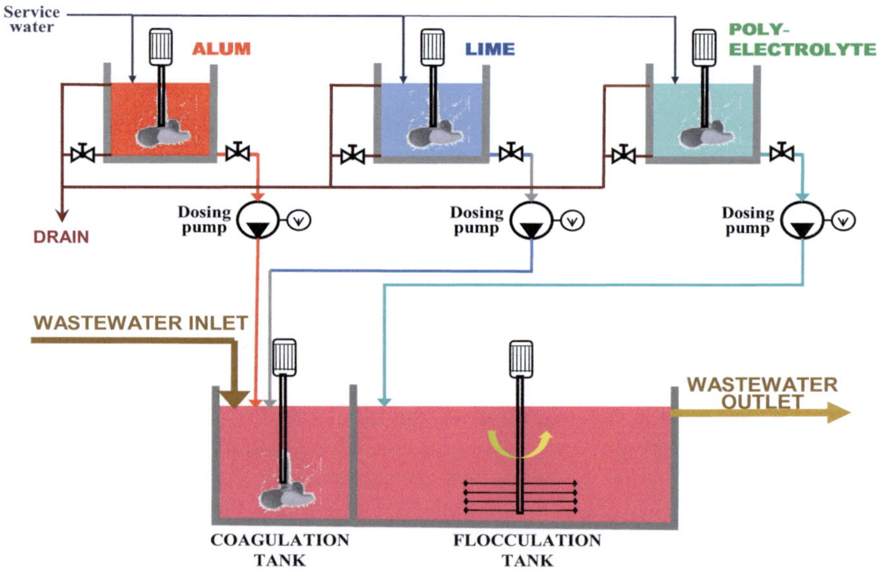

Fig. 13 Schematic view of the coagulation and flocculation system

4.4 Solids Separation—Primary Sedimentation

The main objective is removal of suspended solids; various floatables like fats, waxes, mineral oils, floating nonfatty materials etc. ("grease"), not already removed in the grit and oil chamber (usually positioned between screening and equalization) are also separated here.

Primary settling tanks (clarifiers) are circular (more common) or rectangular with arrangements for continuous, quiet in- and out-flow, grease (scum) removal at the top and sludge removal at the bottom. The key design parameters are detention time (usually 1–2 h), surface hydraulic loading (typically 1–2 m^3/m^2 per hour) and surface solids rate; the latest is expressed in Kg/m^2 indicating the quantity of SS crossing the surface area of the tank over a certain time span (hour, day). Mechanical device (scraper) is necessary in larger settling tanks as shown in the following Fig. 14.

In some cases, mainly due to space shortage, solids are removed by flotation, usually by the Dissolved Air Flotation (DAF) system as shown Fig. 15.

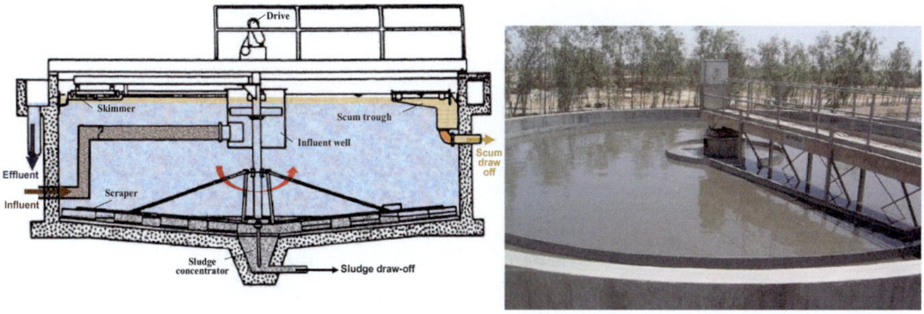

Fig. 14 Schematic cross-section view and external view of a typical circular sedimentation tank

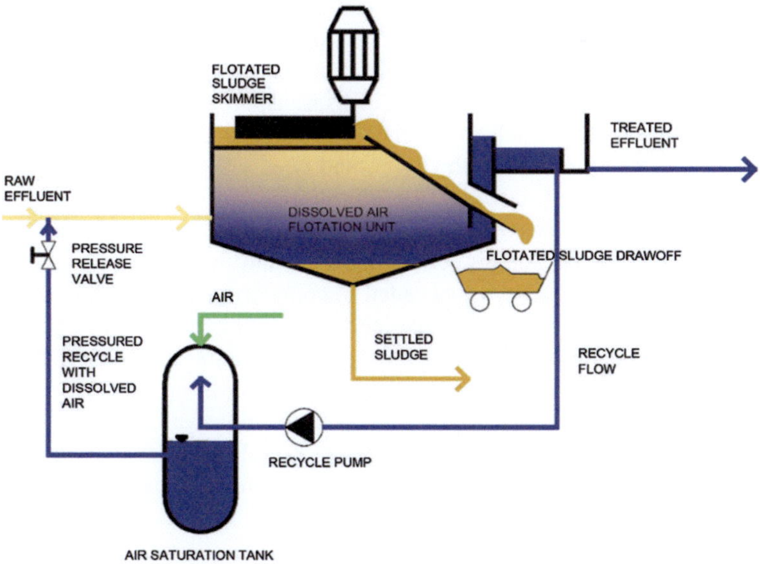

Fig. 15 Schematic view of the dissolved air flotation (DAF) unit

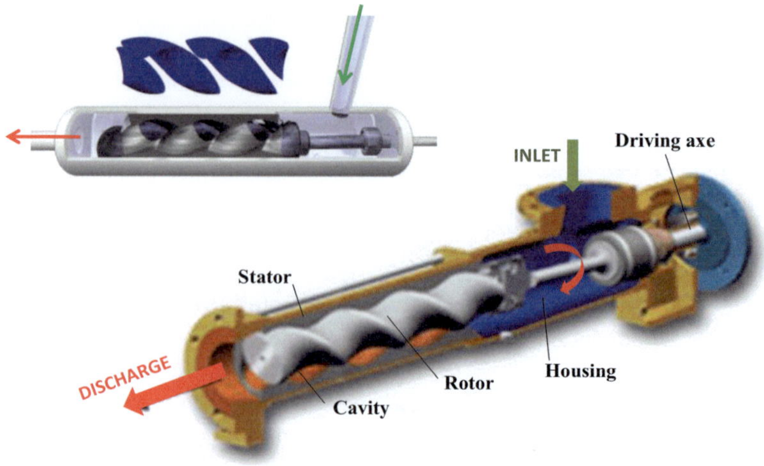

Fig. 16 Schematic cross-section view of a progressing cavity pump type for sludge transfer

4.4.1 Sludge Dewatering[1]

The sludge drawn from the bottom of the tank is in the form of slurry with dry solids (DS) content of only 2–4% (Wastewater Treatment, WPCF - 1985). For its evacuation special, progressing cavity-type pumps as shown in the following figure are used (Fig. 16).

Further reduction of the water content is achieved in sludge thickeners (very much like circular clarifiers) followed by mechanical dewatering in plate filter presses, belt filter presses or decanter centrifuges. In addition to power and chemicals requirements, the key parameter for equipment selection is the achievable dry matter content in the dewatered sludge lowering the cost of its transport and disposal.

The construction of the sludge thickener is in practice identical with that of sedimentator although in some cases the Dortmund type with self-desludging slopes is also used as shown in the following Figs. 17 and 18.

Through increase of dewatering level, improvement of wear and tear resistance (especially in the case of sludges containing fine sand) as well as by lowering the irritating noise, centrifuges have succeeded in conquering a lot of ground in dewatering tannery sludges. The comparative performance characteristics of dewatered sludge from different dewatering equipment are given in the following Tables 6 and 7.

Sludge drying beds, easily constructed with locally available materials were perceived as optimum solution for tanneries in hot climate developing countries. However, they require a lot of area, the output during rainy seasons drops considerably, there is the problem of malodour, they are not easy to clean and made ready

[1] *The main part of sludge stems from the physical–chemical (primary) treatment and that is why it is described at this point.*

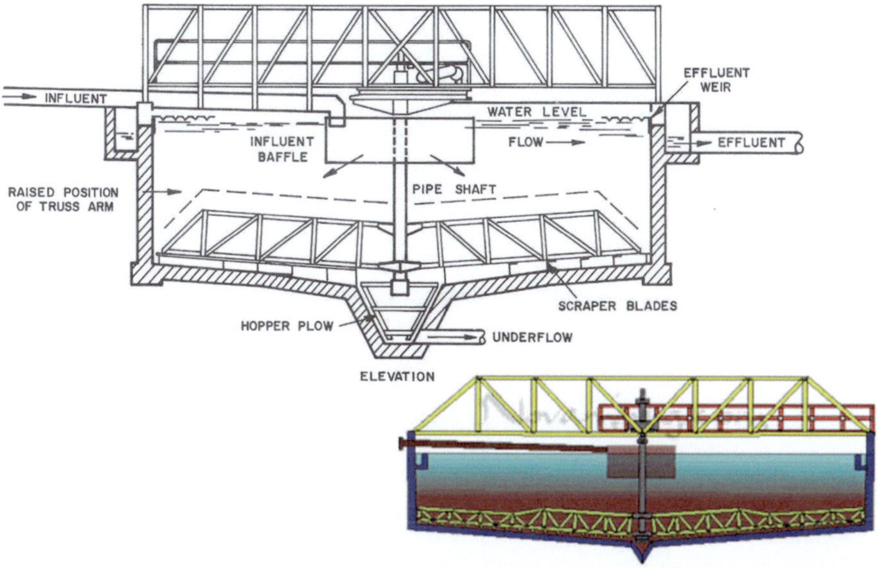

Fig. 17 Schematic cross-section view of the gravity thickener

Fig. 18 Recessed plate filter press with container for dewatered sludge

Table 6 The main performance characteristics of sludge dewatering equipment

	Decanter centrifuge	Belt press	Plate filter press
Way of operation	*Continuous*	*Continuous*	*Batch*
Sludge conditioning	Required (1)	Required (1)	Not required (2)
Washing water	Not required	Required (3)	Not required (4)
Labour	Only supervision	Only supervision	Required during cake discharge
Sensitive to sludge variability	Very sensitive	Very sensitive	Less sensitive
Energy consumption	High	Medium	Low
Maintenance	Sophisticated	Medium	Low

Notes

(1) *Polyelectrolyte (usually cationic): 2–4* g/kg *of DS*

(2) *Sludge conditioning with inorganic chemicals (Iron salts and lime) is not strictly necessary, but recommendable for enhancing filtration rate and regular performances.*

(3) *About 10* m^3/h *of clean water at 4 Bars per meter of belt width is required for continuous belt washing.*

(4) *Periodical cleaning of filtering clothes is required (minimum once per week).*

Table 7 Dry matter content in sludge depending on the process stage and/or type of dewatering equipment

Treatment stage, equipment	DS content%
Sludge after primary clarifier	3–5
Thickened mixed primary and secondary sludge (at specific load of about 60 kg $DS/m^2.d$	6–8
Chamber filter press with chemical conditioning	35–45
Belt filter press with use of PE	20–25
Centrifuge—decanter	20–30
Sludge after stabilization with CaO (mostly after belt filter press or centrifuge)	60–90

* Average values for mixed (primary + secondary) tannery sludges

for the next batch etc. For these reasons, sludge drying beds are still used mainly by small tanneries not close to residential areas and/or as a short-term backup for larger units (Fig. 19).

The products of the primary treatment are:

- The primary treated effluent (the overflow from the primary settling tank- with only residual amounts of chrome and sulphide and significantly reduced BOD, COD and SS content).
- Dewatered sludge (approx. 40% DS, its chrome content depending on the type and efficiency of the chrome management system applied).

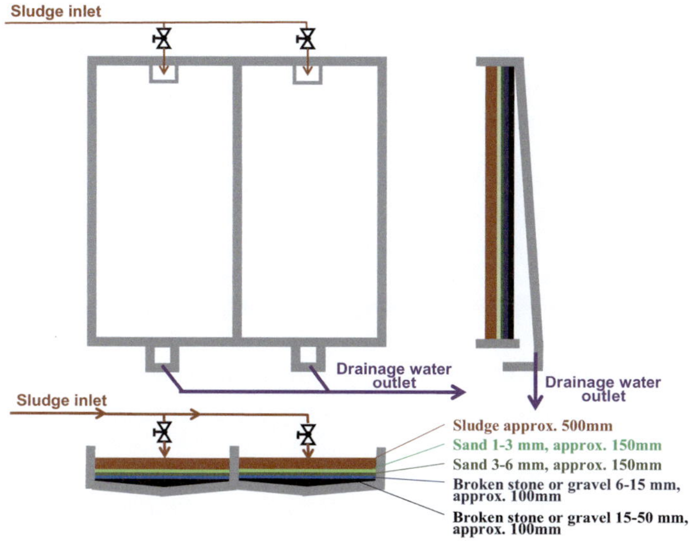

Fig. 19 Schematic views of sludge drying beds

For the fortunate ones, i.e. tanneries coupled to the sewage system, the on-site treatment often ends here—the biological treatment takes place together with urban wastewaters in large WWW. Usually, the industrial effluents make only a small fraction of the total volume so that salinity—TDS they bring along does not represent any problem. However, this situation is still rather rare in developing countries.

5 Biological (Secondary) Treatment

The main objective of this phase is to further reduce the amount of organic and other substances still present in the effluent after the primary treatment stage and thereby satisfy the standards/limits for discharge into the water recipient. Biologically degradable organic substances are converted to bacterial cells, and the latter are removed from the wastewater.

The biological treatment stage duplicates the processes that take place in nature but under controlled conditions and, especially, at highly accelerated pace; however, the efficiency of this treatment stage largely depends on the biodegradability of the polluting substrate, i.e. its inherent capacity to decompose by the biological process. The remaining suspended and colloidal solids are removed by flocculation and adsorption.

While in principle the biological treatment may be aerobic, facultative or anaerobic (or some combination thereof) in practice, almost only aerobic systems are used;

exceptionally, in countries with hot climate and where a lot of land is available, facultative (or preferably aerated/facultative) lagoons are also used.

Due to inherent characteristics of tannery effluents, primarily sulphide/sulphate content and related malodour problem, anaerobic treatment is in practice limited to sludge digestion.

Among many variations of the aerobic processes, the most widely used method is (complete mix) the activated sludge treatment with extended aeration; despite some very interesting features, membrane bioreactors (MBRs) have not made significant inroad in the tanning sector (Eckenfelder & Grau, 1992).

The activated sludge process can be summarized in the following manner:

$$\text{Inert matter} + \text{organic matter} + \text{oxygen} + \text{nutrients} + \text{microorganisms} \rightarrow\rightarrow$$
$$\rightarrow\rightarrow\rightarrow \text{New microorganisms} + CO_2 + H_2O + \text{additional inert matter}$$

Simply said, the microorganisms are stimulated to convert (eat and digest) harmful, oxygen demanding organic compounds into environmentally more acceptable form (microorganisms) and low energy, stable compounds like water, carbon dioxide and nitrogen gas.

The microbial community that does that job comprises various species of bacteria, fungi, protozoa, sometimes rotifers (multicellular animals only found in very stable activated sludge with long retention times), even nematodes, the composition of the community population depending on a plethora of factors (Eckenfelder & Musterman, 1995). A typical activated sludge process flow diagram is shown (Fig. 20).

The biological stage is the most complex part of the overall effluent treatment process, with highest share of investment and operational costs, its day-to-day running requiring considerable skills and experience.

The main operational parameters

The main operational parameters—expressions important for understanding of the process:

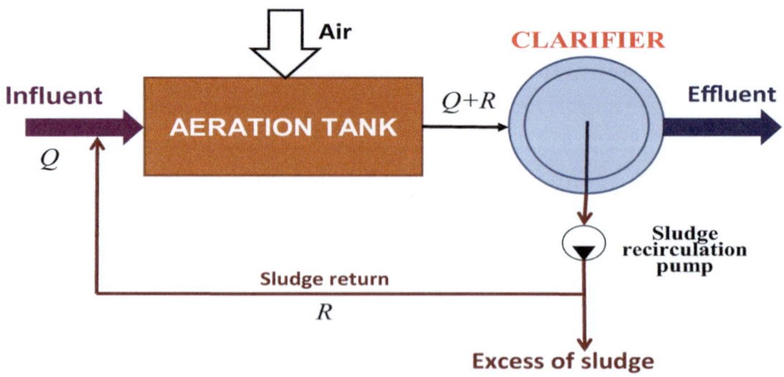

Fig. 20 A simplified flow diagram of the activated sludge process

- **Total Influent Volume, *Q***: Volume of the treated effluent [m^3/day]
- **Tank Volume, *V***: Aeration tank volume [m^3]
- **Organic Loading, *F***: The total BOD_5 applied [kg/day]

$$F = (f * Q)/1000$$

where *f is* the BOD_5 of the influent [mg/l]
- **Mixed Liquor Suspended Solids, *M***:

$$M = (MLSS * V)/1000$$

where MLSS is the concentration of SS in Mixed Liquor in the aeration tank [mg/l]
- **Loading factor, *F/M***:
  BOD_5 kg per day per kg of Mixed Liquor Suspended Solids (*MLSS*) in the aeration tank [mg/l].
- **Hydraulic retention time, *T***:

$$T = (V/Q) * 24$$

where *V is the* aeration tank volume [m^3].

Uniform inlet flow distributed over the entire day provides optimum conditions for absorbing the effect of possible peaks of organic load or toxic substances (shock loads) and enhancing the efficiency of the secondary sedimentation. The BOD here is in practice taken to represent the amount of food provided to the microorganisms contained in the system. Hydraulic retention time is actually the average time (in hours) the influent spends passing through the aeration tank (hours); in extended aeration process, typical for tannery effluents it is usually more than 24 h.

Extended aeration plants are characterized by introduction of wastewater directly into the reactor basin, long aeration, high sludge return ratio, low sludge wastage and high MLSS: the F/M (Kg BOD/Kg MLSS per day) ratio is only $\leq$0.05–0.1 in contrast with conventional (0.15–0.4) or high load type (0.4–1.0) as shown in Table 8.

The other closely monitored parameters are:

Dissolved Oxygen (DO), the content of molecular oxygen in the aeration tank (mg/l); extended aeration units are usually operated in such a way as to keep the DO of the mixed liquor at about 2 mg/l.

Table 8 Typical F/M and O_2 to BOD ratios for different types of treatment

Process	F/M ratio	Kg O_2/Kg BOD
Extended aeration	0.1	2.0
Conventional	0.3	1.2
High rate	1.0	0.8

pH, its optimum normally between 7.0 and 7.5 with an effective process range of 6–9. Reaction between the CO_2 produced from respiration neutralizes the excess alkali, adjustment with lime or other needed only if pH drops below 6.

Temperature affects the metabolic and growth rates of the microorganisms responsible. While higher temperature accelerates the growth rate increases it negatively affects the water solubility of oxygen and the oxygen transfer rate.

Nutrients: the nutritional balance of an aerobic system is primarily based upon satisfying the requirements of the cell structure produced by the removal of BOD from waste. Efficient and successful biological oxidation requires a minimal quantity of nitrogen and phosphorus. A BOD: N: P ratio of 100:5:1 in the waste usually insures adequate nutrition. The tannery effluent is very rich in nitrogen and sometimes poor in phosphorus.

Sludge Volume Index (SVI) is the volume occupied by 1 g of activated sludge after settling the aerated mixed liquor in a 1,000 mL graduated cylinder or Imhoff cone for 30 min. It provides good indication of sludge compacting characteristics and it is very helpful in determining return sludge pumping requirements.

$$SVI = \frac{\text{Settled sludge volume} \cdot 1000}{\text{Suspended solids}} \quad \left[\frac{\text{ml/l}}{\text{mg/l}}\right]$$

Well-settleable and mineralized sludges have SVI 100.

5.1 *Nitrification, Denitrification*

In order to satisfy the legal limits for nitrogen (ammonia and TKN), very often nitrification and denitrification stages need to be introduced into the biological system. Nitrification requires extensive aeration as well as low F:M ratio (<0,1) to facilitate conversion of nitrogen containing organic matter into nitrite (NO_2^-) and nitrate (NO_3^-) salts. The crucial role plays very specific *Nitrosomonas* and *Nitrobacter* bacteria.

Denitrification is the final step of the conversion where the nitrate is reduced to gaseous nitrogen. It occurs while oxygen levels are depleted and nitrate becomes the primary oxygen source for denitrifying (facultative heterotrophic) bacteria. For the conversion, they need sufficient organic carbon as food, and oxygen which is taken from the oxygen contained in nitrate because there is no other source (thus anoxic conditions must be maintained). In this way, nitrate is firstly reduced to nitrous oxide, and then into nitrogen gas which escapes into the atmosphere as gas bubbles during aeration in aerobic tanks. In India, the nitrification and denitrification aspects are not considered seriously especially in CETPs treating effluent from semi-finish to finishing process.

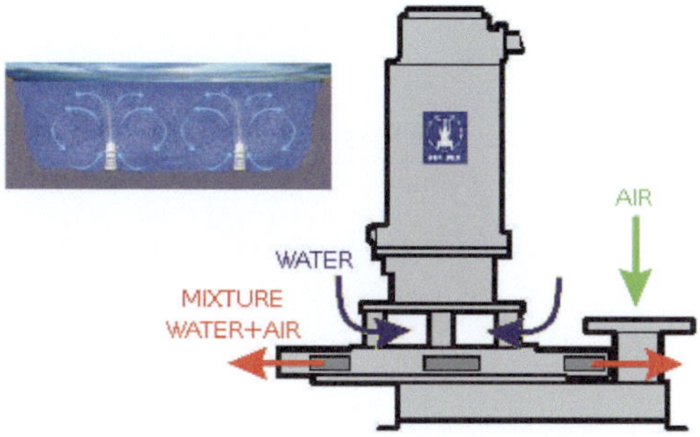

Fig. 21 Forced air, submersed turbine aerator

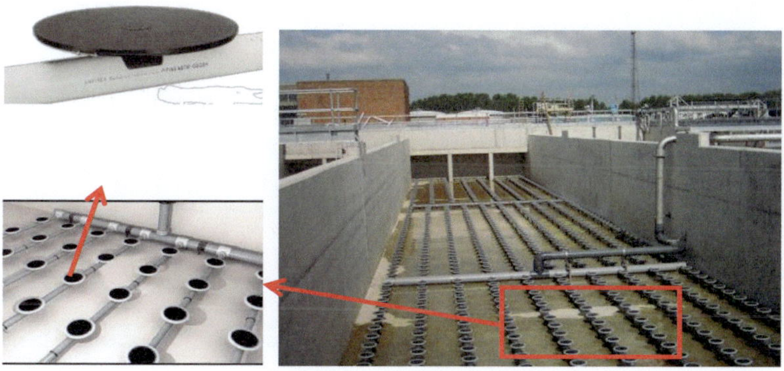

Fig. 22 Bottom air diffuser grid with one section and fine bubble dome diffuser

5.2 Aeration Devices

Water (effluent) aeration is an important business with wide range of models, designs. In addition to cost, reliability etc. the key criterion is the amount of air (oxygen) transfer per KW installed. As mentioned earlier, despite easy maintenance, due to lower transfer efficiency and problems with aerosols and malodour, surface aerators are more and more abandoned in favour submerged turbines, various Venturi-type jets and fine bubbles bottom diffusers. Brush rotors are used in oxidation ditches. A few selected aeration systems are shown (Figs. 21 and 22).

Reportedly, new and more efficient aeration devices such as jet aspirators fixed with mini compressors are being developed and introduced.

5.3 Aeration Basins, Oxidation Ditch

Possibly the best system for biological treatment of tannery effluents is the oxidation ditch, a racetrack-shaped aeration basin, with rotary brush or vertical rotors

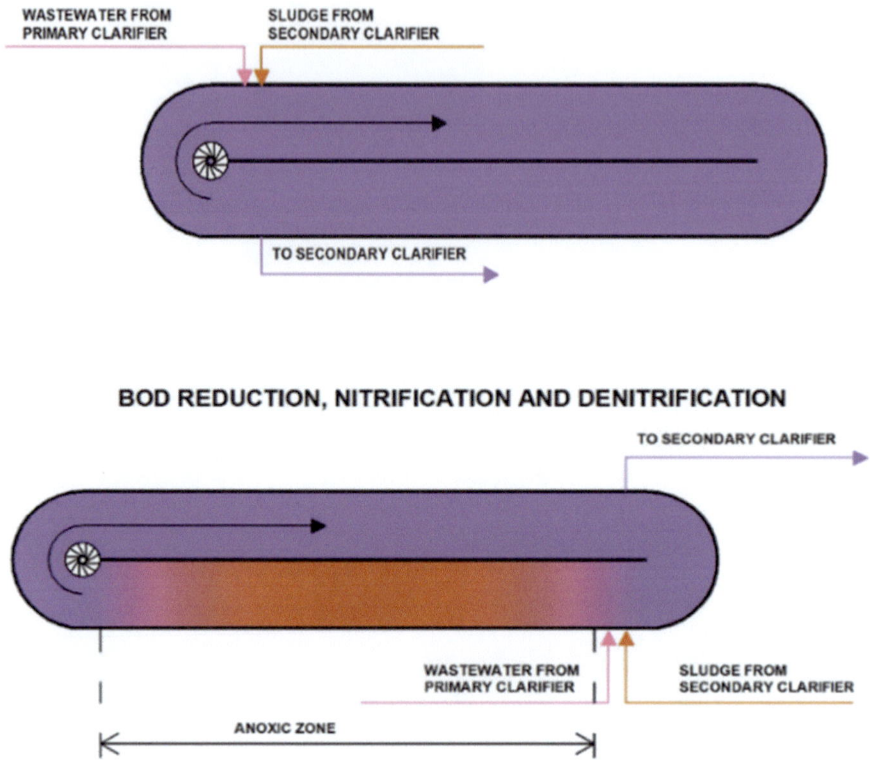

Fig. 23 Schematic diagram of oxidation ditches with BOD removal, nitrification and denitrification

(carrousel) aerators that extend across the width of the ditch as shown in the following Fig. 23.

In addition to simple construction and easy maintenance, the main advantage of the OD is its resilience to variations in flow, pollution load, including shock loads. It is even possible to combine several ovals and within them maintain different aeration regimes suitable for nitrification and denitrification (Figs. 24 and 25).

Owing to the lack of space, many aeration tanks (for example in CETPs in India) are rectangular, with extra depth and appropriate equipment to ensure proper aeration.

5.4 *Secondary Sedimentators*

Their design is very similar to those of primary sedimentators, but the operational conditions differ. In addition, the excess—wastage sludge evacuated at the tank bottom is normally bulkier and more difficult to dewater. Important design/performance parameters are:

Fig. 24 Submersed turbine aerators in the oxidation ditch

Fig. 25 View of anoxidation ditch in operation

Surface Hydraulic Loading or Surface Overflow Rate (SOR): vertical velocity of the influent in the secondary sedimentation tank (m^3/m^2 of tank surface per hour or m/h). SOR of approx. 0.5 m^3/m^2 per hour is usually adopted for secondary sedimentation of tannery effluents—i.e. less than for primary clarifiers.

Surface Solid Rate (SSR): quantity per hour of MLSS (kg) crossing the surface area of the secondary sedimentation tank (kg/m^2 of tank surface per hour) (see primary sedimentation). SSR values between 2.0 and 3.0 kg/m^2 per hour are usually adopted for secondary sedimentation of tannery effluents.

The overflow from the secondary clarifiers represents the fully treated effluent usually fit for discharge into municipal sewer or even the final water recipient—a large lake or a river with adequate flow.

5.5 *Membrane Bioreactors (MBR)*

MBR biological treatment is essentially a version of the conventional activated sludge system with membrane filtering instead of secondary clarifier for solid/liquid separation. Membranes are semi-permeable filters attached to a support matrix, which have a defined pore size. Membranes are operated in a cross-flow modus by pumping a feed solution through the membrane module under pressure and at high velocity. The result is water (permeate) of consistent quality and free of suspended solids. Different configurations of perm-selective membranes can be integrated in the system to achieve micro-, ultra or nanofiltration. The process flow diagrams of conventional and membrane bio-reactor systems are shown Fig. 26.

This very compact and auto-controlled system can be adopted in a modular way and it produces effluent of high quality. The disadvantages are rather high investment, necessity of nitrification and denitrification, decrease in overall operational capacity of treatment plant due to the fast-declining performance of the membranes and high operational costs depending in particular on membranes working life and energy consumption.

It is argued that over the last decade membranes for effluent treatment have improved and investment as well as operational costs have decreased, making the system a potentially attractive alternative to other biological treatments. The purification efficiency at various stages of treatment is given in the following Table 9.

5.6 *Anaerobic Biological Treatment*

As mentioned earlier, despite its higher performance for effluent with high volatile organic content, comparatively lower costs and wide use in some other industries, anaerobic treatment has failed to make any inroad in the tanning industry. The main reason is that the tannery effluent contains high concentration of sulphate which

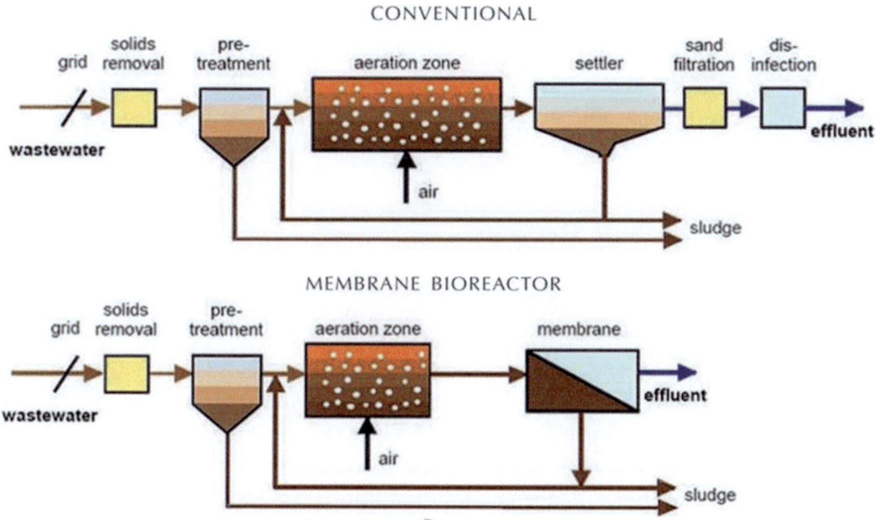

Fig. 26 Simplified schemes of the conventional versus MBR process

Table 9 Purification efficiency of the treatment stages referred to raw effluent

Parameter	Physical–chemical		Biological	
	Reduction %	Approx. value mg/l	Reduction * %	Approx. value mg/l
Suspended solids (SS)	80–90	300–600	95–98	20–50
Biological oxygen demand (BOD_5)	50–65	750–1500	97–99	20–30
Chemical Oxygen Demand (COD)	50–65	1500–3000	87–93	250–400
Total Kjeldahl nitrogen (TKN)	40–50	250–300	60–70	100–200
Chrome	92–97	5–10	99–99.5	<1
Sulphide	80–90	10–20	99–99.5	<1
TDS, mainly chlorides and sulphates	No reduction!!			

* Approximately at the load level of 0.30 kg BOD_5/kg MLSS, oxygen requirement of 2.0 kg O_2/kg BOD_5 and MLSS 3300 g/m^3

during anaerobic digestion develops highly toxic, corrosive and flammable hydrogen sulphide gas (H_2S) (Anaerobic sludge digestion, WPCF - 1992). This aspect as well as limitations during winter period and other O&M difficulties prevail over other, positive features of anaerobic treatment.

5.7 Other Treatment Methods

The overall performance and/or specific requirements have limited the use of other treatment methods to some particular situations.

Lagooning may appear a low-cost technology but large land areas far from residential zones are needed and there is a serious problem of sludge accumulation and removal. Trickling filters might be suitable for a smaller unit in a congested area whereas constructed wetlands (reed beds) are excellent for the final polishing before discharge into the ultimate water recipient but not for full-scale treatment of larger volumes. Depending on the land configuration, cascades can serve a similar purpose.

6 Advanced (Tertiary) Treatment

Tertiary and/or advanced wastewater treatment is employed to reduce the residual COD load and/or when specific wastewater constituents are not removed by the previous treatment stages. Drastic reduction of TDS (colloquially salinity) by Reverse Osmosis (RO) supplemented by Multi Effect Evaporation (MEE) also falls into this category.

6.1 Fenton Process

The Fenton process is an established and very effective oxidation process widely adopted in industrialized countries, which reduces residual pollution by chemically oxidizing recalcitrant COD, i.e. the matter that remains after the biological treatment because the microorganisms did not find it digestible.

The Fenton treatment is based on the formation of hydroxyl radicals that have a strong oxidation capacity and can fully oxidize refractory organic pollutants. These powerful free radicals are generated by the hydrogen peroxide (H_2O_2) in the presence of suitable catalyst, usually the ferrous sulphate ($FeSO_4$). The stages of Fenton process are given (Fig. 27).

Though the Fenton process is an efficient tertiary treatment in reducing non-degradable COD, it requires close monitoring with skilled manpower. The generation of additional sludge and high O&M cost are other limitations.

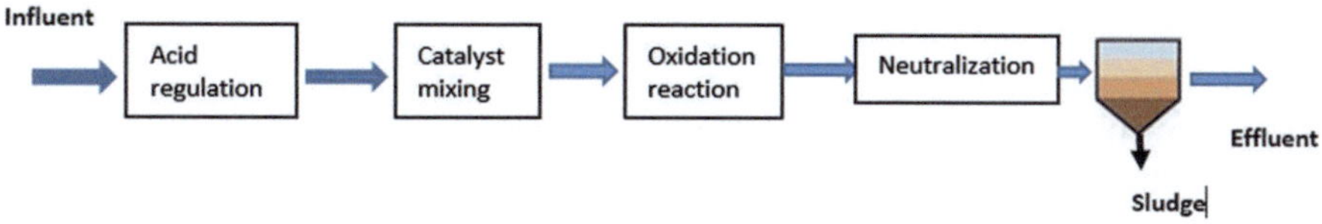

Fig. 27 Stages of the Fenton process

6.2 Ozone Treatment

Advanced oxidation treatment using ozone is being adopted in some of the CETPs to reduce the non-degradable COD and for colour removal. This process has high capital cost and energy requirements.

7 Sludge Disposal

It is important to realize that in addition to purified effluent, ETPs and CETPs also produce large volumes of sludge that has to be dealt with: ideally reutilized or disposed of in safe and environmentally acceptable manner (Richard, 1991).

In comparison with sludges from domestic wastewater treatment plants, the sludge from tannery effluent treatment plants has higher inorganic matter and higher heavy metal content, and, especially, higher chromium and sulphur compound content (Feikes, 1990; Reich, 2000). However, the main stumbling block is the chromium content, legislation and practice in various countries differing a lot.

In any case, handling, dewatering, storage and transport of sludge and solid wastes from ETPs & CETPs should also be safe and not contaminate the surrounding; thus, for example, the collection points should be protected against weather (e.g. rain).

Untanned and tanned wastes generated by the tanning industry are not classified as hazardous waste in industrialized or developing countries. In future, European legislation will prohibit the disposal in landfills of solid waste containing more than 10% organic matter. Currently, the permissible limit in India is 20% organic matter for the disposal of dewatered sludge in secured landfills (SLF). This limit is likely to be reduced in the future.

Chromium content in tannery sludge is the main constraint on its use or safe disposal. Given the high toxicity of hexavalent chromium, many environmental standards also limit total chromium. This limitation arises for want of insufficient knowledge of the difference in properties between trivalent and hexavalent chromium or for fear that trivalent chromium may be transformed into hexavalent chromium during waste disposal/re-use or chlorination of drinking water.

Landfilling in ordinary or special landfills remains the prevailing disposal method. Regrettably, in many developing countries, properly designed, constructed and/or maintained landfills are not available; tannery sludge is just dumped without proper control (Post et al., 2001).

Other, more advanced sludge handling and disposal methods are chemical conditioning with quick lime, thus increasing the sludge dry solids content and improving its appearance and smell, composting (usually aerobic, in either open or closed systems, the compost applied in cultivation of inedible plants) (Epstein, 1997) and thermal treatment (drying, vitrification, pyrolysis–gasification etc.) resulting in lightweight aggregates/oil. The decentralized secured landfill system has been developed and adopted in India and other Asian countries. The typical cross-section of SLF systems, bottom layers, laying of HDPE sheet, side walls, etc. are shown in the following Fig. 28.

It is important to have a leachate collection arrangement, the leachate is to be sent for treatment or, where applicable, to be sprinkled over the top of the landfill.

As per the regulations of the Government of India, sludge that contains less than 5,000 mg/kg of total chromium and less than 50 mg/kg of hexavalent chromium is not considered hazardous. However, all category of tannery sludge is expected to be disposed in the Secured Land Fill (SLF) irrespective of the concentration of chromium in the dewatered sludge.

Sludge from tannery effluent treatment plants is not treated as hazardous material in industrialized countries. However, the presence of chromium beyond prescribed levels limits its alternative uses and hence landfill is the preferred mode of disposal. With stricter regulations governing the disposal of sludge with an organic content of more than 10% or with a potential for value recovery, including the recovery of energy, sludge disposal costs in industrialized countries are expected to rise further. Regulations in this regard in developing countries, although strict (many countries treat sludge as a hazardous substance), are rarely enforced. In India, CETPs generally dispose of sludge on their own premises. Alternative utilization options attempted in India have not yet resulted in a popular mode of disposal. Sludge disposal costs in France and Italy are therefore significantly higher than in India.

8 Common Effluent Treatment Plants (CETPs)

A large number of traditional tanning units with inadequate investment potential and technical skills, including a lack of space typical for developing countries, were conducive for setting up CETPs servicing entire tannery clusters. This concept was then followed in relocations and/or establishing new leather centres/clusters; for practical reasons (only gradual increase in volumes of effluents, seasonal fluctuations), new CETPs have been designed in modular way.

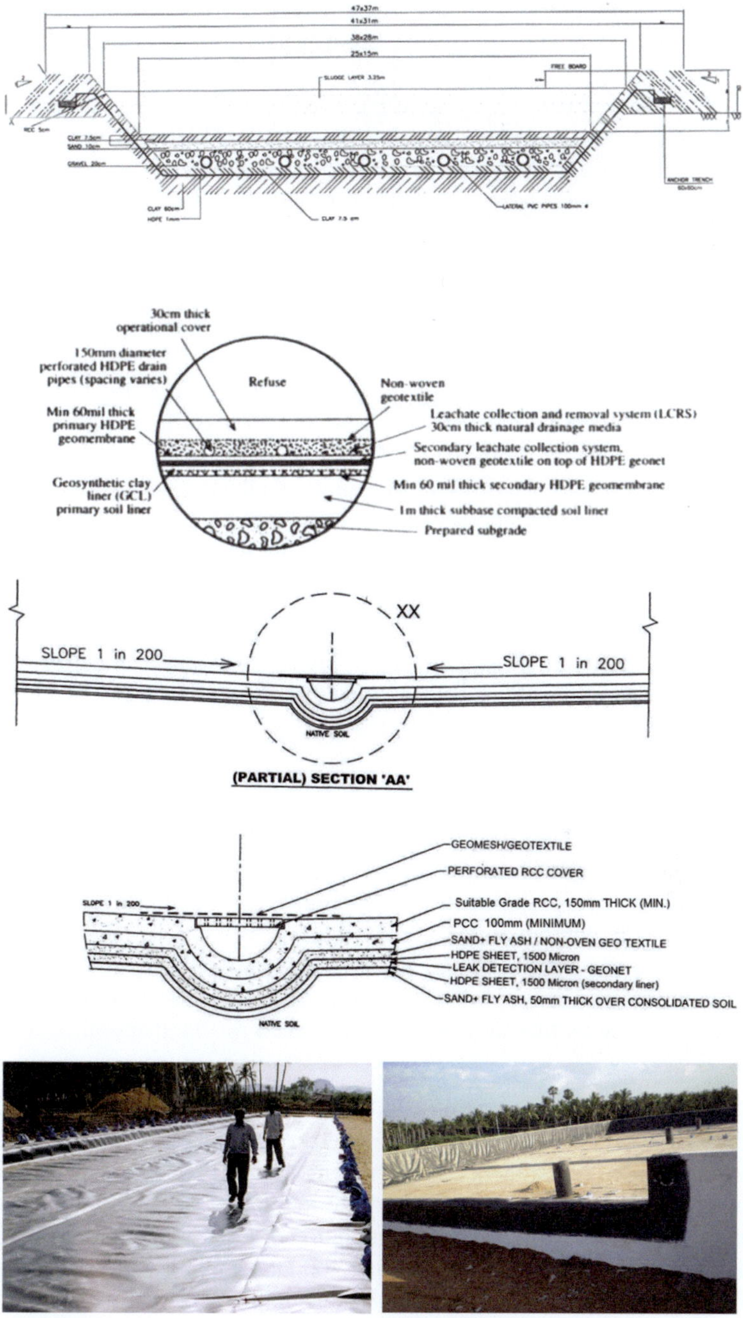

Fig. 28 Landfill: cross-section, bottom layers and laying of the HDPE sheet

The CETPs in Asia have been developed in China, India, Pakistan, Bangladesh, etc., from 1995 onwards following Italy, Turkey and other European countries (Rajamani, 2020).

Estimating effluent volume is one of the most challenging tasks in planning and designing CETPs mainly due to large oscillations in availability of raw hides and skins and market demand. Seasonal peaks associated with religious festivals are also very pronounced in developing countries. Modular approach to some extent helps coping with this challenge.

An important aspect of CETPs operations is the effluent collection network, very often a combination of pressure and gravity line. Scale forming in gravity lines, silting and generation of toxic fumes (H_2S), especially in deep manholes, are the major issues in operation and maintenance of effluent collection and conveyance pipelines. Many countries, including India, started enforcing the TDS parameter in the discharge standards. The safe disposal/utilization of the hazardous category, chrome-containing sludge from the CETPs and non-degradable solid wastes from tanneries such as chrome shavings are the major challenges yet to be fully resolved in many Asian countries (Tanneries and the environment, UNEP-UNIDO - 1991).

The salient features of the CETPs in some leather-producing countries are given below:

Bangladesh: The main tannery cluster in Bangladesh was located in Dhaka city. To overcome the environmental challenges in the city, the tanneries numbering more than 300 units had been relocated in a newly developed industrial estate with Common Effluent Treatment Plant (CETP) of 30,000 m^3/day capacity.

China: Currently, there are about 800 tanneries. As a sustainability measure new licenses are given to tanneries with a processing capacity of more than 3000 tons/year of raw hides and skins. There are about 13 CETPs in operation, some more are under planning; expected to reduce the volume of water usage and pollution load at source through cleaner production programme. The tanneries are permitted to expand the capacity without increase in the water usage. One of the major tanneries has implemented the MBR and RO system for water recovery and reuse. While apparently, there is no specific restriction on the Total Dissolved Solids (TDS) or salinity norms it seems that they will be introduced for the disposal of the treated effluent.

India: There are about 20 operating CETPs mainly located in Tamilnadu, Kolkata, Uttar Pradesh and Punjab catering to the needs of about 1500 tanneries mostly in small scale sector with treatment capacities ranging from 1000 m^3 to 20000 m^3/day. The Government of India provides funding to the tune of about 70% on the capital investment for the establishment and upgradation of CETPs in leather clusters.

A major leather complex with more than 200 tanneries for a processing capacity of 600 tons of raw hides and skins per day is being implemented in Kanpur city (Rajamani, 2020). The CETP with 20,000 m^3/day capacity is being implemented in two phases, the TDS management is done by segregation and separate treatment of saline stream such as soak and chrome bath. The quality salt is recovered from saline stream and chromium recovered in the form of cake or powder. The other sectional discharges with low TDS are collected and treated separately. The treated effluent is further mixed with treated domestic sewage received from the nearby Sewage Treatment Plants (STPs). Zero Liquid Discharge concepts by adopting membrane system for recovery of water from tannery effluent have been implemented in the South Indian tanneries with huge investment of more than 100 million USD.

Disposal of the saline reject stream from Reverse Osmosis (RO), mixed salt generated from Multiple Effect Evaporator (MEE) in absence of scope for disposal into sea or mixing with treated domestic sewage are major technical/environmental challenges. Decentralized Secured Landfill (SLF) system linked with CETPs for leather sector had been implemented in many tannery clusters. (First of its kind in the world). Applied R&D activities are under progress in the area of bio-processing of bio-sludge mixed with fleshing and other degradable wastes, advance oxidation using ozone treatment for the removal of residual non-degradable COD and colour in the treated effluent for disposal into water bodies.

Italy: The second major (about 15%) leather producer in the world following China. There are about eight CETPs catering to the needs of tannery clusters. The sludge disposal represents a major environmental challenge. The CETPs are being upgraded with total aerobic biological oxidation system without use of chemicals for reduction of COD and thus contributing to sludge generation. Thermal treatment of sludge, energy generation from volatile organic matter and overall sludge management are followed. Central chrome recovery and reuse system are being adopted in many tannery clusters in Italy.

Spain: The tannery clusters with CETP are located in Igualada near Barcelona and Lorca near Murcia a coastal town in the southern part of Spain. The CETP in Igualada with a capacity of 9,000 m^3/day has been established with a capital cost of 13 million EUR. Membrane bioreactor with Reverse Osmosis (RO) for water recovery had been established in a CETP near Lorca. The water recovery system from a tannery CETP is first of its kind in the world and was commissioned during 2004–2005. The CETP with ZLD system had faced technical and economical issues in saline water evaporation system in the landlocked area and become unsustainable.

Turkey: There are about 540 tanneries existing in 14 zones. Eight CETPs have been established and are in operation. One of the biggest CETP with a capacity of 36,000 m^3/day had been established at Tuzla with provision for disposal of treated effluent into sea for TDS management. The other major tannery cluster is in Izmir with an integrated CETP. The small-scale tanneries had been resettled in industrial zones. R&D activities on cleaner production and environmental protection are being continued in universities. Sludge disposal is a major problem similar to other countries.

Pakistan: The tanneries in Pakistan are mainly located in Karachi, Lahore, Kasur and Sialkot. A major CETP integrated with domestic wastewater treatment has been established in Karachi for a total capacity of 42,000 m^3/day out of which about

16,000 m^3/day is tannery wastewater and about 26,000 m^3/day of domestic wastewater. The other CETP in tannery cluster is located in Kasur. The CETPs need further upgradation and technological developments to achieve environmental requirements.

9 Reverse Osmosis (RO), Zero Liquid Discharge (ZLD) Concept

This concept has been largely implemented in several traditional tannery clusters in the very arid areas in the state of Tamil Nadu, India where dramatic situation with water and soil pollution along the Palar River together with public and buyers' pressure eventually prompted the state environmental authorities to enforce the discharge limit for Dissolved Solids (inorganic) of 2100 mg/L.

The fully-fledged physical–chemical and biological treatment is followed by tertiary treatment units such as Dual Media Filters, Microfilters, Ultrafiltration (UF), two or three stage Reverse Osmosis (RO) and evaporation of RO rejects of volume using MultiEffect Evaporator (MEE) and Agitated Thin Film Dryer (ATFD) (Emmanuel, 2001 & Rajamani, 2019).

Essentially, the ZLD systems concentrate dissolved solids present in the fully treated effluent by RO and MEE/ATFD until only damp residual remains. Nearly all purified water, the permeate (about 80%) and condensate (close to 20%), is reclaimed and supposedly reused. Reportedly, there are some difficulties in the reuse of condensate due to malodour caused by the presence of ammonia.

The ZLD is actually not so much treatment but rather a salt sequestration and removal system. Due to its complexity, it requires close and continuous monitoring of many parameters. The schematic diagram on the principle of RO system is shown (Fig. 29).

Preparatory, post-CETP "conditioning" steps, in particular water softening, often require dosing of different chemicals, including salts, which is quite a paradox for what is essentially a salt removal system. The Ultrafiltration (UF) system prior to RO system installed in one of the tannery CETP in India is shown (Fig. 30).

During initial years of operations, the estimated yearly average water recovery (permeate) rate for three CETPs with RO + MEE system was about 72–76%.

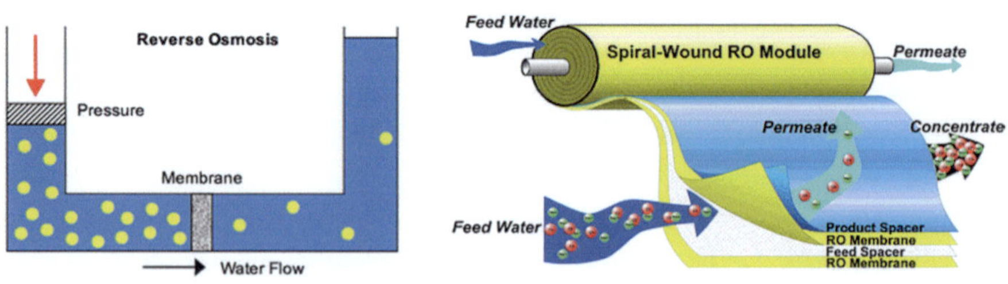

Fig. 29 Reverse osmosis (RO): principle, schematic diagram

Fig. 30 Ultrafiltration (UF), an important step in preparation for RO

However, the consumption and the cost of energy exploded. The salt residue produced and by now accumulated in terms of thousands of tons poses a very serious environmental challenge. Since it is a mixture of mainly chlorides and sulphates and there is no viable technology for their separation, the challenge is to find a way for its utilization and/or safe disposal.

There is no doubt that industrial scale ZLD in treatment of tannery effluents is technically feasible. However, due to very high capital and operation costs (in particular power consumption), including the problem of storage and disposal of solid residue its wider acceptance is quite unlikely (Buljan et al., 2017). A possible viable option might be separate evaporation of only highly concentrated saline streams (soak, pickling) preferably without RO because that requires another, smaller volume but parallel full-scale treatment (Sampathkumar & Viswanathan, 2001).

Potential alternatives for survival of the local tanning industry include a combination of short- and long-term options such as construction of proper sewage systems in large townships facilitating mixing and joint treatment of tannery and municipal wastewater, slaughter of some livestock (buffaloes, goats/sheep) and salt-free preservation, concentration of wet blueing etc.

10 Occupational Safety and Health (OSH) at (C)ETPS

10.1 General

The management should specify and provide appropriate Personal Protective Equipment (PPE) such as boots, gloves, goggles, face shields etc. for tasks where it is required. Similarly, safety (warning, mandatory) signs need to be explained and suitably displayed.

The entire effluent treatment area tends to be slippery, both metal and concrete surfaces. Employees working over or near water tanks where there is a risk of

Fig. 31 A ring buoy conveniently positioned at the treatment plant

drowning should be provided with approved life jackets or buoyant work vest; ring buoys with lines should also be readily available (Hannak et al., 2001) as shown in the following Fig. 31.

10.2 Risks Associated with the Hydrogen Sulphide (H_2S), Malodour Problems

Odours associated with wastewater are difficult to quantify because they are caused by a wide variety of compounds, it is a nuisance that is more qualitative than quantitative: sensitive persons easily detect very low concentration of odoriferous substances in the air (sulphides/other sulphur compounds, ammonia, amines, etc.). Local geographic and climatic conditions such as wind direction, land shape, air humidity, ground and air temperature, etc. Along the treatment line, the main sources of bad smell are:

- Oxidation of sulphide in the beamhouse streams
- Equalization, homogenization
- Sludge thickening
- Biological aeration
- In-plant storage of the dewatered sludge
- (Temporary) sludge disposal site.

Yet, the main source of bad smell remains stripping of hydrogen sulphide; however, it is not the concentration of sulphide per se but lowering of pH: the not disassociated H_2S is present only at pH below 10. Thus, it is crucial to control pH and in case of need alkalis like NaOH or lime to make pH > 9.5–10 are added. More extensive, uninterrupted aeration may help but sometimes rigorous (and expensive) methods like adding hydrogen peroxide or pure oxygen are necessary. Nowadays, in some places, nearly the entire ETP is covered and the air purified.

Despite the fact that outdoor the risk of H_2S is lower, the area should be monitored using fixed, portable and personal detectors, and all concerned workers should be well equipped and trained to deal with emergency situations (Hannak & Jayaraj, 2015).

The tanning industry is not a major air polluter. The main air pollutants are hydrogen sulphide and ammonia emanating from liming, deliming and pickling operations and waste treatment operations as well as volatile organic compounds (VOC) used in the finishing department. This holds especially true for tanneries with poor housekeeping practices, including improper handling of solid waste. Incineration of tanned leather wastes is a health hazard whereas the new sludge treatment technologies (such as thermal drying and pyrolysis) may produce exhaust gases (fumes) that pose a threat to the environment and workforce alike. Standards governing exhaust gases have been set and are relatively strictly enforced in industrialized countries.

10.3 Noise

The main source of noise are air blowers supplying compressed air to aeration systems; they need to be appropriately positioned and dampened. Pumps, screening devices and sludge dewatering equipment can also be rather noisy.

11 Monitoring, Treatment Costs

The extent of monitoring activities largely depends on local legal requirements and (C)ETP size and it includes a wide range of parameters: effluent volume(s), peak loads, pollution loads before and after treatment, chemicals dosing, dissolved oxygen (DO), sludge volume index (SVI), DS content in sludge etc. as one aspect of the overall company's Environmental Management System (EMS). The costs of treatment vary a lot, primarily depending on the local costs of power, chemicals, sludge utilization and disposal options, finances and labour (Buljan, 2005) as shown in the following Fig. 32.

The highest component: power; sludge handling and disposal costs comparatively low. However, if the reverse osmosis (RO) and evaporation (MEE) stages are installed, the cost of energy escalates even further.

In industrialized countries, the largest component, often around 40% of the total costs, is dewatering and safe disposal of solids (sludge).

While working out the costs of an ETP in an individual tannery is quite straightforward, different methods are applied to compute the distribution of the CETP operation costs among individual tanneries. The key parameters used are typically the following:

- Water consumption, m^3/day

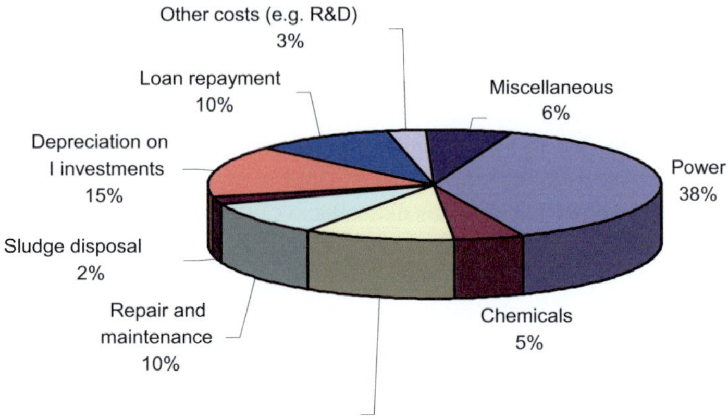

Fig. 32 Typical structure of treatment costs in selected CETPs in India (before ZLD)

- Production capacity, tonnes of wet salted hides/day
- Actual pollution load in terms of COD (even better COD removed), suspended solids, TDS, sulphide and chromium
- Production output, m^2/day of wet blue, crust or finished leather.

Pollution load-based distribution of costs requires appropriate on-/off-line monitoring.

A typical flowchart of a fully-fledged tannery effluent treatment plant is shown in the following Fig. 33.

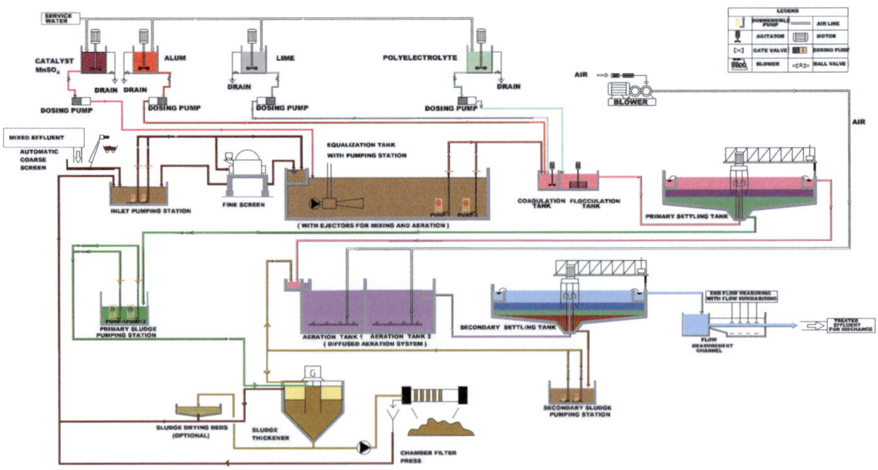

Fig. 33 Simplified flowchart of the fully-fledged tannery effluent treatment plant

12 Challenges and Opportunities

The following are the main immediate and long-term challenges to the tanning industry.

12.1 Science and Technology

- Avoidance of salt preservation, reduction in salt usage in the process.
- Replacement of currently used non-degradable/and hazardous chemicals with more environment-friendly chemicals.
- Significant improvement of efficiency and exhaustion of chemicals used in the process.
- Utilization of non-collagen, i.e. non-leather-making substances and solid waste in general.
- Recycling of residual chemicals, leather and leather products.

12.2 Global Pressures and Public Perception of Leather Under Influence of Misinformed Groups and Lobbies

There is a lot of incorrect and obsolete information in public media linking the leather industry with cruelty toward animals, deforestation (in particular rain forests) and even to greenhouse gases (GHG), especially methane generated by livestock. A few remaining ugly spots in developing countries are presented as representative of the whole industry; some of them in North Africa, in terms of production are irrelevant and actually mainly kept as tourist attractions.

The fact that tanners all over the world use only "instant" powders containing harmless trivalent chromium (Cr^{3+}) salts is disregarded and chrome tanned leather and leather goods made of it are drummed as being carcinogenic due to the presence of hexavalent (Cr^{6+}) and, in the best case, only chrome-free leathers should be used. Paradoxically, even fossil-based synthetic materials are sometime preferred over leather.

Admittedly, it (unnecessarily) took some time for tanners to realize and to accept that under some circumstances (long storage, high temperatures, use of low quality fatliquors etc.*) small amounts of* Cr^{3+} *can be oxidised to* Cr^{6+} *form. Again, the real risk from the* Cr^{6+} *is regularly grossly exaggerated. Furthermore, nowadays this can be easily prevented by using appropriate retanning agents (*e.g. *tara vegetable tannin) and appropriate fatliquors.*

Actually, it is astonishing that it is too often overlooked that tanners convert a waste product from the meat industry into a highly versatile, sophisticated and yet beautiful material. The cell cultivated leather appears to be still pretty far from replicating the fascinating natural structure.

Due to their (in old days) noxious nature, tanning activities and tanners in virtually all cultures all over the globe were assigned to a very low social status, very often entirely outside "normal" society. In the worst-case scenario, such deeply rooted biases compounded by very aggressive campaigns based on false claims and agendas and possibly excessive legislation might jeopardize the future of the tanning industry as we know it (Král et al., 2014).

12.3 Total Dissolved Solids (TDS) and Salinity

Crystal clear, odourless water resulting from a fully-fledged conventional treatment is in arid regions not only useless but actually potentially harmful due to TDS content (colloquially salinity). This is compounded by the fact that currently only a few developing countries have large meat processing plants offering the possibility of skipping the preservation step (salting) and almost immediately proceeding with tanning up to wet blue or even fully finished stage (Buljan, 2004). A new concept of segregation, separate treatment of soak, spent chrome baths and composite streams is shown in the following process flow diagram (Fig. 34).

Attempts with mechanical desalting of hides and skins as well as solar evaporation pans (including accelerated evaporation trials) proved to be of little impact.

More recently, a system where treated general effluent joins municipal wastewater, spent chrome floats are collected from the entire cluster and chrome recovered at the central plant whereas Reverse Osmosis (RO) and Multi Effect Evaporation (MEE) are limited to the soak stream has been proposed.

Traditionally, tanning activities have been associated with the proximity of water. Nowadays, due to TDS issue, the proximity of large urban agglomeration with proper sewage and wastewater treatment plants is additional precondition for survival of tanneries in arid regions. Furthermore, existence of large abattoirs might eventually increase the scope for processing fresh hides and skins. At the same time, relocation of tanning capacities, even if limited to wet processing stage, has its serious social consequences.

12.4 Safe Disposal of (Chrome-Containing) Sludges

A number of solution for utilization and/or safe disposal have been proposed, practised, tested, applied at pilot and industrial scale: landfill, land application, land reclamation, composting, anaerobic digestion, thermal treatment, vitrification, pyrolysis, brick making etc., none of them quite satisfactory. There is certainly no universal

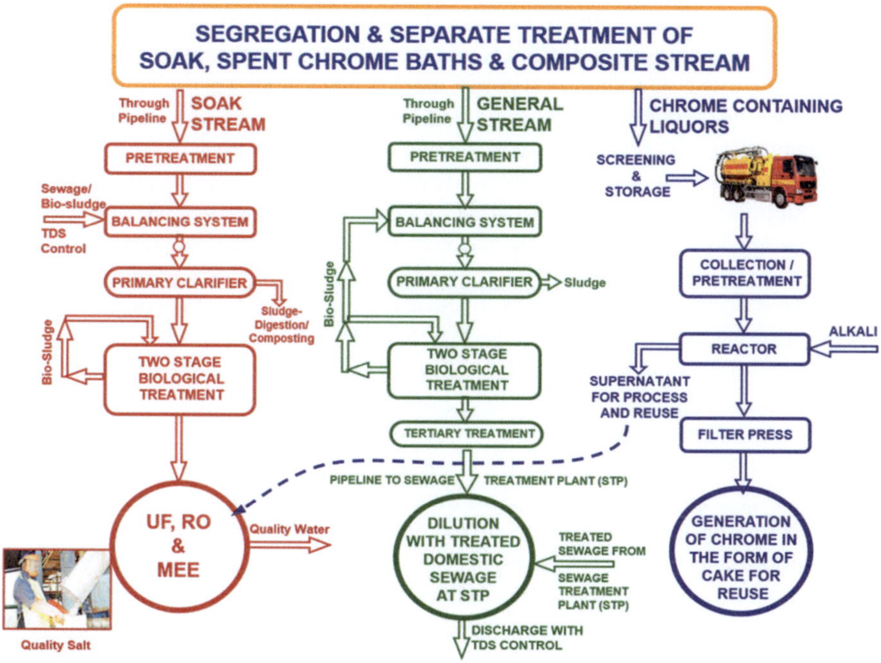

Fig. 34 One proposal for segregation and separate treatment of soak, spent chrome baths and composite streams. * Note purification and treatment of the supernatant from the central chrome recovery

solution for sludge utilization/application. Each (C)ETP produces sludge of specific characteristics and various regions and countries have quite different regulation for sludge utilization. Therefore, prior to any (C)ETP construction detailed assessment of options should be prepared and the most suitable application proposed.

13 Conclusions

The cattle, sheep and goats are reared for their value in dairy products, wool and for their meat, never bred and raised for their skins.

It is very unlikely that the majority of the world population will turn vegetarian and there are no signs of dramatic increase of utilization of raw hides and skins for non-tanning purposes. Thus, the meat industry will continue providing the basic raw material to the leather industry that transforms the natural inter-woven structure into a variety of leather types suitable for very different uses: footwear and apparel manufacture, car industry, equestrian sports, etc.

The leather industry needs to continue its efforts in improving its performance, from water consumption and effluent treatment, exhaustion of chemicals, utilization of solid waste and sludges, energy management, adherence to high Occupational Safety and Health (OSH) standards to implementation of Corporate Social Responsibility (CSR) policies both at the national and global level.

Similarly, it is important to take a proactive role in battling negative publicity and spreading of misleading information by fake green lobbies and pressure groups. Better coordination among various national, regional and global leather organizations, including R&D institutions would make a stronger impact. Continued interaction with the leather industry clusters and dissemination of technology by the national and international institutions are essential.

The expanding leather products sector (footwear, leather goods, apparel), including significant local consumption, points towards competitive and successful overall leather industry in developing countries.

The scope for utilization, possibly even recycling of disposed leather products (apparel, footwear, leather goods, furniture and transport upholstery) might turn vital for the long-term perspective of the entire leather sector. A paradox here is that fashion changes contribute to higher demand but at also the volume of disposed leather items; finally, its longevity is another proof in favour of leather's sustainability.

References

Aloy, M., Folachier, A., & Vuillermet, B. (1976). Tannerie et pollution, Centre Technique du Cuir (CTC).

Anaerobic sludge digestion. (1992). Manual of Practice No. 16.. Taskforce on anaerobic digestion, Water Pollution Control Federation (WPCF).

Best Available Techniques. (2013). Reference Document, BREF, EU.

Bosnić, M., Buljan, J., & Daniels, R. P. (2016). *Pollutants in Tannery effluents (Sources, description, environmental impact).* UNIDO.

Buljan, J., & Rajamani, S. (1996). A system for recovery and reuse of chromium from spent tanning liquor using magnesium oxide and sulphuric acid. TECHPACK/UNIDO/RePO.

Buljan, J., Ludvik, J., & Reich, G. (1997). IULTCS Congress "Mass Balance in Leather Processing".

Buljan, J. (2004). Some considerations about the problem of salinity of tannery effluents. LGR, Germany.

Buljan, J. (2005). Costs of Tannery Waste Treatment, UNIDO Leather Industry Panel meeting.

Buljan, J. (2007). Benchmarking in the tanning industry, UNIDO Leather Industry Panel meeting Gramado.

Buljan, J. (2014). Some considerations about international technical cooperation in the tanning industry, John Arthur Wilson Memorial Lecture.

Buljan, J., Emmanuel, K. V., Viswanathan, M., Bosnić, M, & Král, I., (2017). Analysis of flow and energy aspects of zero liquid discharge (ZLD) technology in treatment of tannery effluents in Tamil Nadu, India, IULTCS Congress Chennai.

Buljan, J., & Král, I., (2019). The framework for sustainable leather manufacture (2nd ed.). UNIDO.

Carré, M. C., Vuillermet, A., & Vuillermet, B. (1983). Tannerie et environnement. Centre Technique du Cuir (CTC).

Money, C. (1999). Clean technology challenges. In *Science and technology for leather into the next millennium,* Tata McGraw Hill Publishing Company, pp. 284–294.

Eckenfelder, W, W., & Grau, P. (1992). *Activated sludge process (Design and control, theory and practice).* Technomic Publishing Company Inc.
Eckenfelder, W. W., & Musterman, J. L. (1995). *Activated sludge treatment of industrial wastewater.* Technomic Publishing Company Inc.
Epstein, E. (1997). *The science of composting*, Technomic Publishing Company Inc.
Feikes, L. (1990). Ökologische Probleme der Lederindustrie, Bibliothek des Leders, Band 8, 2. Auflage.
Hannak, J., Buljan, J., & Sahasranaman, A. (2001). Occupational safety and health aspects of leather manufacture (OSH Manual). CTC &UNIDO.
Hannak, J., & Jayaraj, G. (2015) *Safety Handbook—How to Deal with Hydrogen Sulphide Gas in Tanneries & Effluent Treatment Plants.*
Knoedler, J. (2014). Management of water-use in tanneries. Retrieved from http://www.nothinghide.org/LeatherFacts/Water:reduce,reuse,recycle/Summary
Kràl, I., Schmèl, F., & Buljan, J. (2014). The future for leather. UNIDO. Retrieved from www.nothing-to-hide.org, www.leathernaturally.org, mike@internationalleathermaker.com; and series of columns/articles Leather naturally! and Nothing to Hide in International Leather Manufacturer (ILM), and World Leather (WL) respectively.
Leder, Fabrik., Melvisharam., Emmanuel, K, V. (2001). Operation of pilot Reverse Osmosis (RO) unit at ATH, UNIDO.
Mosiewicz, J., & Forestal. (1976). *The modern rational pit-drum tannage of vegetable sole leather.* International Limited.
Post, V., Swaminathan, R., Aloy, M., & Poncet, M. (2001). Manual on landfill for tannery sludge. UNIDO.
Rajamani, S. (2019). Innovative technologies on cleaner production and waste management in tanneries—Paper presented in 5th International leather engineering congress innovative aspects for leather industry.
Rajamani, S. (2022). New and innovative cleaner production process for sustainable development in leather process and environmental protection. In *10th Freiberg Leather Days in Chemnitz.*
Reich, G. (2000). Ökologische Aspekte wichtiger Gerbverfahren (Eine vergleichende Betrachtung der Chromgerbung und chromfreier Gerbverfahren).
Richard, M. (1991). Activated sludge microbiology. *The Water Pollution Control Federation (WPCF).*
Sampathkumar, S., & Viswanathan, M. (2001). Accelerated evaporation of saline streams in solar pans. UNIDO.
Spellman, F. R. (1996). *Stream ecology and self-purification (An introduction for wastewater and water specialists).* Technomic Publishing Company Inc.
Tanneries and the environment. (1991). A Technical Guide, UNEP-UNIDO.
Rajamani, S. (2020). Sustainable environmental management in Indian & Asian Leather Industries. In *Paper presented in XVI International Scientific-Practical Conference Leather and Fur in the XXI Century.*
Wastewater Treatment (Sludge Conditioning, Thickening and Dewatering) Volume A. (1985). Water pollution control federation (WPCF) www.leatherworkinggroup.com, LWG Environmental Audit Protocol, Issue 6.0.

Current Trends in Leather and Leather Products

Cost and Process Control in the Leather Product Supply Chain

Phil Shaw and Lee Lythgoe

1 Introduction

The global supply chains that serve the leather product markets are complex and rely on a large number of people, processes, infrastructures and systems to ensure that the finished product is available to the customer whenever and wherever they want them, and at an acceptable price. In this chapter, we focus on footwear, but the principles will apply to other leather products equally.

With up to 30 different materials needed to make some types of shoes, sourced from almost as many countries, and with some products manufactured half a world away from their final market, it is critical that the control of the events throughout the chain is carefully managed and maintained. Not only is this important to meet delivery and quality expectations but also it is critical to also ensure that the costs incurred throughout the entire process are within budget.

If we consider the supply chain as a series of discrete, yet interconnected, phases, then it is apparent that when each one works to the agreed plan, all aspects of the delivery will be intended. However, it takes only one of these phases to breakdown to affect the entire chain. Since few companies control every phase of the supply chain, any failure can significantly impact on the overall costs and delivery. As we have seen in recent times (highlighted below), a major problem can impact delivery in both the short and long term.

The Suez Canal plays an important role in the global economy by facilitating world shipping and trade. This waterway allows huge container ships to navigate a relatively direct route from Asia to Europe, rather than being forced to take a 3,500-mile diversion around Africa.

Around 12% of global trade passes through the Suez Canal, carrying goods worth well over USD 1 trillion (GBP 844 billion) each year. Approximately 50 ships a

P. Shaw (✉) · L. Lythgoe
SATRA Technology Centre, Northamptonshire NN16 8SD, UK
e-mail: bulletin@satra.com

S. K. Janardhanan and L. A. Zugno (eds.), *Emerging Trends in Leather Science and Technology*, https://doi.org/10.1007/978-981-99-9754-1_6

day, carrying cargo valued at USD3 to 9 billion (GBP 2.5–7.6 billion), will pass north or south through the canal between Suez Port and Port Said. According to the Suez Canal Authority (SCA), nearly 19,000 ships, with a net tonnage of 1.17 billion tonnes, used the canal in 2020.

The grounding of the 'Ever Given' container ship in 2021 amply demonstrated potential vulnerabilities. Therefore, any delay or problem in the supply chain is likely to have an immediate effect on the cost. Thus, the method of controlling each aspect of the cost build-up must be carefully and continuously monitored.

In order to understand the process of cost control, we should consider two parts: (i) establishing realistic and achievable cost standards and (ii) monitoring and maintaining actual costs.

There is no point in accepting product orders based on the price that the customer requires. The only valid cost is the figure that can be reliably and consistently achieved by the supplier and that will ensure an adequate profit.

2 Result and Discussion

2.1 *Designing to Control Costs*

At the initial design stage of a new product, there is the opportunity to consider the likely effect on costs. While it is not the primary responsibility of designers to consider the eventual costs, they can, working with development personnel, review the materials and constructions.

Typically, the cost of manufacturing footwear in the main shoemaking areas of the world will be less than 20% of the final price paid by the customer, so there is limited scope to significantly influence this final figure. However, every effort should be made to select materials, components and processes that are cost-effective. Ideally, using existing materials and components (which have an established cost record) will help, but the dictates of fashion can often require novel materials and processes. The development of each new style in preparation for manufacturing is also an opportunity to reduce any excess costs (for example, by minimising unused materials during cutting), as well as balancing the investment in labour-saving equipment against the cost of using manual labour. Materials inadvertently chosen that are not suitable for shoemaking will require additional processing, and thus will incur additional cost—such as textiles that fray when die cut.

2.2 *Manufacturing Costs*

Considerable work has been carried out by SATRA to determine the costs of materials and labour needed to manufacture the shoe, and most companies have their own

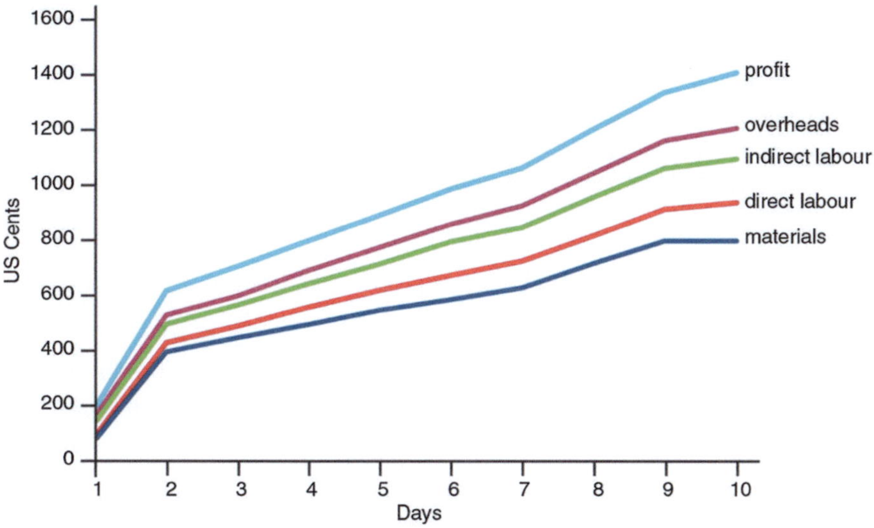

Fig. 1 An example of the build-up of manufacturing costs

methodology for adding the relevant indirect costs and overheads to these (see Fig. 1). It is worth investing considerable effort to control each category of cost, but often little is known about the actual overhead costs incurred by each style.

The existing premise is that the more complex the style (both in terms of materials and the processes needed), the greater the overhead costs that should be attributed to this complexity. Generally speaking, this is a valid point in that footwear that requires more operations, and thus takes longer to make, will utilise more space, energy and other resources. However, this may not necessarily be true for constructions that require specialised equipment or need precisely controlled process or storage conditions. From work carried out by SATRA, it is clear that the allocation of the different categories of cost will be spread unevenly across the various phases of the manufacturing phase of the supply chain. For instance, most of the cost of materials will be incurred during the actual assembly operations, as will the direct labour. In contrast, indirect costs and overheads can be the result of inspections, storage and delays, which may occur spasmodically and at unplanned times.

2.3 *Breakdowns, Problems and Delays*

Even the most carefully planned sequence of operations can be disrupted by unforeseen difficulties. Few companies can claim to achieve 100% efficiency of their expected, and therefore costed, performance, and any disruption will almost certainly add cost to the operation. While it is not possible to precisely calculate a contingency to allow for the unexpected, it is feasible to use sensible estimates and historical data to build some additional allowance into the product cost.

In terms of cost control, it is essential to have adequate levels of materials and components, as well as correctly maintained equipment to minimise any chance of delay. A planned schedule of servicing key equipment, together with trained personnel, can eliminate most types of breakdowns.

2.4 Process Control

Increasingly, the use of a wide variety of materials can require differing process control conditions to meet manufacturing specification, which will require constant checking to ensure that both of the required conditions are set and maintained throughout the production phase. Although this will incur an additional overhead cost, it will be more than balanced by the reduction in the cost of any incorrectly made products.

In simple terms, 'process control' requires that an appropriate degree of control is exercised over all steps in the manufacture of a product.

An ideal starting point is to list all steps or processes involved in the design, manufacture and dispatch of the product. Every process will receive an input (in the form of information or a product) from the previous process and will output that information or product to the next process in the sequence. Each process should be examined, and the following questions asked:

- has the required input to this process been made clear?
- how can it be ensured that the input is to the agreed quality standard?
- is the required output from this process understood?
- does the process provide the desired quality of output?
- how is the quality of the output to be judged?
- are all the necessary personnel trained, and do they understand what the process involves?
- is there a procedure that employees can follow?
- do employees know the expected level of quality?
- how is the completion of each process recorded (if required), so that the next process can begin?
- is any monitoring equipment regularly maintained and calibrated?
- are the necessary processing conditions known—such as temperature and time?
- is it necessary to inspect the output against clear quality standards—for instance, correct back heights or colour matching?
- do employees know the action to take in the event of faulty products, equipment or materials?

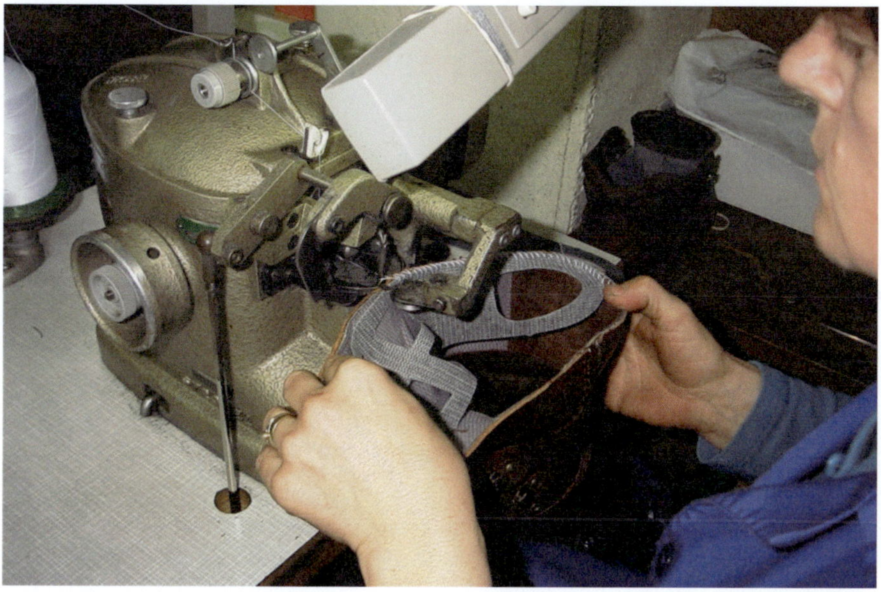

Fig. 2 With many different processes being carried out on each production line within a factory, control of each process is essential

2.5 Managing Procedures

Process control is about the management of procedures, people and equipment. It is, therefore, a basic requirement of manufacture. All manufacturers exercise process control in one way or another, but what distinguishes the successful company from the not-so-successful one—labour costs being equal—is the effectiveness of the process control used (Fig. 2).

2.6 Monitoring of Processes

All processes should be checked to ensure that they have been carried out correctly and meet the company's needs. Therefore, it is important to consider how best to monitor each one and to ensure that procedures make monitoring possible.

The procedures must generate evidence that action has been taken. In many cases, this will only require individuals to confirm that they have completed a specific task, such as signing against that operation on the work documentation. However, where measurements are made (for example, in manufacturing processes), these should be recorded to generate feedback and traceability, in case of quality problems. This objective evidence is then examined, often by carrying out an audit–the management tool that is used in order to assess quality.

2.7 Process Parameters

There are many shoemaking processes that require control. However, not all of the processes can be measured directly or easily. Consider the force exerted by a forepart lasting machine. In cases such as this, the alternative is to assess the result of the process, such as the width of the lasted margin. If the margin is too wide or too narrow, it suggests that something is wrong. It could be that too little force has been applied, or the patterns may be wrong for the materials being used. A rule can be used for such measurements.

In addition, for operations such as forepart lasting, the effectiveness of the process is dependent on the operative's skill. Humans cannot be programmed to perform a task precisely the same way every time. However, this does not prevent in-process inspection in place of measurement.

In the case of forepart lasting, the key areas would be:

1. Ensuring that the facings are straight
2. Keeping straight any vamp medallion or cap that is used
3. Checking that the lasting allowance is to the agreed measurement
4. Making sure that the upper material is 'down to the last'
5. Confirming that the leather is not cracked or has any dimpling (often called 'orange peel effect').

Process control may involve changing a procedure, based on the results of process monitoring. Once the process monitoring tools have detected an 'out of control situation', the person responsible makes a change to bring the process back into control. A typical programme might incorporate the following aspects.

'Out of control action plans' detail the action to be taken once an out-of-control situation is detected. A specific flowchart that leads the person responsible for taking remedial action through the corrective procedure may be provided for each unique process.

'Advanced process control loops' are automated changes to the process that are programmed to correct for the size of the out-of-control measurement.

If the process is out of control, the person responsible looks for the cause by following the out-of-control action plan associated with the control chart.

Control charts are used to routinely monitor quality. Depending on the number of process characteristics to be monitored (for example, bond strength or temperatures on heat setters), there are two basic types of control charts. The first, referred to as a 'univariate control chart', is a graphical display (chart) of one quality characteristic, such as critical process temperatures.

The second, referred to as a 'multivariate control chart', is a graphical display of a statistic that summarises or represents more than one quality characteristic (for instance, temperature and the length of time during which bonding pressure is exerted).

2.8 Indicators on Machinery

Many machines have indicator gauges for temperature, pressure and time. These gauges give useful information on the control of the process but must be properly maintained and checked for accuracy using independent, calibrated equipment. Machines are becoming more sophisticated and are fitted with comprehensive monitoring facilities—either as standard or optional extras.

In some cases, highly visible warning systems (such as audible alarms or flashing lights) are also built in, so that any malfunction is immediately apparent. However, it is important to remember that the monitoring systems only monitor, and it is the control system that implements actions. For example, in the case of back part moulding or box toe application, the temperatures indicated cannot easily be related to bond line temperatures without carrying out the additional in-process monitoring trials described.

2.9 Process Monitoring Tools

Accurate control of time and temperature plays an important part in footwear manufacture, and a variety of low-cost tools are available.

Raw materials used or products manufactured throughout each stage of the production process may also require testing within an in-house laboratory. In-house testing facilities enable manufacturers to conduct scientific analyses of raw materials and processes against product specifications.

Testing is typically conducted to agreed assessments such as SATRA test methods, which stipulate that a controlled atmosphere is required for testing. The temperature and humidity of an environment can significantly affect the results achieved when testing certain types of footwear materials, often resulting in the need for a dedicated testing area.

SATRA provides expert consultancy and Laboratory Accreditation services to global brand owners and manufacturers.

2.10 Time, Temperature and Light Levels

Throughput times and temperatures are key parameters for many shoemaking operations, and accurate measurement (within the specified parameters laid down by the quality control department) and control of both are essential. Fortunately, there are many measurement tools available, from low-cost, low-tech apparatus, through to far more sophisticated electronic equipment (Fig. 3).

A simple stopwatch is the preferred time measurement option, and this must not be underestimated as an important part of process control. All the time controllers

Fig. 3 Time, temperature and other process parameters used in production should be specified on the machine so that the operator can easily see if it is 'out of control'

on processing equipment should be checked, and a correction chart posted alongside the equipment. If no time controller is present, speed settings must be specified and posted on the machine.

With respect to temperature measurement (as briefly mentioned above) perhaps the most versatile equipment is a hand-held digital thermometer with a range of thermocouple probes to suit particular applications. Three types of probes are recommended, as follows:

A fine wire thermocouple for measuring air temperatures. This should be long enough to pass along the whole length of any heating or cooling tunnels. This probe can also be used to determine bond temperatures by inserting it between materials during lamination.

A spring-loaded flat probe that is used for taking measurements of material surface temperatures.

A needle probe for measuring the temperature of hot liquids, such as molten hot-melt adhesive.

Although the hand-held digital thermometer is one of the most versatile devices available, alternatives include heat-sensitive crayons, labels and paper strips. The crayons remain the best choice for measuring adhesive re-activation temperatures, and heat-sensitive labels can be helpful when attached to hot surfaces to give a continuous indication of operating temperature. A number of these change colour at specified temperatures, and some show a reversible colour change on cooling.

In addition, an infrared thermometer, sometimes referred to as a 'non-contact temperature gun' can be used for measuring surface temperatures. Normally equipped with a laser for accurate positioning, they can be extremely useful especially if the surface is too hot for a contact probe or contact may damage the surface, such as adhesive activation temperatures on box toes. However, care must be taken to ensure that the 'gun' is properly targeted so that it is not influenced by any other sources of infrared radiation from devices that are located nearby.

For operations such as the inspection of leather or finished goods, adequate lighting is essential, and there are published guidelines for each type of process. Electronic light meters are readily available to ensure the level of light is correct, of the right type (for instance, as produced by 'daylight' tubes) and pointing in the correct direction.

2.11 A Permanent Record

With heat setters, tunnel heaters, dryers and coolers now becoming key components of many assembly lines, an ideal monitoring alternative to a long-wire thermocouple and a stopwatch is a compact data logger. This is a robust data logging and recording device, which is small enough to be mounted in a last to pass through the tunnel, thus simulating a part-made shoe. The sensors record a temperature/time/humidity profile of the process (Fig. 4).

The data from the sensor can then be downloaded into a computer for examination in a readily understandable format. Specially written software also creates a tamper-proof record of the conditions measured.

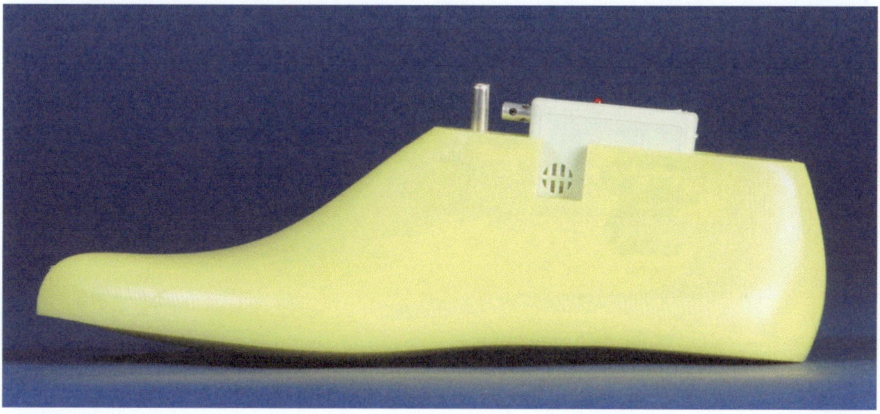

Fig. 4 A data logger (shown here fitted in a last) records a temperature/time/humidity profile of the process

2.12 Pressure Measurement

There are several ways to check the pressure applied in various shoe manufacturing processes.

Manufacturers should check the uniformity of pressure and ensure that adequate pressure has been applied. SATRA Foil is an embossed aluminium foil in which the embossing collapses under a known force—shown in use in Fig. 5. When this is sandwiched between two materials which are then pressed together, an indication is given of both the uniformity of the pressure applied, and whether or not the pressure was above a given threshold value. This procedure can be stated in the in-factory process control documentation that describes the frequency at which the process should be carried out.

Another product (SATRA pressure pack) can assess the uniformity of sole bonding pressure. In this instance, a pimple rubber—similar to table-tennis bat rubber—and carbon paper create an imprint on a sheet of paper that is situated on the sole unit. Areas showing a lack of imprint are not receiving sufficient pressure during the operation.

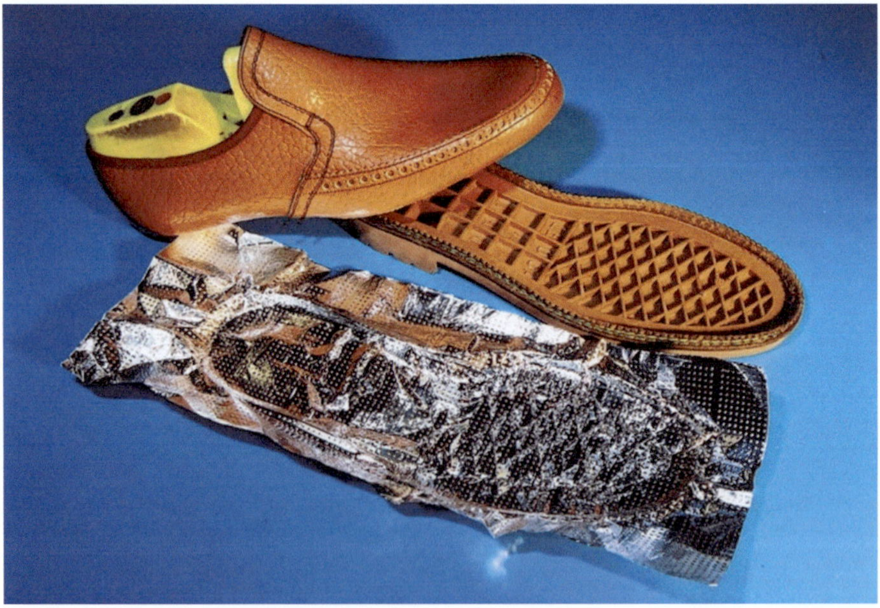

Fig. 5 The assessment of the uniformity of sole bonding pressure

2.13 Calibration and Maintenance of Tools

Whatever tools are used for process monitoring, they must be regularly maintained and calibrated, as it is pointless measuring process conditions with an inaccurate tool. It is essential that any devices used to check process conditions are calibrated by a competent provider and that calibration certificates are obtained. SATRA offers a fully traceable calibration service for physical testing and verification equipment.

2.14 Process Monitoring Records

Where standalone process monitoring record sheets are used, these should be designed to ensure that they contain all relevant information and can be quickly completed by the person who is taking the measurements.

It is best to have charts for each process, since this allows out of control conditions to be quickly identified. Action limits can be drawn on the charts so that if manufacturing conditions stray outside established parameters, they become immediately apparent and corrective action can be taken.

For a number of processes, part of the monitoring procedure will involve calibrating temperature or time settings. Therefore, the records must be referred to during production, to ensure that operatives use correct settings. Monitoring records should be prominently displayed alongside the process for easy reference. If the settings used by operatives are not displayed on the individual processes, they should be recorded on the work tickets in order to provide traceability back to the monitoring exercise.

2.15 Available Technology

Internet-based data monitoring brought new capabilities and unprecedented access to process measurement, and 'remote monitoring' took on a new meaning with cloud-based systems. With the availability of such instrumentation, remote access can be made from anywhere in the world that a smartphone is able to receive a signal.

Standard process sensors—such as thermocouples or resistance temperature detectors (RTDs) for temperature, pressure transducers or flowmeters—can be monitored, controlled or logged in almost any location, whether in the office next door or in another country.

From the simplest application, viewing data through a web browser on a cell phone or laptop, to more sophisticated uses (such as sending a text or email message when an alarm occurs, or transmitting a data log file over the internet from a remote location to a central office), a user can gain access at any time and from any location that has internet access.

The 'passive monitoring' that is built into internet-connected systems also brings a fresh understanding of 'process control'. The user has the knowledge that the process measurement and control system—no matter how large—is functioning optimally and can be confirmed by a glance at a web browser. Historical process information is now often cloud-based for remote viewing.

As an example of the capability of these devices, consider the monitoring of the environment in a warehouse or storage facility where conditions are critical. Perhaps temperature, humidity or maybe even the security need to be maintained. Thanks to standalone units with built-in sensors, temperature, humidity and dew point can be maintained without needing anyone onsite. Pre-defined alarms can provide alerts if the conditions exceed the stated parameters. The system can send an email to the responsible person, as well as to a distribution list. A text message can also be sent to any cell phone. If an IP camera or webcam is available, the scene can also be viewed on a computer.

The so-called 'fourth industrial revolution', often referred to as 'Industry 4.0' includes the 'Internet of Things' ('IoT'). The Internet of Things is a network of physical objects that are connected to the internet so that they can exchange data and information in order to improve productivity, efficiency, services, and more. The protocol allows for short-range wireless communication from device to device and ultimately to the internet. Devices that can communicate with each other can be aware of the status of each.

For process monitoring, this can even mean that a device can alert the need for attention and service via very low energy backup systems, even if it has failed or become disconnected. Therefore, a device that requires consumables can send an alert about an impending need to be refilled and direct that need to the appropriate resource—whether an onsite maintenance person or a supplier in a completely different country.

2.16 Ensuring Consistent High Quality

Process control is a vital part of modern shoemaking. Done correctly, it aids the selection of the correct materials, eliminates product variability and reduces factory rejects and wear failures. Ultimately, it should ensure that the customer receives a consistently high-quality product. Further, incorrect process and machine conditions can also lead to inefficiency and breakdown, increased cost and waste.

If properly organised, effective process control is not a costly exercise. An effective process control system can be a well-targeted use of quality management time and effort, that produces significant savings. The most involved aspect of process control is process monitoring. However, as identified here, only low-cost tools are needed, and little time is taken in taking the necessary measurements. Initially, the number of measurements may seem large but, as experience of each process grows, the number of these can be reduced.

2.17 Quality Control

Despite recent developments in lean manufacturing, the majority of footwear is still produced in batches to meet the customer's demand. When a quality issue arises, one faulty product can delay a large quantity of production and the dilemma can be to either hold up the entire production batch until the reject has been rectified or replaced, or to allow the incomplete batch to proceed to completion and delivery.

Of course, there is no simple answer to this—it is a question of calculating if the financial implications caused by a delay to the customer's batch and, therefore, late delivery is greater than the cost of manufacturing a small batch of additional products to make up the deficit. Research by SATRA has indicated that a small batch of work made as an urgent order can increase the actual manufacturing costs by 100–150%.

The provision of a dedicated, multi-skilled team to provide excess capacity whenever required can help to minimise this type of problem, as well as allowing the build-up of production skills to new operatives. Companies which have invested in this type of team to complement to main production environment have added the cost of the team to the overheads.

An additional cost issue related to quality control is the use of non-standard materials or components. Difficulties arise when materials are not as specified, so it is essential to have authorised and approved alternative suppliers as a fallback. However, since the cost of bulk production will be based on known materials and components, any changes can lead to both delays (which, in turn, will lead to additional costs) and possible changes in the price of these alternatives. The automotive industry has led the way in reducing lead times and levels of materials and components to drive down manufacturing costs, but these are often linked to severe financial penalties where deliveries do not meet the agreed standards.

2.18 Is It a Runner?

When considering the likely cost of each style, one important consideration is how often the specific design will be manufactured. Traditionally, we think of footwear as one of three types—'runners', 'repeaters' and 'strangers'.

'Runners' are those styles which are made consistently and comprise the bedrock of any production plan. Styles such as everyday fashion or school shoes would be runners, as they tend not to change in either design or materials. Hence, the production personnel become used to these and can achieve a high level of efficiency. Runners should be easy to predict and to control the cost.

'Repeaters' can be slightly more problematic, as they are generally produced in reasonable quantities, although they are not required regularly or consistently. Being more difficult to predict, they may incur some additional costs due to their disruptive nature.

'Strangers' are unexpected, infrequent or even one-off orders, and are best treated as 'made to order'. Inevitably, the difficulty that they can impose on the scheduling and cost control of production will lead to significant excess cost.

Therefore, while a general approach to identifying likely costs (such as materials and direct labour) is still necessary, each of the three types above will require different overhead and indirect labour costs. It is not possible to provide exact numerical guidance as to the effect of each, as the individual circumstances will vary enormously.

However, research by SATRA has indicated that repeaters can add between 8 and 12% to the manufacturing cost of a runner, whereas strangers can increase the runners cost by 10–17%.

Some companies will produce a scale of cost for each style based on a variable level of orders and frequency and will use this to benchmark the effect of the three types.

2.19 Average Size

Many companies calculate the expected cost of a new style (and use this to project the selling price) based on the midpoint of the size range. This has traditionally been an acceptable idea since they had a reasonably clear idea of their market and the likely range of sizes that would be needed.

However, SATRA's global foot survey has highlighted the major variations between the average size for different genders and countries (see Figs. 6 and 7). Much of the footwear currently being produced is sold internationally, and thus this significant difference between the average size can impact on the actual cost.

For instance, if the average size of an adult's shoe that is purchased varies by only one size (and when this happens, it is mostly to be larger than expected), the extra materials needed could be over 5%. In the example shown earlier, this would add at least USD 0.50 (GBP 0.42) to the manufacturing cost of each pair. There is less variation in the cost of labour of all kinds.

2.20 Demand Amplification

Because footwear is, in the main, a fashion product, the volume of sales of individual styles will fluctuate with the dictates of the market. This can have serious implications for controlling the costs throughout the supply chain. While sudden increases in demand for a best-selling style can represent a successful sales campaign, it can impose significant stress on the control of costs.

Some studies have concluded that an unexpected 10% increase in sales can result in an up to 40% increase in inventory throughout the manufacturing and distribution chain, with consequent increase in costs. Since the costing model is generally based

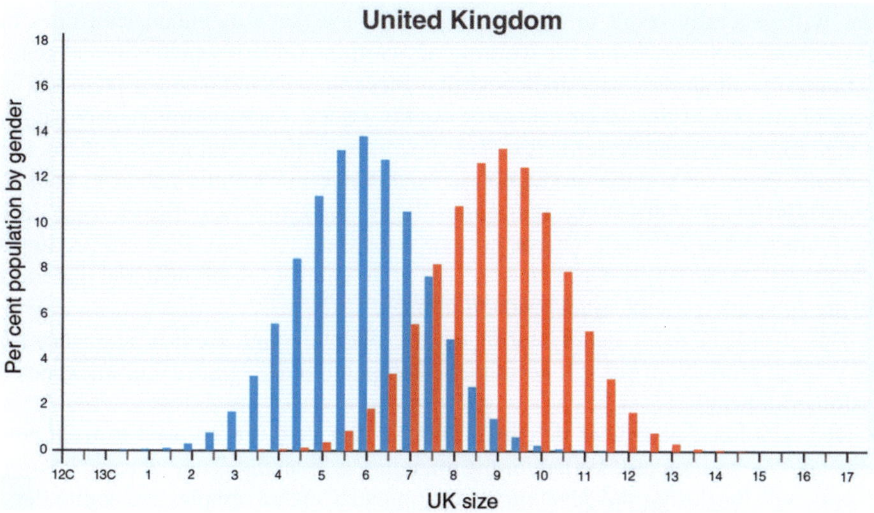

Fig. 6 UK women's (blue) and men's (red) population foot size distribution

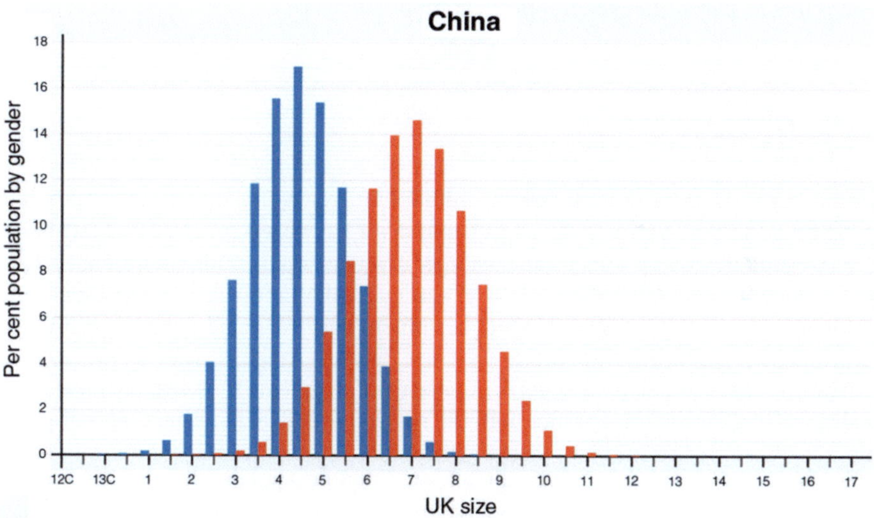

Fig. 7 China women's (blue) and men's (red) population foot size distribution

on a known quantity or output, any additional demand can create problems. Some materials or components may not be available within the required timescale from the usual supplier, and experience has shown that change to alternative suppliers will inevitably lead to excess quality control costs. Similarly, adding extra manufacturing time will generally result in either lower efficiency (such as unbalanced lines or inexperienced operatives) or the working of overtime, both of which will add cost.

Some recent studies have indicated that these unexpected, sudden increases in demand are one of the major causes of quality issues. One solution that is widely used in footwear manufacturing is to have flexibility in the manufacturing phase. For example, expected volume production is carried out on carefully balanced lines to create high levels of efficiency, thus controlling cost. Rather than disturb these lines when unexpected or small, late or custom orders are received, they are produced using autonomous teams which have access to the relevant equipment and materials but are separate from the main production lines.

This allows the main manufacturing team to maintain high levels of efficiency for the volume production but still permits the company to meet unusual or unexpected customer demand.

One additional benefit of the small team is that they can be used to train new personnel to the standard needed for the main production line. They can also be used to work with the design and development personnel to create samples and pathfinders.

Of course, there is an additional cost for this type of team, as it cannot ever reach the high levels of efficiency—and therefore, cost control—of the main production units. However, the most widely used method of assigning a cost to the team is to regard this as part of the general production overheads.

2.21 Transportation

Once completed, it is essential that the products are transported as quickly as possible to the final market. At this stage, significant costs have been incurred and any delay in converting the final product into customer revenue simply adds additional costs in the form of indirect labour (warehouse, storage and administration personnel) and overheads. Also, the finished product is vulnerable to damage from handling and incorrect storage during shipping.

The correct choice of packaging is important to protect the footwear but increasing environmental issues can conflict with the desire for adequate protection. All packaging represents an additional cost over and above the manufacturing figure, and very little of this is truly reusable. Although it is tempting to standardise packaging to save money and storage, it is important that the actual packaging is designed and selected specifically to protect the product—which will vary in size, weight and fragility, depending on the style. The cost of packaging will vary considerably but can be as much as USD 1.50 (GBP 1.27) per pair. It is therefore essential to consider that the packaging will not only protect the footwear when transported from the factory to the point of sale but increasingly also when the footwear is mailed to the customer (when purchased remotely, such as on the internet) and also returned by the customer.

The majority of footwear is currently produced in countries that are not where the final sale will take place, leading to a huge demand for safe, efficient and cost-effective transportation.

There is a vast difference in the cost of transporting finished goods, depending on the method selected. For example, if the footwear is transported by sea (usually in a container), this will typically cost between USD 1.00 and 3.00 (GBP 0.84–2.53) per kilogram, compared to USD 6.00 and 8.00 (GBP 5.06–6.75) per kilogram if the same products are sent by air. Of course, there is also a significant difference in the time taken. By air, the products can arrive in 2 or 3 days rather than taking anything up to 30 days by sea. However, this is purely the time taken for the actual movement of the products—there can be significant delays at the ports where custom inspection and checking the relevant documentation can add a number of additional days to these figures.

When considering the costs of transportation, it must be remembered that the prices quoted can be either 'Free on Board' (FOB) or 'Ex Works' (EXW). FOB is usually only available for sea freight. In a nutshell, FOB means the seller takes full responsibility for the package to the point that it is onboard and ready for transportation. The responsibility for successful delivery is ultimately split more equally between the merchant and the consumer. EXW indicates the seller has a responsibility to make sure the buyer can access and pick up the cargo at the seller's place of business.

Transportation cost and associated risks are no longer a burden for the seller under the EXW option, and this favours the shipper. Free on Board and Ex Works are both international trade terms known as 'Incoterms' (an acronym standing for international commercial terms). These dictate the responsibilities of buyers and sellers, including which parties are required to cover all costs and arrangements related to the shipping of goods. In addition to these costs, there may be others such as insurance, security and specialised storage for some types of footwear.

2.22 Retail

At the end of the supply chain are, of course, the customers. This is where all the hard work leads to, and it should result in sales to satisfied clients who will not only pay the full selling price but also return again to the brand when further purchases are required. In addition, such customers may recommend to their family and friends that they should do the same.

In terms of cost control, if all the issues previously highlighted have been successfully dealt with, the product should be not only available but also represent value for money. As so many footwear purchases are now made by the customer without the opportunity to discuss and receive advice from a retail expert, then the labelling and product properties claimed are the key information that will be used by the customer when choosing.

Many sports or fashion products will be linked to, or endorsed by, celebrities. This is generally achieved at a significant cost and must represent a viable investment. Celebrities are regularly in the public eye through the media and social networks, and we have seen many times how bad news about a chosen celebrity can cause a significant loss of goodwill from the public, inevitably leading to a loss of sales.

The companies which place the products on the market—whether they are traditional retailers with a high street presence, or internet-based—have the difficult task of maintaining not only adequate levels of stock to meet whatever the customer requires but also access to rapid replacement of products that are sold. The public has become used to fast (often next day) delivery from the major online selling companies, together with easy and usually free return for those products found to be unsatisfactory.

To put this in context—the number of adults buying online has risen from 25% in 2003 to 50% in 2008, and at least 75% in 2018—the most recent 'normal' pre-COVID-19 year. Between 20 and 30% of this latest figure involves the use of a mobile ('cell') telephone. However, this move into ecommerce is not without its financial problems.

A recent survey has indicated that the cost of picking, packing and dispatching an order worth USD 23.69 (GBP 20) is the same as an order of USD 118.45 (GBP 100). This has been confirmed when one UK department store is said to have told suppliers that any online order under the value of USD 35.52(GBP 30) is unprofitable. Additionally, a very high percentage of fashion orders placed will be returned. The chief executive of one high street brand owner is reported to have said: 'This varies by subcategories but, in the fashion category, a retailer could have a returns rate of anywhere between 30 percent and 50%'. Another well-known ecommerce fashion trader has estimated that every British pound that moved from stores to online last year cost an additional 6pence in overheads, such as warehouse picking and delivery. In its 2018/19 results, the retailer commented: 'The digital evolution has not been easy, and nor is it without cost'. Nevertheless, the current mix of store and internet retailing will continue to develop, and most commentators believe that there is no way back to the traditional pattern of high street purchasing.

2.23 Using the Correct Technology

Many of the examples of cost issues already mentioned in this article can be addressed by using relevant and appropriate communication. In most cases, this means using modern technology to ensure both rapid and secure transmission of information, and online meetings to discuss and resolve problems (see Fig. 8). There is little firm or reliable data to show the financial impact of delays due to poor communication or wrong decisions, but several recent studies have recorded a reduction in profitability of up to 50% for each week lost from the promised delivery date.

The key point is that the effect and cost of poor communication are magnified the closer to the selling date in the supply chain that the delay occurs. As the footwear progresses from the manufacturer to the customer, its value (and therefore cost) increases. Any delays when the footwear is at its final (maximum) value

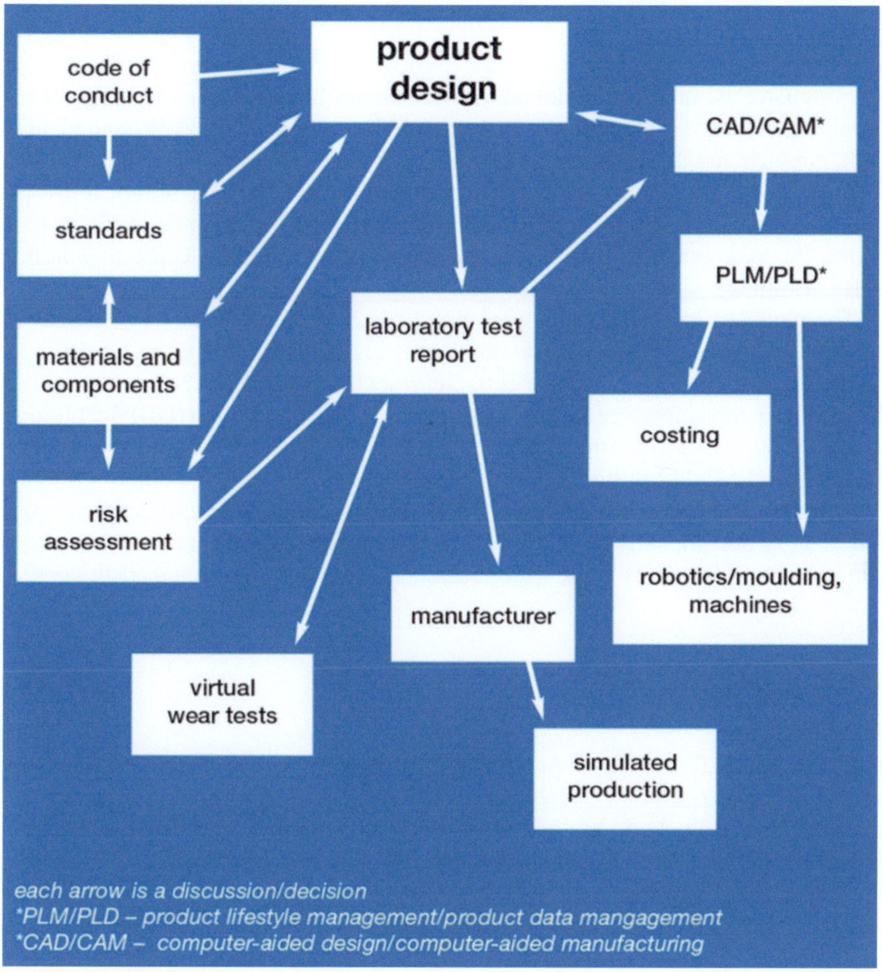

Fig. 8 Connecting existing 'islands of technology'

will inevitably cost more as the capital represented by the completed shoes could have been usefully used elsewhere.

Companies are already reaping the benefits of using technology to speed up communication. This is particularly true at the design and development stage, where historically there have been extended times needed to collect, process and discuss the complex and interlocking information about each new style. The use of shared information between manufacturers, transportation companies and those in the eventual distribution and retail part of the supply chain will, undoubtedly, lead to similar and perhaps even greater benefits.

2.24 *Fraud*

The extensive use of internet sales has provided many benefits for both customers and suppliers, but there is a downside. Remote sales are at risk of fraud through identity theft, copying and theft of charge cards and websites selling counterfeit products.

In 2019, the total value of all UK e-commerce was USD 821 billion (GBP693 billion), of which around 9% (USD 74 billion/GBP 62 billion) was in the fashion sector. At least one-third of these sales were online, totalling USD 24.9 billion (GBP21 billion).

The recorded level of e-commerce fraud for 2019 in all sectors was USD 733 million (GBP619 million)—just under 1%. It is likely that this same level occurs within the footwear and fashion sectors, representing some USD 249 million (GBP 210 million). Since footwear accounted for about USD 11.2 billion (GBP9.5 billion), then it is reasonable to assume that the same level of fraud (1%) will occur—costing the industry USD 112 million (GBP 95 million).

There is a wide range of fraud-prevention software which will monitor transactions and identify potentially false or misleading issues. However, the most secure way to minimise fraud is to only work with known and trusted suppliers and customers.

3 Conclusion

The fragmented formation of many global supply chains means that establishing, monitoring and controlling individual costs require serious effort and sophisticated systems. The whole world has been affected by the events of the COVID pandemic, and many of the established patterns of trade and relationships have been damaged—some irreparably. As our industry re-establishes new ways of working, this is an opportunity to put in place stronger, more durable and more informative systems to control costs.

Leather and The World of Fashion

Jayne Simone Estève Curé and Mohamed Sadiq

1 Introduction

1.1 Leather and Clothing: An Ancient Love Affair

Leather & the story of our civilization are intricately intertwined and have been for thousands of years… this material considered to be precious, noble, and practical since the dawn of mankind has become part of our daily lives. Leather has been and still is, one of our favourite materials for inventing and making different types of clothing and accessories. Beyond the importance of practicality in the world of Fashion are aesthetics, emotion, and desirability. Leather has secured itself a key place within the system of Fashion and trends continuously proving itself to be relevant, attractive, and popular.

J. S. E. Curé (✉)
Jayne Esteve Cure Fashion Consultancy & Training, 75003 Paris, France
e-mail: jayne@jayneestevecure.com

M. Sadiq
CSIR-Central Leather Research Institute, Adyar, Chennai 600020, India
e-mail: mdsadiq@clri.res.in

S. K. Janardhanan and L. A. Zugno (eds.), *Emerging Trends in Leather Science and Technology*, https://doi.org/10.1007/978-981-99-9754-1_7

Part I

2 A Journey from Clothing to Fashion

2.1 The Different Functions of Clothing and Leather

The naked bodies humans are born with need constant protection from many things, including weather conditions. Thus, clothing has a fundamental meaning for Humanity which is functional. However, dress has progressively taken on other dimensions that are less and less materialised and practical but more and more symbolic and immaterial such as adornment or social, professional, or cultural identification. Exploring these different dimensions will take us from the most functional and tangible elements to the most symbolic meanings, including the notions of aesthetics and fashion trends.

2.2 Protection and Utility

Clothing ancestrally protects humans from weather conditions, including the sun, the cold, rain & wind. However, it can also be utilitarian, for example, protecting workers for specific jobs or providing certain characteristics for particular activities such as sports.

2.3 Identification—Social and Cultural Aspects

Within our past or contemporary societies, social & cultural identification is made possible through the type of attire people wear. It can be about gender differentiation & professional, cultural, or religious identification. The type of garment worn, the material used, its shape, volume, length, and details, as well as the use of colour and patterns, can symbolise a rank, a function and a role within Society.

Clothing and specific materials employed have often been used to indicate someone's status or position. For some ancient societies such as Rome, leather is used to make army uniforms and clad the outfits of their belligerent leaders. It can be remarked that the sturdiness, yet suppleness of leather contributes to its' popularity, notably for its' protective qualities stemming from its' numerous technical proprieties. Also, on another level, it can be said that marking a social status or rank can be associated with the choice of rare luxury materials and finishings to conceive and adorn an item of clothing in an elaborate & sophisticated way.

2.3.1 Modesty

Since the beginning of Time, most cultures across the world have considered, scrutinised & judged the human body through the lenses of modesty. Often generally hiding, but sometimes exposing, the male/female body according to different societal eras & moments based on evolving beliefs and rulings. Historically, the Design of clothing is usually an activity under the epoch's relationship to the notion of modesty, directly affecting the type, shape, detail and material used to conceive clothing.

2.3.2 Status

As introduced previously, clothing can symbolise or reflect a person's status. It can be in tune with a particular lifestyle; typify cultural interests or affiliations & social or religious identifications, including vocational/professional differentiations. This is a compelling and universal mode of communication symbolised by specific attire where again, types of clothing & accessories through shape, colour & materials are used to differentiate the symbolism of clothing. Typically, the uniform for school children or nurses illustrates this dimension of apparel.

2.4 Adornment/Ornamentation

Clothing, solely designed for functional purposes, can take on another dimension when enhanced by adornment and ornamentation. These are employed to heighten product attractiveness, expressing uniqueness and creativity.

2.4.1 Fashion

Clothing, notably Fashion clothing, mirrors our identity, making it possible to assert ourselves as individuals. The great Christian Dior said, "Men and women have an instinctive desire to show themselves to others…" & Fashion allows us to assert our social status, affirm our values and beliefs and visually declare something about our individuality, our personality, and the way we want to be seen by others and the World.

Fashion can also be described as an aesthetic awareness or demonstration of oneself. This can be the longevity of self-expression in tune with the ever-evolving Times; here, we can call it "style" to the extent that it goes beyond the notion of Fashion… The following famous words of Gabrielle "Coco" Chanel perfectly cinch this idea: "Fashion goes out of fashion, style never does". Most importantly, to understand Fashion, one has to consider its' ephemeral quality as it inevitably goes out of Fashion.

Fashion is a sociological reflection of a given moment in Time, inexorably linked to the notion of the "Spirit or Mood of the Times" also recurrently referred to by the Fashion industry by the French term "Air du Temps". This phenomenon is perpetually evolving and transforming under various influences such as climate, social, political, economic, artistic or technological evolutions, to name a few. This is key to comprehending the fashion system which has prevailed for several centuries. The French Sun King, Louis the 14th, not only understood this but used it to assert his power and authority. He influenced how people dressed, largely magnifying his geopolitical role & place within what could be called the European countries of that Time. He understood and incorporated in his style & method of ruling the importance of sophisticated attire. He extensively enacted on the evolution of French creative & industrial know-how, enabling & promoting the development of high-level techniques and pushing for refinement & sophistication. Industries elaborating sumptuous fabrics such as silk or intricately beautiful embroidery or lace became highly successful, such as the use of feathers and other exquisite ornamentations.

In the context of Fashion, leather has become a constant feature inspiring the world's most renowned designers and a whole global industry and market. With its ever-evolving cycles, it is continually desirable and everlastingly used to design, produce and sell fashion products worldwide, maintaining its strong position at the heart of trends. Leather is at the heart of the creative chain that drives the fashion industry. Thus, it is intrinsically related to the creative forces driving Fashion. Our next topic shall endeavour to dissect and explain the notions behind the innovative sources & processes driving Fashion.

3 The Creative Sources Driving Fashion

3.1 What is Creativity?

Creativity is a complex, intricate, manifold term encompassing several distinctive notions & features. To explain this term simply is challenging. We can say that it defines the will and capacity for humans to conceive, build and enforce something innovative or new. This can be intangible such as an idea or solution to a problem, or tangible, taking the shape of an object or any other materialised element.

Delving deeper into this notion reveals a process of imagination & creation of new concepts and ideas. Pushing further… creativity is about conceptualising thought, imagination, and ideas. It can therefore be materialized, staying in idea form or thought process or becoming an actual object.

To conclude, creativity, with its multiple dimensions of innovation and newness, manifests itself in the world of Fashion at different levels, differing significantly from one type of fashion House/Brand/Distributor to another to meet specific market needs.

3.2 The Act of Designing (Fashion Designer)

In English, the one who creates, or designs clothing and fashion accessories is named a Fashion Designer. This terminology is associated with the concept of "creativity" (previously seen above). The act of creating Fashion is aligned with the concept of Design found in the terminology used to define this process. What is Design in general? It is a creative manifestation of a concept by a designer. It is the act or process of intentionally creating.

The word Design is by no means only linked to the world of Fashion as many different types of products can be "designed". A design produced by a person or group entails conceptualising an idea from the imagination that brings an innovative quality. Imagination with intention is used within a process; thus, fashion design is about an intentional creating process for the designing of Fashion.

3.3 Distinguishing Different Levels of Fashion Designing

In English, we tend to put different types of fashion design jobs/levels under the same denomination: "Fashion Designer". However, this term covers several different levels and types of fashion design types that need to be differentiated to understand or topic better. The French language offers various terms that cover distinct definitions, as listed below.

Haute Couture

What is *Haute Couture*?

French by nature, it exists within an official system that resembles, for example, the production of Champagne in France…it is possible to use a similar winemaking Champagne method in different parts of the world. It can only be named "Champagne" if produced within the specific Champagne region of France and, most notably, if officially recognised as a label *Appellation d'Origine Controlée* (AOC) that names, protects and preserves the Champagne designation.

Haute Couture functions under a similar system that legally designates & protects Parisian Couture Houses since 1945. However, the overall system began a little after the 1st Couture House was launched in Paris by the British Charles Frederick Worth in 1858.

First and foremost, it is a designation protected by the French Federation of Haute Couture. A commission establishes yearly, a list of companies officially recognised as *Haute Couture* Houses.

Strict specific rules regulate the practice of the unique French system of "Haute Couture", such as making each Design solely by hand. This implies a high level of technical excellence with exceedingly qualified ancestral Arts & Crafts techniques. Haute Couture is bespoke and made to measure to the client's body measurements, which means that there are no two identical dresses made. There is no stock in advance, so it is impossible to go into a store and leave with an Haute Couture design/product.

So, this is very specific to Parisian-based Fashion design, although there are also invited Members from other countries who are officially recognised by the French Haute Couture syndicate. This terminology is quite specific to French Fashion, as the official *Haute Couture* fashion show weeks only take place in Paris, encompassing the highest possible level of artistic creativity and technical excellence.

3.3.1 Fashion Designer

After one hundred years of ruling the Fashion System of global trends, the Haute Couture system came to its' paroxysm in the late 50s, making way for the explosion of ready-to-wear in the 1960s. This development of ready-to-wear and fashion distribution was driven by a new youthful generation named the "Baby Boomers" impacted by the substantial societal & cultural changes taking place at that time.

This new era, mainly in the 70s and 80s, gave birth to new type of "Fashion Designers" such as Jean Paul Gaultier, Thierry Mugler and Kenzo. They had unbounded creativity, a solid individualised & recognisable identity, and an acute commercial sense. This new generation of fashion designers create global and original concepts for their signature ready-to-wear brand or other existing brands. Design is executed in an innovatively creative way without producing *Haute Couture* designs… although several of these designers did move on to do *Haute Couture* at the turn of the millennial.

Styliste de Mode

To conclude, this series of definitions of types of Fashion Designers found in today's fashion system, is the "*styliste de mode*". This type of designer creates fashion products for a brand collection(s) (or for several brands if the designer is a freelancer). The "styliste de mode" is asked to respect the brands' DNA/creative identity/Market positioning rather than Design under his own creative identity and name. The "*styliste de mode*" can design a whole collection or specialise in a specific product for a brand such as dress, knitwear, or leather goods.

3.4 Fashion as a Form of Art Related to the Zeitgeist

Fashion is driven by change, change that reflects ongoing societal transformations. Change is, synonymous with constant renewal and creativity. One of the principal driving forces of change that impulses Fashion is generated by the artistic approach of Fashion Designers to their creative output.

What is Art?

Here we encounter again the notions of creative skills linked to imagination & innovation expressed here through different Art forms, Fashion being one of them.

So, what is Art? This is a vast question. Art expresses a strong vision, will and capacity to create and innovate in a visual manner through the creation of paintings, sculptures, architecture & other Art forms.

It is important to stress that the production of Art through the ages has aimed to create an effect on the beholder (public), conveying messages & feelings that mostly hold an emotional power materialised aesthetically often through what can be called beauty. This emotional power conveyed through Art is fundamental to our topic of Fashion and is a driving force for success. It is also an essential influence that feeds the Trend System. It is about saying something to the surrounding world, creating a dialogue between the Artist and those who behold (or wear) the artistic production. As the great Art historian, Ernst Gombrich, said "*The painter must leave the beholder something to guess*".

We shall explore different dimensions of what drives Art and how it is in tune with the fashion system through quotes from great Artists across Time.

Michelangelo, one of the greatest artists of all time, said "*If people knew how hard I worked to get my mastery, it would not seem so wonderful at all*". Of course, knowing that it took time and tremendous effort to achieve such a level of genius/ mastery is more than wonderful, leaving us even more in awe of his Art. The fact is that, often, to exist and be impactful, Art needs not only inspiration but also technical excellency. Something very elaborate and difficult to achieve can look very simply when obtained through the mastery of specific techniques. This is also very true for Fashion. A convincing example of this is Cristobal Balenciaga, the great Spanish *Haute Couture* fashion designer who was based in Paris after the Spanish civil war. He strove to reach a pure form of Design, developing a simplification process to achieve an aesthetic that he was looking for during his world-renowned career. When he felt that he had achieved it, he retired and stopped designing because he had obtained his goal and achieved his aim. To this day, he is universally considered to be the Master of all the talented Couturiers and Fashion designers. In this vein of thought regarding Fashion, one must mention the great know-how and savoir-faire of the *Haute Couture Ateliers* (workshops) who translate the *Couturier's Design*, materialising it with brio and talent.

The following quote by the painter Paul Gauguin, "*Art is a mad search for individualism,*" strikes a chord with how Fashion is a way to express our individuality through clothing.

We can observe the following three major types of relationships between Art and Fashion:

1. Art as an inspiration for Fashion
2. Art & Fashion collaborations
3. Fashion, a form of Art.

3.5 Art is an Inspiration for Fashion

Yves Saint Laurent designed the legendary Mondrian-inspired dress collection in 1965 as a homage to Pietr Mondrian's abstract Art pieces of 1935. The deep connection Y.S.L. had with Mondrian's Artwork resonated with the highly innovative and dynamic 60s era, where shortened and straight dresses were trendy at that time and the perfect shape for Yves Saint Laurent to turn Mondrian's Artwork into wearable Art. The collection became a substantial planetary success and is considered, to this day, to be one of Yves Saint Laurent's signature designs. Throughout his career, Art was a constant inspiration for Yves Saint Laurent. He interpreted and translated into fashion pieces celebrated major Artworks by painters such as Van Gogh, Matisse and Picasso.

3.6 Art and Fashion Collaborations

The relationship between Art & Fashion has resulted in successful creative collaborations between contemporary artists and Fashion Designers. Marc Jacobs, who launched the very first ready-to-wear fashion collection for Louis Vuitton in 1997, initiated & developed several highly fruitful collaborations with artists such as Stephen Sprouse and his graffiti version of the Louis Vuitton monogram, Takashi Murakami and his multicoloured LV monogram with cartoon eyes or again Richard Prince with his paint-splattered canvas coated handbags that soon became best sellers.

More recently, Simon Porte Jacquemus collaborated with the Canadian, New York-based artist Chloe Wise. The communication campaign of the Spring Summer 2019 collection was entrusted to the Artist whose paintings showcased the collection.

3.7 Fashion, a Form of Art

Designing Fashion in a singular personalised way, driven by a powerful, recognisable aesthetic in tune with the *Zeitgeïst*, can be considered a form of Art. In the same line of thought, creative Luxury Fashion Houses whose original designer is no longer head of Design, such as Dior, Balenciaga or Saint Laurent, appoint an Artistic Director to lead the overall creation of the House. Today at Dior, it's Maria Grazia Chiuri, Balenciaga is driven by Demma Gvasalia and Saint Laurent by Anthony Vaccarello.

What is an Artistic Direction? It is about applying the designer's identity, vision, and feeling to the whole Brand top down, including retail, artistic values, communication & company culture. The definition of an artistic or creative direction can vary significantly from one House to another. The Artistic Director is asked to bring cohesion and cohesiveness to the brand to ensure a global identity perceptibly in tune with the DNA House.

Lastly, the classical composer Sibelius said, "*Art is a signal of civilisation*". A parallel can be drawn between this notion and the forces driving Fashion's evolution that mirror the *Zeitgeist, Air du Temps* or Sign of the times. This phenomenon points to a period in Time defined by civilisation expressing itself through Art.

3.8 Fashion and l'Air du Temps or Zeitgeist

The Zeitgeist or Air du Temps designates a spirit or mood that defines a specific period… subsequently Time, always in motion, is a succession of these periods. Its perpetual changing nature has an ephemeral quality, making it hard to predict durably. However, the Artist perceives the *Zeitgeist* and how it reflects Society at a given moment… often "feeling" what is coming next. This perception is conceptualised and materialised through the act of creation, perceptible in the work of Art. A Fashion Designer with a singular creative vision, signature, and identity draws from his perception and concept of "*l'Air du temps*" with an artistic approach to his work.

This awareness and vision of the *Zeitgeist* are translated into his creativity, particularly perceptible in the Design of fashion collections, communication campaigns & other marketing strategies. This phenomenon explains why common ideas spring up simultaneously in different collections by different designers/brands at the same time… they feel, see, and perceive the same things simultaneously yet translate them in different ways expressing their specific creative identity. Consequently, Fashion can be observed as a sort of reflection of Society at a given moment. This idea brings depth and complexity to how Fashion can be defined and perceived… it is not just artificial, "bling," or shallow… it springs from Society and reflects the world we live in and the questions we can ask ourselves. It is a mirror effect of our ever-evolving Society at different moments of its history.

This notion of perceiving and encapsulating *l'Air du Temps* in Creative Design at the right moment for the market to find desirable is essential for a brand's performance. The brand's identity is composed of equilibrium between how it expresses its history/DNA, the singularity of the Creative Director, who is at the head of the House at a given moment, how it fits with l'Air du Temps and, very importantly, how the brand wishes to develop its business strategically.

3.9 Identity and Singularity

In Psychology, identity is the consciousness, and we could say awareness one has of oneself. How do I perceive myself, what consciousness do I have, and how do people distinguish me and recognise my specific identity?

We can equate identity with uniqueness which is very much linked to the idea of singularity, which can also be collective when an identifiable element is shared. Singularity is about being unique; the singular character of something or someone that is distinguishable; recognisable! In the realm of Fashion Design, it is the way a creative person leaves an imprint in the world through an act of creativity which is innovative, singular, recognisable and desirable! To summarise, Fashion Houses & Designers strive to create and nurture a solid seductive identity that necessitates having an identifiable vision and creative singularity that evolves with external stimuli related to current issues.

3.10 Creative Identity

Creative identity involves & evolves with stimuli related to current issues. In tune with the Creative/Artistic Director or Designer's vision and creative singularity of the House/brand, it can be defined in the following ways.

Having a Vision

Cultivating a personal vision of "*l'Air du temps*" and translating it into desirable creative work.

Having a Voice

Personal expression & communication to voice an idea/opinion with Fashion Design as a mode of expression. It is not just about being mainstream and following the general trends but also about having a voice and using Fashion to say something; a way of expressing one's singularity/vision—also linked to storytelling and creating a narrative between the designer/product & final customer.

3.11 Singular Creative Universe and Brand Identity

Drawing on specific sources of inspiration & being in tune with *Air du Temps.* As seen previously, having a singular creative universe is essential for a House/brand to obtain recognition. The designer at the helm of a success story cultivates a solid creative universe to develop the brand. Sources of inspiration enrich this journey, personality, a specific way of interacting with the world and the capacity to create innovative content for storytelling. A creative signature leads to brand identity.

Codes & references

Codes & references make up specific design vocabulary that identifies a Designer and/or Brand. Brands and fashion Houses aim to have distinctive, seductive and recognisable visual codes and references that are instantaneously identifiable. Jean-Paul Gaultier, Chanel, Chloé, Jacquemus, Marine Serre and Gucci are good examples of this, visually recognisable in their creative, inspirational universe and style of Design.

Creative signature

Most of these designers strive to have a distinct creative universe enriched by their personality, life journey, encounters & muses, recurring sources of inspiration, way of looking at the world and need to express themselves via their creativity. Storytelling is also essential to define a creative signature, usually based on a skilful mix of brand heritage and contemporary storytelling, such as the world of travel at Hermes, which is one of the leitmotifs of the brand. Brands strive hard to have, keep and develop a solid desirable creative signature, the basis of brand identity.

4 Leather and Fashion—Iconic Wardrobe Pieces

4.1 Iconic Leather Pieces and the Timeless Wardrobe

Leather designs, mainly fashion accessories, are amongst the most desirable iconic fashion pieces of all Time, putting leather at the heart of Fashion. They are timeless in their Design and very recognisable in their aesthetic & technical conception and production; they symbolise a solid creative brand identity based on Heritage and brand DNA. The timeless quality of these designs is fertile ground for constant reinterpretation, according to the changing *Zeitgeist*, so both timeless and fashionable qualities prevail. Iconic pieces have a strong identity signature that makes them instantly recognisable and coveted. Iconic is a powerful terminology usually attributed to a person or object worthy of veneration. Such is the commercial power of iconic leather fashion designs that consumers revere.

Some iconic pieces

Here are some examples of the most loved iconic leather fashion pieces:

Chanel 2.55 Handbag

Gabrielle Chanel designed and developed in February 1955 what was to become one of the most iconic, timeless, and desirable handbags of all time: the 2.55. Its Design is simple; a rectangular shape with a metal clasp and signature flat chain for the handle, making it easy to wear in everyday life. What is revolutionary in this Design is that Gabrielle Chanel chose to make the 2.55 in soft calfskin leather with a quilted aspect inspired by her passion for the equestrian world.

Gucci wide leather logo belt

The brand's founder Guccio Gucci and his double G initials inspired the famous and instantly identifiable two intertwined Gs logo that his son, Aldo, created in 1933. Gucci's wide leather logo belt is an iconic must-have piece.

Louis Vuitton leather Luggage & Handbags

Over the past 150 years, Louis Vuitton has been designing Luxury leather-made luggage and handbags, making the company the most successful luxury brand in the world.

Hermès Kelly & Birkin Bag

Probably the most iconic handbags of all time, the Hermès Kelly and Birkin have played significant roles in transforming Hermès into the luxury Fashion House it is today from an equestrian saddle maker of the 19th century. The Kelly handbag was initially inspired by a functional large leather bag that was reduced in size and feminised in its shape, becoming a huge worldwide success when worn by Grace Kelly of Monaco. Similarly, the Birkin bag was made functional and aesthetically desirable and named after Jane Birkin in 1984, subsequent to a creative conversation she sparked off with the C.E.O. of Hermès, Jean-Louis Dumas, about her ideal dream handbag.

4.2 The Creative and Marketing Ingredients that Make an Iconic Product

Six analytical criteria can be studied to understand the complexity of what makes a product an iconic success. The iconic Trench coat and leather Stiletto Court shoe will be used below to illustrate this approach.

#1. First and foremost, one must consider the product's historical & storytelling dimensions, including when, why and how it was designed and developed.

Trench coat illustration: The first world war, the British soldiers in the trenches, a functional revolutionary coat & fabric design, espionage films and 1950's glamour movie stars… here are some of the storytelling ingredients for the Trench Coat evolving from purely functional to glamourous.

Leather Stiletto Court shoe illustration: Heels became popular in the 16th Century with Venetian women wearing chopines, a large platform shoe used to elevate their skirts from the mud. In the 18th century, Louis XIV, the French "Sun King", wore red heels to symbolise his authority. Men and women wore Heeled shoes at Court to show their power, opulence and wealth. Thin stiletto-type heels were first seen in the late 19th century, as evidenced by numerous fetish drawings. The Stiletto, as we know it today, debuted in the 40s with the shoe designer André Perugia. He worked with great fashion designers such as Paul Poiret, Jacques Fath and Hubert de Givenchy. The 50s popularised the stiletto court shoe for women, notably with designs made by Roger Vivier for Christian Diors' New look silhouette. The 80s decade saw a massive rise in the trend for this type of shoe associated with Power dressing. This leather shoe type is now a classic staple of modern-day wardrobe.

#2. This leads to by whom the iconic product was designed and what reference brands are involved and assigned to it, for example, Burberry for the Trench coat. References for the Leather Stiletto Court shoe include André Perugia (one of the first shoe designers to experiment with height and slimness), Roger Vivier (Christian Dior designs & Famous customers such as Queen Elizabeth 2nd or Marilyn Monroe) and more recently, Christian Louboutin (known for his famous red-soled Stilettos) or Jimmy Choo (renowned for glamourous heels with luxurious embellishments such as crystals or feathers).

#3. An iconic fashion piece often embodies several cultural references encountered during its lifespan. These can come from various artistic fields such as the cinema or pop stars or be referenced in a historical moment. For example, Humphrey Bogart and Lauren Bacall wear trench coats in the epic film Casablanca. There are multiple cultural references to the leather stiletto, such as the Ferragamo pair Marilyn Monroe wore in the 50s, helping to make her seductive walk famous or the Manolo Blahnik shoes loved by Carrie Bradshaw in Sex and the City T.V. series.

#4. The technical aspect of the product is also essential to understand its iconic dimension, how, where and by whom it is made and what materials are used, i.e. the Trench coat has the following features: double-breasted with buttons, buckle sleeve straps and also the gabardine fabric and nova check lining. The leather stiletto shoe has a long, thin, high heel that exceeds 8 cm and a pointed, rounded or peep toe. The leather finishings, such as full grain, patent or exotic leather, can vary greatly.

#5. Fashion trends & styling drive the renewal of iconic pieces enthusing freshness and a trendy contemporary dimension to stay modern and desirable. The Trench coat is permanently up-to-date by designers and brands that play on volume, fabric types, finishings, colour & print. The Stiletto is a universally acclaimed "must-have" item that designer brands in tune with trends constantly reinterpret. It can be styled for a dressy fit with a cocktail dress or skirt/pant suit or worn in a more streetwear style with jeans or jogging bottoms.

#6. Lastly, one can study how contemporary designers and brands reinterpret and reappropriate iconic pieces by using them out of context, like a sunny summer Trench coat or a hybrid stiletto version such as the Balenciaga Stiletto Croc shoe.

5 The Basics of Fashion Marketing for Trend Forecasting

5.1 Nature of Fashion

Fashion is a prevailing style, custom, usage or mode at a given time. Fashion is a dynamic mechanism of change. There is a psychology behind Fashion, a reason for its existence, constant change and constant appeal. To those who do not understand it, Fashion is an elusive mystique. To those who do, Fashion is a precise science.

What is Fashion?

Fashion plays a significant role in the consumer product sector of the global economy. It permeates innumerable products of human consumption, from footwear to automobiles. Apparel is the most predominant area where Fashion rules high over times immemorial, and a visible example is the global appeal and acceptance of denim pants, fade-outs, wash-outs, stone wash, etcetera for over a decade. In the complete set, overall personality concept fashion and colour developments in clothing influence the Fashion in footwear. Humans wear hats, leather garments, belts and footwear and is susceptible to frequent fashion changes. It is reported that in the U.S. alone, about 20,000 new footwear styles are being introduced annually, though many may have only marginal variations. Fashion need not necessarily be western and the privilege of advanced countries. Many developed countries are increasingly becoming sources of inspiration from the culture, architecture, and temples that portray the advanced fashion trends exhibited in apparel. It is observed that Fashion is not necessarily the brainchild of the designers, manufacturers or consumers. However, Fashion is evolved, keeping a sense of continuity with the prevailing trend in the well-defined seasons. The Fashion within reach should be attractive, appropriate and affordable. Fashion, like humour or music, is a social reflection.

Man is the only animal to embellish and modify his natural shape. The human body is the basis of all Fashion. People wear footwear—for whatever makes them feel better, younger, thinner, more affluent, prettier, sleeker, more artistic, intellectual, rebellious or whatever images they wish to adopt or promote. The old saying that you cannot judge a book by its cover does not hold for people. One can tell a lot about people by how they dress, cut their hair, wear jewelry or make up. Fashion is a language of signs—a nonverbal system of communication. Fashion is body language and clothes are the vocabularies. In the broadest sense, Fashion is man's oldest Art. It is also one of his most outstanding examples of collective achievement, for it is a culture created by the convergence of individual expression and popular consensus. It does reflect the taste, mood and tone of the Society. In the last fifty years, Fashion has emerged from the salon and entered the marketplace, becoming accessible to and almost as diverse as its ever-growing audience. It is more relevant in respect of apparel. Fashion reflects our world. However, Fashion, like other amenities, can only be one man or woman's secret, or it can transcend personality and speak for a large and diverse community. The term fashion may be applied to a consumer product, and it can also be used in describing a pattern of consumer response towards new forms of consumer product. For example, one can say that buying or wearing sports shoes is a fashion. Fashion may be analysed by directing attention to the physical characteristics of the products. Alternatively, an examination may be made of the process whereby new fashions emerge and are adopted by consumers.

Role of Fashion

Fashion plays a significant role in the consumer product sector of the global economy. Apparel is the most predominant area where fashion rules, for example, global appeal and acceptance. The overall personality concept, fashion and colour developments in clothing influence fashion in footwear.

Fashion culture

Fashion is found in every culture from the most primitive to the most sophisticated.

Fashion and Style

Style is a pattern or model. Fashion is a prevailing style at a given time. Style should not be confused with Fashion because Fashion is variable, whereas style is not.

High Fashion

High-priced, exclusively designed, bonded styles accepted by fashion leaders who are a limited number of persons above normal means.

Mass Fashion

It refers to Fashions that have achieved considerable popularity and are therefore mass produced and sold in large volumes in various price ranges and retail outlets. In the U.S.A., athleisure footwear represents more than 70% of all footwear used.

- Classics: Styles that have long-term popularity, a long period of slow initial growth in sales followed by a higher and longer peak of acceptance and a period of gradual decline.
- Colour: Colour is one of the important ingredients of fashion products. International organisations offer comprehensive colour forecasting twice a year (fashion seasons: Spring/Summer and Autumn/Winter).
- What are new fashions: New fashions are related to evolving from their predecessors in continuous but gradual progress. A fashion has a necessarily temporal linkage with the preceding fashion form.
- Historical continuity, incidence and duration.

Volume fashion

It is designed and priced for mass market and mass consumption. Volume fashion with mass-produced Fashion is churned by frequent turnover. Volume fashion began after the end of the second World War. Women's ankle-length skirts were cut down to calf high, scandalously knee high. So women's feet and ankles came into prominence. They had to be dressed in a glamorised manner giving rise to the requirement of "FASHION SHOE". Women's fashion shoes with lots of high heels, instep straps, decorative bows, and buckles with a good combination of leather (colour) were formed. Similarly, men's shoe styles sought a breakdown from traditional gentleman's shoes. Young College men took the lead, followed by others. The new era of Fashion gave rise to new demand for creative shoe designers. Volume fashion is believed to be normally evolved due to social and economic changes. The very nature of the footwear industry today is the child of this memorable era.

Fashion and Fad

- Fashion is a particular version of basic style that appeals to contemporary or popular taste. F.A.D. can be defined as a practice followed for a period of time with exaggerated zeal or craze.
- F.A.D. gains extremely extravagant popularity of specific style for a short time, different from true Fashion in having something unexpected. It is a pattern of a craze on a spectacular rise which would undergo sudden demise. In some instances, F.A.D. may turn out to be Fashion. Fashion products show a more gradual growth in sales followed by a plateau in popularity and a continuous period of sales decline. They were quickly accepted but soon became a declining style. Like a firefly, a F.A.D. has its moment of brilliant flash, then instantly fades and disappears. Styles and fashions have reasons; F.A.D.'s do not. Footwear F.A.D.s remain one of human nature's unresolvable mysteries, never to be understood, yet a reality we learn to live with.

6 The Cycles of Fashion & The Industry's Trend System

Fashion Cycles

Fashion is a dynamic mechanism of change through which a potential object is transmitted from its point of creation to general introduction, discernable public acceptance and eventual obsolescence. Fashion is more or less than the prevailing style at any given time.

Ripples & waves

Fashions change but basic styles never. The type of a Derby remains the same though its fashion cycle may oscillate between pointed or square toe, low or high heels etc. Here cycles may be referred to as waves. A significant fashion change in a cycle or a wave involves a radical shift in direction, e.g., thick to thin soles, from heavy to elegant looks. New fashions are introduced every season and every year. These are not cycles but ripples within the same cycle; e.g., an embroidered vamp may be considered as a ripple in a court shoe, and so is a fur tongue in a boot style. It must be remembered that ripples do not have a lasting impact as waves do.

Why Does Fashion Change?

The need for fashion change is due to psychological satisfaction rather than physical benefits. People take a fashion due to market saturation, boredom or desire for novelty. Style, colour and quality reflect the wearer's status, association and image. It is this which pushes people to be turned to the fashion world.

Who are the architects of fashion change?

There is a persistent and widespread myth that manufacturers or designers are the instigators of significant fashion changes and stimulate sales by "forcing" people to buy the new fashions. Both these views are absurd. The movement of significant style is beyond their control. Manufacturers and Designers do not make the fashion weather. They only forecast it by researching the past.

Fashion cycle changes occur due to a combination of political, social and economic factors establishing a new lifestyle. These are the result of a mass shift in Fashion. Change is a consequence and not a cause by itself.

The length of a fashion cycle depends on the following factors:

- The extent to which the Fashion meets the customer's needs
- The consistency that Fashion has with other prevalent trends.
- In America, the average duration of a fashion cycle was approximately nine years. A significant new cycle goes into the ascending for Two years, peaks for approximately five years, then moves through a descendance over the next two years as a new cycle moves upward.

Trickle-down theory

- Here a fashion product is followed first by the higher ranges of Society. Therefore, as it gains popularity among the lower strata, it is mass-produced and becomes a fashion statement.

Bubble up theory

- Some fashion products are popularised by lower strata of Society and then climb up the social ladder. Denim pants, perhaps the biggest fashion statement of the century, are a perfect example of a fashion product which had its roots in the masses and therefore gained popularity among the elite.

The adopters of the Fashion cycle can be classified as:

- Innovators: A small group which wants to distinguish itself.
- Early adopters: They have above-average exposure to Fashion and are more than the innovators in number.
- Early majority: Through not leaders, they are active members of the fashion change.
- Laggards: They belong to people who prefer not to be left behind and therefore join the fashion change.

The stages involved in the adoption of the fashion cycle are:

- Knowledge: An awareness of the existence of the innovation.
- Persuasion: An evaluation of an invention that may or may not be favourable.
- Decision: Leads to acceptance or rejection of the fashion statement.
- Confirmation: Adopter joins the fashion bandwagon.

The fashion cycle consists of four stages:

- Distinctiveness stage: Some customers take an interest in something new to set themselves apart from other customers.
- Emulation stage: Other customers take an interest to emulate the fashion leaders, which results in additional production.

- Mass Fashion stage: Fashion becomes extremely popular, and the product is mass-produced.
- Declining stage: Customer begins to move to another fashion trend.

Sources and inspirations for Design

Designer collections, Architecture, Stylists/Modelists, Nature, International Shop Reports, Forecast Magazines, Fabric Forecasts, Lifestyles, Museums, Film/Media, Footwear Schools, Travels, Club & Street Fashion, Footwear Magazines, Art, Music, Footwear Fairs, Footwear Buyers, Fashion Bulletins, Fantasy, Tanners, Culture, Last Makers, Design Movements.

7 Fashion Design and Style

Individual interpretations or versions of a style are called designs. Styles are forms that tend towards uniformity, and fashions prefer continual change. Design is a unique combination of characteristics within a given class of products distinguishing one item from another. Shoes, sandals, house wear may each be seen as an individual class or product with an infinite number of design possibilities within each class. Differences can exist on silhouette, construction, leather used, colour, surface pattern, accessories used and texture. Each separate Design exists as a highly individualistic creation.

Style

- Characteristic mode of presentation that typifies several similar objects of the same class. A product that has one or more objects is of the same class.
- A product that has one or more specific characteristics that distinguish it from other products of the same category.

Individual interpretations or versions of a style are called designs. Styles are forms that tend towards uniformity, and fashions tend towards continual change. Style is a characteristic or distinctive method of expression, presentation or conception. A style of footwear that is temporarily adopted by a discernable proportion of social group members because that chosen style is perceived to be socially appropriate for the Time and situation. Among the numerous styles, only a small number will be accepted by Fashion conscious consumers at any given time. These accepted styles are deemed fashions. Styles do not become Fashion until it achieves popularity and remains so till it maintains popularity and remains so till it maintains popularity. Fashion is believed to normally evolved due to socio-economic changes. However, there are two types of fashions viz. attributed to being consumer-created and producer-created. The length of a particular fashion cycle depends on the extent to which the Fashion meets genuine needs is consistent with other trends in Society, satisfies society norms and values and does not meet technological limits as it develops. There is yet another description of Fashion though the basic concepts are the same. Every Society above the subsistence economy exhibits changing styles in its pure and utilitarian art form. The rate of change is more significant in the last six to seven decades than at any time in history. Changes occur due to functional forces, and fashion useful forces may be defined as those resulting from our requirements by technological advances of every sort and socio-economic developments which allow for improvement or diversification of what has gone before.

Why does fashion change

In affluent economics, attitudes toward consumption tend to be related more to psychological satisfaction than purely physical benefit. People take to fashion due to market saturation, boredom or desire for novelty. Footwear is not desired to cover one's feet or confer warmth and protection. Style, colour and quality reflect the wearer's status, associations, self-image and attitudes, as well as his more apparent characteristics like age and social class. Fashionable behaviour is stimulated by a range of desires such as personal recognition, self-assertion, attraction to the opposite sex, non-conformity in the face of convention, improvement of appearance, the desire for ownership, display of wealth, fear of ridicule or social disapproval, personal work, and physical wellbeing. It is a non-verbal form of communication. Man is a complex creature, and besides economic activities, one should take more of psychological makeup, the Society in which he lives, and the cultural background to study the buying behaviour and fashion consciousness.

7.1 Interrelated Concepts & Fashion Trends

The interrelated concepts to study of fashion trends are perception, social class, collective behaviour and culture.

Perception

A process through which one makes sense of the world—perception through which an individual receives stimuli and gives them meaning—perception is subjective, limited to existing beliefs and attitudes, and selective in response. All promotional messages do not appeal to all. Perception is further divided into:

- Awareness
- Interest
- Knowledgeability
- Perceived risk.

Consumer risks include loss of status, self-esteem, and social disapproval by making incorrect buying decisions. Many consumers follow rather than lead fashion purchases to avoid financial and social risks.

Awareness

- Primary level: consciousness of particular features such as silhouettes and colour of footwear
- Secondary level: physical characteristics and construction of footwear
- Tertiary level: individual physical and social needs—size, shape and colour—special footwear features and requirements.

Awareness of fashion footwear trends increases with higher levels of social class, income, education, number of organisational membership and magazine subscription. Young men usually are more aware of fashion footwear, particularly those who subscribe to fashion magazines.

Interest

The proportion of people usually interested in fashion footwear.

Knowledgeability

Through Fashion and other magazines, cinema, T.V., Newspapers observations—window shopping, other consumers.

8 The 5 Basic Elements Needed for Shoe Design

In this chapter, we understand that shoe fashion consists of five essential elements which the shoemaker analyses before it can become a popular fashion. Anatomy, style and trends are studied before designing the shoe; this consists of five essential elements the shoemaker analyses to produce a popular fashion shoe product.

Shape

This is the basic architecture of all Fashion, determined by the last. Last is the closest resemblance of the human foot with its curves and arches. Key elements of the shoe shape are toe shape, heel height, side wall shape and edge slant. In all, Fashion begins with form.

The shapes can be pointed, rounded, square or flat. Similarly, heel heights could vary from low, medium, to high fashion heels. The toe shape is an essential aspect of Fashion. Edges slant for the classic or may have a walled shape for the casual shoes.

Design

It is the Design that helps to establish the shoes' basic styles. The Design is composed of the upper patterns, which can take any endless forms. It is primarily here that the designers' creative talent is applied, and the shoes' look becomes so vividly visible to give it its distinctive character. The designs establish the proportion of the shoe. For example, based on the Design, the opening neckline differs in an oxford and a pantafola model. The style lines clearly distinguish the gender of a shoe.

Materials

A given shape and Design assurance attains other distinctive characteristics with the materials used. Shoes of identical shape and Design will look different using different materials. Qualities of materials are:

- The texture is the feel of the materials, which gives it its fashion character.
- Weight is the thickness of the materials, and due to this, different materials are used in various types of footwear for other fashion effects.
- Surface grain and finish give the material its look, which is vital in fashion footwear.

Colour

Colour illuminates the whole illusion of fashion shape, Design and materials. All colours are variations of primary colours but throw light on the enormous range of shades and other subtleties of colour in Fashion. There are different reflections of colour, which is why we use colour terms like bright, pastels, flat or soft, dark or translucent. Colour preferences are very much personalised by customers, and they call it to taste. The selection of colours also reflects moods and personality.

Ornamentation and treatment

This includes the subtle touches that bring together all the other elements of Fashion and presents a fashionable product. Buckles, trims, eyelets, lace ends or fancy buttons give the necessary finish to the product. The French binds, decorative tassels, perforations and similar treatments complete the product for exhibiting to the customer.

9 Fashion Forecasting System

Research and observation are critically important in the fashion business. By watching for directions in designer collections, studying markets, reading the best fashion publications, and observing fashion leaders, manufacturers and retailers can predict what most customers will want in the foreseeable future. To keep up with the changing world of Fashion, one must constantly read trade and customer periodicals. Researching design sources such as museums should also become routine. Fashion awareness should become second nature for anyone serious about a career in Design or merchandising. Fashion forecasters have to identify and analyse signs indicating a change. They base their judgement on experience, awareness of trend-setting designers, fashion cycles, and economic, social and technological influences on the leather industry and the consumer.

Fashion forecasters must do all of the following

- Decide which fashions are prophetic

- Estimate which segment of the market will accept a particular fashion
- Determine at what Time these fashions will be acceptable to target customers.

Fashion forecasters analyse the designer collections for what they call "prophetic styles' fresh ideas that capture the mood of the times".

Targeted Customers

- A noted economist has said that "the central function of the entrepreneur in the fashion industry is far less than an efficient organisation of the production of a given commodity and much more the shrewd anticipation of the changing preferences in his numerically restricted clientele—his small niche in the great neighbourhood of women".
- The adaptation of a trend depends on the particular group of people who are potential customers. Sources of influence and adaptation Fashion varies with the consumers' age range, income level, lifestyle and fashion preferences.
- What are they buying this season?
- What are the activities and occasions for which they need clothes?
- What are their priorities?
- Are they innovators or imitators?
- What people, periodicals, or other influences will affect their choice?

How does location affect timing?

The timing of fashion-trend adoption and adaptation depends on the geographical location of potential customers. Manufacturers and retailers determine how long it takes for their customers to accept new Fashion; by knowing the difference between origin and acceptance, they can predict when their future products will be accepted.

Analysing customer preferences

Market Research

Manufacturers and retailers "shop the market to see what is selling well and why". They compare the styling, price, fit and quality of lines that compete with their own. They also analyse lines at higher price levels to get a feeling for directions that may soon affect their price levels. Market information sources include analyses of new collections, sales records, and customer surveys.

Sources of Information

Concerning the firm's targeted customer group, the fashion forecaster calls on the following.

Careful observation of current events that have captured or are likely to capture the imagination of customers and affect the styles they will prefer. Awareness of the current lifestyles and dress of those men and women most likely to influence what the firm's own customers will ultimately adopt. Study of sales trends in various sections of the country, not only for the forecaster's own company but for competing companies to whatever extent is possible.

Sources of Information

An intimate knowledge of the fashion opinions of their sources of supply.

Familiarity with professional sources of information such as fashion reporting services, fashion periodicals, consultants' opinions, analyses offered by resident buying offices and the like.

- Exchange of information with non-competing concerns
- Understanding of and constant awareness of the inevitable and evolutionary nature of changes in Fashion.

Importance of awareness

Designers surround themselves with photographs of ideas, fabric swatches, and anything else they will stimulate creativity. They leave their studies to shop, visit museums, study nature, attend the theatre or people-watch. Awareness is the key to developing good Design. Designers must study almost constantly to capture the subtle nuances of the environment. Merchandises, too should be open to everything around them that affects the clothing industry: economics, politics, Art. Good taste and a good sense of timing come from exposure. Constant exposure to beautiful merchandise helps a buyer distinguish genuine beauty and quality from pads and mediocrity.

Research and observation are critically important in the fashion business. By watching for directions in designer collections, studying markets, reading the best fashion publications, and observing fashion leaders, manufacturers and retailers can predict what most customers will want in the foreseeable future. To keep up with the changing world of Fashion, one must constantly read trade and customer periodicals. Researching design sources such as museums should also become routine. For anyone serious about a career in Design or merchandising, fashion awareness should become second nature.

10 A Look at Tomorrow's Trends

Anything goes! There are no limits—for colours, materials, shoe styles or designs. 'No limits—for no-one & nobody!' One can feel something big happening here. Thinking out of the box, daring to do something new while retreating into all things proven and familiar at the same time. Though every key here is the interplay of colour and the related materials. This symbiosis creates new possibilities and looks that become a creative challenge for all of us! Add to this the ever stronger competition "Leather versus High-Tech". The positive impact of high-intensity colours is undisputed, and the importance of colours—be they loud, neutral or feminine and delicate—is pivotal for changing seasonal Fashion. This positive impact of high-intensity colours, materials, shoe styles or designs remains undisputed and essential for seasonal shoe trend styling.

Joie de vivre, energy and power also having visions for the future, is the message of the changing seasonal trends! Social media increasingly characterise our everyday lives. Here, however, fakes are permitted. Inspired by modern paintings, L.E.D.s and screenshots or nocturnal club life, psychedelic effects are produced full of extravagance and glamour. Education, Heritage, Art and Culture, Sculptures, Museums and Architecture, Ateliers, Research Labs, and Technology also fit the picture.

Examples

It can be noticed that in certain seasons, there are excesses in shoe fashion, colour and materials. Prints become even more fancy, wild and extreme. Thanks to digital technologies, authentic motifs are even harder to identify and categorise. Furthermore, the world of bits, bytes and pixels brings about image transitions that our eyes have never seen before. Artificial and shiny fabrics dominate the material selection and technical finishes—be they vinyl or patent, metallic, mirroring or transplant—anything goes! The mix of modernity and classic elements is often also seen in the materials as part of the game.

Through the 60s: Shoes were sold, but back then, it was not a necessity—it was a luxury. People wanted to wear them to look different. In the eighties, there was some colour to the shoes being sold. People wanted to wear shoes and move on. In the early 2000 shops started taking shape, and we all remember shops on high streets, and some malls selling shoes. Shoe Designers need to constantly understand the Upward Trends and Sliding Trends for the season in address.

Studying constantly the behaviour, for example: What are the Rising Trends—are there Shiny patent and polished leather, Silver sports and formals, influences from the 40s, 60s, and 80s, is there a larger percentage of boots to shoes, is Fashion influenced by military boots & soccer shoes, narrow toes, plainer dress uppers, sports shoes with all-over prints, green Fashion, soft unstructured uppers and skinny flexible soles or the opposite with chunky soles. Likewise, a study of the Falling Trends could be—Matt Leathers, inspiration from the 50s, a larger percentage of shoes to boots, bulky square toes, fussy brogues and frayed edges, vintage, stiff uppers and thick inflexible soles. A lot of attention is paid to honing 'Design' capabilities, introducing technologically advanced tools to aid in the 'design process and developing' 'merchandising' expertise amongst our product design teams in the footwear companies.

11 Conclusion

In a globalizing economy, design is perceived as a new engine of economic and industrial growth with breakthrough innovations playing a pivotal role in positioning leather in the global arena by value addition and enhancing competitiveness. Leather is undoubtedly a fashion product for which colour, texture and other highlights add to the fashion value of creatively designed leather products. To emerge as a strong global player in the world leather trade, all efforts to take proactive measures to be ready with the fashion leathers when Fashion does emerge is crucial. Success in working with high-end customers requires building design capabilities that enable us to offer exciting collections that can compete internationally. Much attention is needed to be paid to honing 'Design' capabilities, introducing technologically advanced tools to aid in the 'design process and developing 'merchandising' expertise amongst our product design teams in the footwear companies. Leather has been used by humanity for clothing & accessories since the dawn of time, covering and protecting the human body against external aggressions whilst also facilitating daily life with accessories such as shoes, belts or the purse, the ancestor of the handbag. Their primary function, solely utilitarian, soon bore other deeper symbolic dimensions that came into being, including different levels of identification and adornment. Leather has become centre stage within today's fashion realm. It is the primary material for making world-renowned iconic luxury products such as the Hermès Kelly bag or Christian Louboutin red sole pump. It is also a contemporary key direction for future trends that run the fashion system.

Leather in an Innovative World of Materials—A Classification

Michael Meyer and Sascha Dietrich

1 Introduction

Leather is able to protect against wind, heat, cold, sun and mechanical injury, it is a long-lasting material with a high wearing comfort. It becomes heavy when it gets wet, though it remains a stable material, and it is not easy to clean without the application of a top coat finish. Leathers are made from hides, but an impressively broad material's property spectrum can be adjusted ranging from hard to flexible depending on the manufacturing technology. While leather had been used historically for many different applications, it is today mainly used for shoe, upholstery, clothes and accessories and its natural surface structure and appearance plays an important role on sale.

Textiles are of lower weight. They cannot protect against wind and rain. However, appropriate coating compensates this disadvantage. Today, the disadvantages of both material groups have been overcome by the generation of increasingly complex designs of materials and material combinations in final products. To achieve a highly functional performance, shoes and outdoor clothing are combinations of dozens of materials, textiles, leather, synthetic membranes, rubber and synthetic polymers. Most of these functions can only be achieved by the combination of several layers of different materials. In recent years, the materials are enriched with additional functions as electric heating, electronic components and data processing.

To be used for upholstery, clothing, shoe, and accessory the materials or their combinations have to be flexible to some extent, and they must be processable. For many applications materials have to be tight against fluid water and wind stopping, open for water vapour, they should absorb more humidity the nearer the material is to

M. Meyer (✉) · S. Dietrich
FILK Freiberg Institute gGmbH, 09599 Freiberg, Germany
e-mail: michael.meyer@filkfreiberg.de

S. Dietrich
e-mail: sascha.dietrich@filkfreiberg.de

S. K. Janardhanan and L. A. Zugno (eds.), *Emerging Trends in Leather Science and Technology*, https://doi.org/10.1007/978-981-99-9754-1_8

the body. This property correlates with the wearing comfort. Furthermore, materials have to be robust enough to achieve a useful life span which is common for the aimed application scenario.

Today, materials that are successful on the market have to achieve a long list of technical properties which are determined in regard of their final use. The highest requirements are needed in aviation, public transport and automotive. Lists of more than hundred parameters to be tested are not unusual. They comprise physical properties, chemical properties, residues of chemicals and contaminants, stability under climate changing conditions, biochemical properties, and many more. In fashion the technical requirements are usually not as high as in the transport sector and special designs and haptics become more important.

A further very important requirement is that materials have to be available in industrial scale at prices which can compete with material prices in their segments, e.g. fashion, transport, upholstery, shoe or accessories, unless a final article is only sold in limited numbers and the manufacturing process is only manual production. This means that raw materials for the production of these materials have to be available in sufficient amounts.

Due to increasing effects on the ecosystems along the materials production, their life and their disposal additional parameters have to be addressed apart from the technical properties of materials. The materials or products made of them shall be part of a cycle which is now often investigated by life cycle assessments. Part of these assessments are an evaluation of whether the materials are bio-based, fossil-free, biodegradable, that parts of the material stem from recycled raw materials or by-products of other processes (e.g. polyethylene terephthalate (PET), fruit pomace, fruit plant fibres, cork), or that the composite material itself is completely recyclable. Finally, this list culminates in the requirement that the materials have to follow a vegan mindset. These materials have been summarized as next-gen materials, which claim to show high performance, to be animal-free and more sustainable (Siu, 2022).

In contrast to leather, the synthetic or bio-based alternative materials are not exactly defined. These materials comprise a broad number of different compositions, which are usually adapted to a function as part of a final article. Fashion as well as automotive and furniture companies are continuously asking for new materials besides leather, artificial leather and textiles for their use cases. During the last decade a plethora of new materials has been proposed and prepared in lab scale to be applied in shoes, accessories, and upholstery. Some of them appeared on the market and have been investigated in a comparative study for their technical performance (Meyer et al., 2021). The investigated materials comprised a shoe upper leather, a PUR coated artificial leather, and the materials Pinatex®, Kombucha, Vegea®, Desserto®, AppleSkin®, Teak Leaf®, SnapPap® and MuSkin®. However, these are just examples for many other materials which show similar compositions. Other materials are available for the same or similar applications, e.g. ELeather, leather board, or textiles coated with cork and wood veneer. Beside leather, the principal composition of most materials consists of several different layers, of which each has one or several different functions.

2 Historical Review

The use of animal hides as raw material intensively correlates with human evolution. As long as mankind lived as hunter and gatherer, humans used all materials, which nature could them provide, mineral as well as plant and animal based materials. Hides as parts of hunted animals were used to prepare tools, shoes, cloths, arms and wineskin. Plant-based materials e.g. grass served for insulation (Dickson et al., 2003) and the Ice man already used fungi-based insulation materials as part of his hat. Net making and braided baskets were known in the late Paleolithic but first weavings on looms were found in the early Neolithic (Grömer, 2016). 5000-year old paintings in Egyptian graves showed tools used in tanneries which allow to reproduce the technological steps of leather making. At this time, to mummify deceased Kings textiles were soaked with natural resins, waxes and pitch, the first coated textiles. For thousands of years no progress was observed neither in the development of new materials nor in that of new technologies. Only industrialisation, a synonym for replacement of labour by machines, led to the invention of textile machines to harvest fibre plant, to separate fibres from capsules, bark or non-fibrous components, to spin threads, and to weave and knit textiles. The prices dropped down. In the leather industry the raw material could not be improved, but the labour intensive and time consuming processing steps of mixing and soaking were improved by rotating drums, and machines were invented for the mechanical treatments as fleshing, splitting, shaving and embossing. Both branches benefitted from the success of the chemical industry in the late nineteenth century which synthesised pure chemicals as basic chemicals, enzymes, fatting agents, surfactants, and dyes, and tanning agents for the leather industry. The use of coatings based on oil, waxes, paraffin and linseed oil to manufacture coated textiles, e.g. oil clothes, chintz, linoleum, remained state of the art until the breakthrough of the industrial polymer synthesis. The discovery of the polymers polyvinylchloride (PVC) and polyurethane (PU) in the mid of the twentieth century and their use to coat textiles led to a breakthrough in function and output. The prices for these so called synthetic leathers amounted to only one tenth of leather. However, these advantages were gained by the disadvantage of a poor wearing comfort. The synthetic polymers do not absorb water and they make the materials completely tight against water vapour.

3 Materials Structures—Schemes and Microscopic Pictures

The technical requirements for leather and leather-like materials to be used in fashion are lower than in most technical applications, e.g. in the automotive or the public transport sector. Three groups of parameters are important, which comprise physical, chemical and comfort parameters. The commonly measured physical parameters are tensile strength, tear resistance and flexural resistance. Comfort parameters can to some extent be reflected by the water vapour permeability and water absorption.

The drawings (Fig. 1) reflect the schemes of the microscopic pictures and show leather and the different compositions of the materials used as leather substitutes.

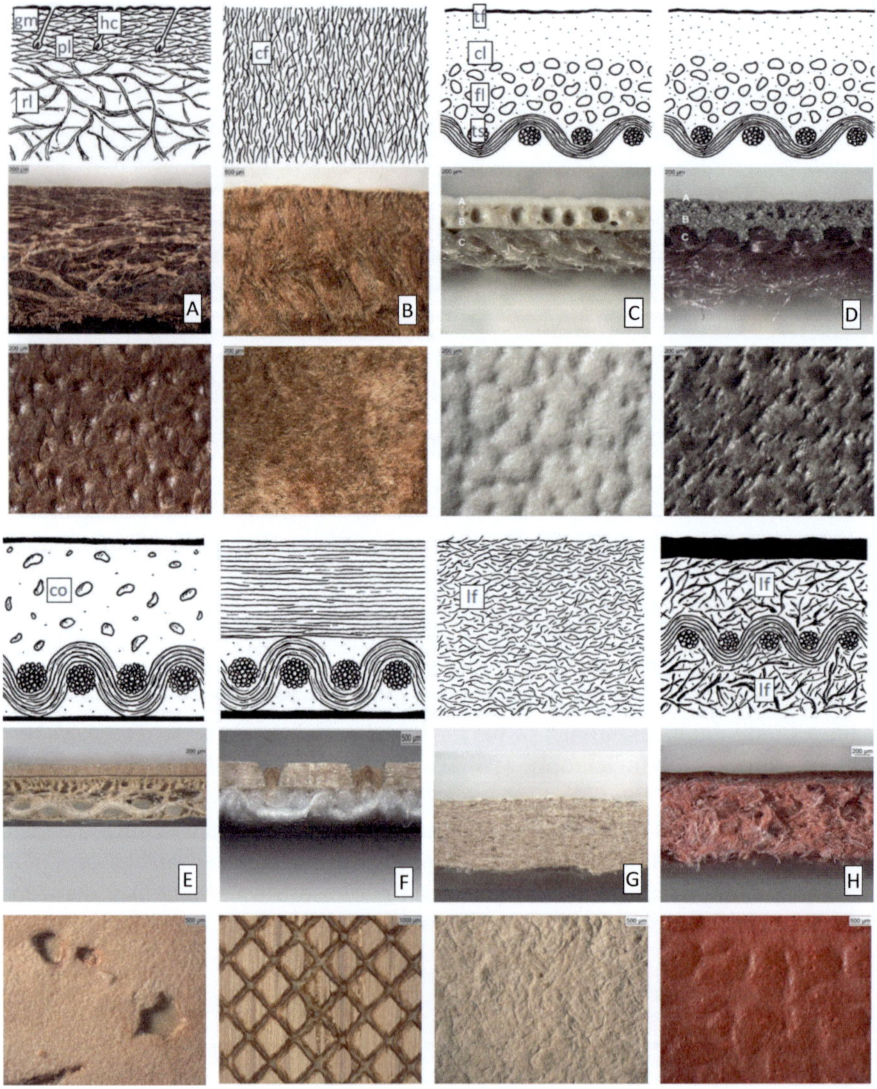

Fig. 1 Microscopic structures (cross-section middle; surface below) and schemes (top) of leather (**a**), fungi material (**b**), artificial leather with (**c**) and without (**d**) fruit-based additives, cork (**e**) and veneer (**f**) coated textile, leather board (**g**) and ELeather (**f**). The drawings show the principal compositions. (rl..reticulary layer; pl..papillary layer; gm..grain membrane; hc..hair channel; ts.. textile support; fl.foamed layer; cl..compact layer; tf..top finish; cf..chitosan fibres; lf.. leather fibres; co..cork layer. More details see text. Drawings by Enno Klüver, FILK)

An animal hide which is chemically cross-linked is called leather, when its original fibrous structure, as it is grown, is still preserved. Leather had been manufactured in the past for many different applications that needed a flexible material with high tensile strength and robustness. It shows a gradient of collagen fibers from the backside to the top with a very fine layer of fibres on top appearing as slightly glossy surface (Fig. 1a).

The hair channels are responsible for the typical appearance of the surface (grain). Therefore, leather preserves principally the structure of the skin with some of its biological functions as heat and water diffusion control and protection against mechanical injury. Leathers in the transport sector are usually additionally finished by a final coating with synthetic polymers e.g. polyurethane or acrylic finishes. If the surface is additionally embossed it appears not anymore as grown structure but as a technical grain, and the coating affects the comfort behaviour.

Artificial leathers (Fig. 1d) made from Polyurethane or PVC consist of a flexible fabric which is coated with synthetic polymers often with three or four layers. The top finish is again a polyurethane or an acrylic layer similar to coated leathers with a leather imitating structure embossed. The subsequent layers consist of a compact layer of soft PVC or polyurethane followed by a second porous layer from the same polymer. Often the textile is glued on the porous layer by an adhesive. Therefore, artificial leather is a multilayer system which combines different functions of leather by different separate layers.

The synthetic films usually contain fillers to reduce the costs per area. For this purpose low-cost inorganic materials as e.g. limestone is used but as well cellulose particles. The materials Apple skin® or Desserto® contain by-products of the food chain (fruit pomace) and plant components, resp. which are expected to be available in sufficient amounts without competing with a use in the food chain (Meyer et al., 2021). The brown inclusions in Fig. 1c reflect some of these filler particles.

The fungi based material (Fig. 1b) consists of grown very fine hyphae. Currently, most of these materials are offered as single-layer materials without any textile support and without a topcoat but it is expected, that these materials will have to be supported and coated to achieve a usability which will be accepted on the market. The fungi layer consists partly of chitin, a structural microfibrillar polysaccharide (monomer N-acetylglucosamin). However, the cell walls of fungi are composed of many other components accompanying the structural part as other saccharides, proteins, lipids and pigments. The compositions change depending on the species, age, nutrient composition and supply and the environmental conditions. The biological function of these hyphae is to resist against the different internal and external osmotic pressures, and to harmful physical, chemical, and biological aggression, provision for the shape of the cell, recognition of external surfaces, inert or biological, the reception of external stimuli, selective permeability to large molecules and the accumulation of molecules important to the physiology of the cell, including nutrition (Ruiz-Herrera, 2012).

The porous material is composed of fine brown fibres with a fibre orientation perpendicular to the surface. These fibres appear at the surface as a fine lawn without a distinct structure. The material which has been investigated (Meyer et al., 2021) is grown in nature as wood and tree fungi called trama and is only available in limited amounts. To make higher amounts of fungi based materials available, recently other fungi species have been investigated by several research groups, which are growing faster and which can be cultivated biotechnologically (Cerimi et al., 2019; Gandia et al., 2021; Haneef et al., 2017; Jones et al., 2020; Vandelook et al., 2021).

Additional biotechnologically produced materials (not shown), e.g. bacterial cellulose, which had been proposed as well as an alternative material consisting of cellulose fibres which are partly glued together. Upon oven drying, these materials build compact layers. They show multilayer structures only on a microscopic scale and fibres appear only at the electron microscopic level at high resolution. The flexibility is reduced upon oven drying and the materials appear paperlike without sufficient drapability (Illa et al., 2019; Scionti, 2010; Zhang et al., 2011).

Leather board has no distinct structure (Fig. 1g). It mainly consists of natural rubber which is filled with leather fibres that have been intensively ground. In Europe it is required, that the content of leather fibres is higher than 50% to be named as "leather board" according to the European standard EN 15,987. The binding system has to apply to the fibres to achieve a homogenous board-like material with sufficient mechanical properties.

The cork material (Fig. 1e) reflects systems of textiles which have been covered by nature materials as cork, leaves or wood veneer (Fig. 1f). Flexibility can be fairly achieved by special additives, coatings or in case of veneer by laser-cutting of rhombs when the multi-layer system is already established. The impressive surfaces of the cork material and wood veneer are mainly design driven. Specific other functions as high technical performance or higher comfort are missing.

Finally, "engineered leather" (ELeather; Fig. 1h) reflects a textile support which is flocked on both sides with larger leather fibres than used for leather board. These fibres are hydro-entangled with a low binder content and the composite is then coated by a polyurethane finish.

4 Manufacturing Technologies

Principally, the structure of the skin is preserved during the leather making process. Skins are by-products of the meat and milk industry and not manufactured with the aim to be tanned. They are single pieces which have to be handled as such and they have to be treated batchwise. The surface of the skin is not uniform, nor its thickness and its structure.

Technologies to manufacture artificial leathers aim to overcome the mentioned disadvantages of leather. Therefore, many advanced technologies on the industrial scale comprise roll-to-roll processes which combine several layers to mimick the functions of leather.

More recent developments include biotechnological processes using bacteria, fungi and their mixtures to produce cellulose, proteins and chitin / chitosan for applications for which leather is used today. Yet, these processes work on a batch basis and a continuous production is not in sight.

4.1 Leather

Leather is made by unhairing of skins, the separation of the collagen fibres by chemical crosslinking and lubrication, and a mechanical treatment to adjust the final properties. The manufacturing technologies of leather are manifold and comprehensively described by outstanding textbooks (Covington, 2009; Heidemann, 1993; Stather, 1957).

The final properties of leather are adjusted by the choice of the raw material and the technology of leather making e.g. the kind of tanning and retanning agent, the kind of fat and the fat content and the intensity of mechanical treatment. Variations of thickness, surface and structure are homogenised by mechanical means e.g. fleshing, splitting, shaving, re-tanning and embossing. The final finishing of several microns with synthetic polymers is necessary in applications which require very high technical performance, e.g. fastnesses in the transport sector. Resulting leathers cover a very broad spectrum of properties. They can be board like as sole leather up to soft and flexible as chamois leather. All materials consisting of tanned hides are named leather, however.

4.2 Artificial Leathers

The description of artificial leathers is much less clear. Since several decades, artificial leather mainly consists of synthetic polymers commonly produced by coating of flexible or drapable supports, mostly textiles, with synthetic polymers. The textiles are usually made of woven or non-woven polyesters (PET) and sometimes knitted fabrics. Other synthetic fabrics and also cotton, viscose and other bio-based textiles and their blends with synthetic fibres are used as well. The technologies to manufacture artificial leathers differ fundamentally, depending on the polymer used for the coating layer and its properties. Commonly polyvinylchloride (PVC) or polyurethane (PU) are most important, but recent developments comprise e.g. silicones, polylactic acid, polyhydroxyalkanoates and others. Two different manufacturing principles are commonly used, which are thermoplastic and wet coating, respectively. Further, coating can be manufactured directly or as reverse coating (Träubel, 1999).

4.3 Polyvinylchloride (PVC)

PVC is a thermoplastic polymer, basing on vinylchloride, which is polymerized e.g. by emulsion (E-PVC) or suspension (S-PVC) polymerization. S-PVC granules are highly porous while E-PVC granules are more compact. To manufacture artificial leather the hard and brittle PVC has to be added by substantial amounts of plasticiser to achieve soft PVC. Plasticisers are high molecular esters of phtalic acid (e.g. dioctylphtalate—DOP), cyclohexanedicarbonicacid (diisononlyester—DINCH), citric acid or adipic acid which are added up to 60% of the whole soft PVC system (Albano et al., 2020; Hosney et al., 2018; Jia et al., 2018). If bio-based plasticizers are used, they account for a comparably high bio-based content in soft PVC.

Most commonly, a plastisol is prepared by using E-PVC, which is mixed with plasticiser, stabilisers, fillers, colouring agents and other additives. Plastisols are viscous fluids which consist of PVC granules, plasticiser and additives as inhomogenous mixture on the microscopic level. Preparations are stable up to months at room temperature. To manufacture films, the plastisol is spread by a knife on a transfer paper and heated in a heating tunnel (for technical drawings see (Smith, 2010)). By heating the wet film the plasticisers increasingly diffuse into the granules acting as external plasticisers which separate the polymer chains. A microscopically highly viscous, but homogenous solution of PVC polymer in plasticizer which contains all other additives is achieved at temperatures >130 °C. This mixture solidifies by cooling.

Foam coats are prepared by addition of the blowing agent e.g. azodicarbonamide or sodium-bicarbonate to the plastisol and this preparation is coated as the last layer before the textile support is laminated into the wet compound. Finally, the composition of different layers is heated at a temperature higher than the decomposition temperature of the blowing agent. Bubbles develop and a foam layer is formed. Alternatively, the composition is cooled down and a third adhesion coat is added which glues the textile support to the PVC-film. The surface of the material can be adjusted by a final embossing step.

Artificial leathers made from PVC are sold for low prices (2022: upholstery 5 €/ m^2). The price is mainly determined by the costs of the raw materials and connected to the naphta prices. “Green” soft PVC, synthesised from bio-ethanol, using renewable energy along the whole process and added with bio-based plasticizers is possible (Alvarenga et al., 2013a, 2013b; Martinz & Quadros, 2008) but is much more expensive than fossil-based PVC. “Green” PVC does not differ from fossil-based PVC, neither in its chemical structure nor in its performance. It differs in regard to the used raw material and the production process of the PVC.

4.4 Polyurethane-Coating

Polyurethane based artificial leathers are as well manufactured by a reverse process, but for selected other applications direct coating is possible. The polyurethanes are applied as solution in organic solvents, as high solid, or as aqueous dispersion. Polyurethanes are synthesized from polyols and isocyanates. Polyoles are basing on polyesters or polyethers, and they are used with varying molecular weights and additional side chains. Therefore, the chemical variety of polyurethanes and the variation of the resulting properties are very high.

Direct coating of PUR solution is usually performed onto the substrate by knife on air, knife over roll or knife over rubber blanket application (Smith, 2010) and the penetration of the polymer into the substrate is adjusted by the kind of this process. A deeper penetration increases on one hand the hardness of the final article, and on the other hand adhesion between the support and the coating layer is improved. These technologies are used to coat fabrics, which are not necessarily sold as artificial leathers, because the resulting materials appear more rigid than leather.

To manufacture artificial leathers the indirect coating process requires the construction of the complete system according to a reverse principle similar to the PVC coating with plastisols. The top layer is applied on a transfer paper which is often coated with silicone, the compact layer is coated on the dried film with the same technology as described above. A foam layer is either applied as the compact layer and foamed by chemical agents or the PU-dispersion is physically whipped and then coated on the previous layer. Finally, the support is added into the wet polymer films. To achieve sufficient final stability, systems are applied which are composed of two components, polyoles and monomer, pre-polymer, or masked isocyante respectively. Demasking of masked isocyanates and the reaction of the isocyanate component to urethane groups is achieved by heating the material preparation higher than 165 °C in drying and heating channels.

Today, polyols as one component of polyurethanes can completely be produced from bio-based raw materials. The used precursors comprise e.g. glycerol, triglycerides, fatty acids and (poly)saccharides and the commercial availability is high (Furtwengler & Avérous, 2018; Singh et al., 2020). In contrast, the supply of isocyanates which are completely made from bio-based raw materials is possible as well, but currently is not economic.

Recently, several PU-based materials came on the market which contain by-products of the food industry (e.g. AppleSkin®) or other plant residues (Desertex®, Vegea®). These materials act as filling agents. Micro-cellulose of common PU-formulations is exchanged by these bio-based agricultural by-products.

4.5 Silicone Coatings

Silicones are used as low molecular oils, as higher molecular sealing grease, and as very high molecular resins and rubbers. The use of silicone oils in the textile industry has a long tradition especially as hydrophobing and antisoiling agent. Silicone rubbers are used to impregnate and coat textiles to achieve closed and tight surfaces. They show high mechanical stability which is improved by the addition of silica, and a temperature stability up to 250 °C. Their cold flexibility is high at temperatures lower than zero centigrade (Cuthbertson, 1956). Recently, artificial leathers have been developed as well on the basis of silicone rubbers, firstly of two compact layers and subsequently by coating of textiles with a foamed silicone layer and a compact coating. Foaming can be achieved by nitrogen-free blowing agents and by the addition of polymer based microspheres. A certain variety of different silicone monomers and pre-polymers allows a flexible silicone chemistry but not comparable to that of polyurethanes.

Resulting artificial leathers are highly flexible over a broad temperature range, they are hydrophobic, flame retardant, they do not contain plasticiser, they show high fastnesses against oils, soil, UV-irradiation, and sweat, and they are very stable against hydrolysis. However, artificial leathers made of silicone are costly due to the high raw material costs and the silicone-like haptic of the surface has still to be improved.

4.6 Polymers from Bio-based Monomers and Intermediates

Common coating systems for flexible substrates (artificial leathers) are often based on different polymers for the substrate and the coating, and sometimes even the coating layer consists of different polymers. For the substrate mostly synthetic polyesters are used and the coatings consist of PVC or polyurethanes with a top finish of acrylates. Material recycling of these systems is only possible by chemical methods. Mostly they are disposed or incinerated, however (Maisels et al., 2021). In artificial leathers and coated fabrics the intensive adhesion of the components does not allow a mechanical separation, and recent developments aim to use monomaterial systems. PVC and polyurethane supports are not available for these purposes.

Therefore, it is aimed to develop bio-based and biodegradable thermoplastic polymer systems e.g. polylactide (PLA), polyhydroxyalcanoates (PHA), polyhydroxybutyrate (PHB), or polybutylenesuccinate (PBS) to manufacture artificial leather. The monomers or the polymers itself are manufactured by bacteria in biotechnological processes (Auras et al., 2022; Patti & Acierno, 2022).

Currently, poly(lactic acid) (PLA) is the most commercially successful bio-based polymer. The monomer lactic acid is produced through fermentation from starch, sugar, or pre-treated leaves, stems and stalks from corn fibre, corn stover, sugarcane bagasse, rice hulls, woody crops, and forest residues (Singhvi & Gokhale, 2013).The global production capacity achieved around 300 kT/a in 2020 (Jem & Tan, 2020). While fibres and textiles can be manufactured from PLA (Siegmund et al., 2009; Yang et al., 2021), albeit with poor thermal, mechanical and rheological properties, coating is currently not possible with the available PLA derivatives and formulations. The PLA based coatings appear brittle and do not achieve the necessary physical and mechanical parameters. Intensive investigations are performed regarding plasticising, to adjust the glass temperature and to improve the mechanical properties especially of the coating.

4.7 Leather-Board and Vegetable Substitutes

Leather-board is a material which consists of leather fibres which are embedded in natural rubber and sometimes additional synthetic polymers as e.g. acrylates. The name leather board was already coined in the late nineteenth century for board like materials which contain leather fibres or dust (Zaehnsdorf, 1890). Leather board is manufactured similar to the paper process. The solid components of an aqueous dispersion, which contains leather fibres, additives and natural rubber as solidifying and binding agent are precipitated with aluminium sulphate, the precipitate is pumped on the continuous sieving belt of a Foudrinier machine (Zhang & Yang, 2020) and the water is removed along the process first by sequential mechanical processes (gravitation, suction, pressing), then by thermal drying and finally compacted on a calender. The board-like material is wound up on rolls at the end of the process. Embossing and a final finishing with different lacquers are common processes to adjust the final appearance (Träubel, 1999).

The process of leather-board manufacture was firstly adapted from the paper and board process to reuse leather fibres as by-product of the leather industry. However, plant based fibres can be bound as well by natural rubber, e.g. recycled cotton from the textile industry, sisal, hemp residues or plant fibres which are by-products of the food industry. Recently, dried fruit pomaces, extracted coffee grounds and other plant based agricultural by-products are options to manufacture animal-free products with properties similar to leather board with the same technology. However, the materials show low water vapour absorption, the materials appear stiff and flexibility is limited.

4.8 Biotechnologically Produced Materials

Kombucha

Kombucha tea is a beverage obtained from a symbiotic culture of acetic and lactic acid bacteria and several yeasts. The sour taste results from the fermentation of sugar, which is added to the tea and which leads to a mixture of several organic acids. Furthermore, a biofilm is produced which consists mainly of bacterial cellulose and which binds about 200 times of its own weight of water (Villarreal-Soto et al., 2018). Drying of this bacterial cellulose film results in a paper-like material, which has to be washed intensively in advance of a use, due to the many acids and aromas which are still part of the biofilm. A production as continuous film seems currently no option, because the main product Kombucha is a beverage and the solid film is only a by-product.

4.9 Fungi Based Materials

While Muskin® is trama, a part of the parasitic wood fungus *Phellinus ellipsoideus*, Mylo®, ecovative® and MycoWorks are examples for fungi based materials or companies which manufacture and sell biotechnologically manufactured fungi materials. These fungal materials are produced in semi technical scale by liquid or solid state fermentation. While the former requires liquid nutrition media containing low molecular weight sugars e.g. molasses in a strictly controlled biotechnological process, in solid state fermentation the fungi grow e.g. on saw dust enriched with easily digestible components like sugars, fats and proteins. The formation of stipes, caps and spores has to be prevented by a strictly controlled environment adjusting especially the humidity and oxygen content. The formed mats have to be washed, stabilised mechanically and chemically, dehydrated, and enriched with humectant agents. If the harvested mats are split, they appear as velour like material. If they are embossed a desired final structure can be achieved, but the resulting materials appear paper-like, because the basic structure chitin-chitosan-fibres behave similar to cellulose fibres upon drying (Jones et al., 2020). Up to now manufacture is still a batch process, but recent research proposes a continuous process in future (Vandelook et al., 2021). Upstream processing needs a sterilisation of the raw materials and nutrients and downstream processing requires this treatment step as well. Drawbacks of the materials are their hydrophilic properties and their low mechanical stability. This has been tried to improve by intensive chemical treatment as crosslinking and addition of humectants and only intensive dying leads to attractive colours for fashion. Therefore, a couple of technical an economic aspects of processing have to be solved before these materials will compete with leather and synthetic materials (Gandia et al., 2021).

4.10 Bacterial Cellulose

Bacterial cellulose (BC) has been studied intensively over the last 130 years (Mühlethaler, 1949). After the discovery of cellulose producing bacteria *Acetobacter xylinum* in the late nineteenth century the structure of this kind of cellulose, many cellulose producing organisms, and the metabolism pathway of their cellulose production have been investigated. BC shows long nanofibers with highly parallel ordered cellulose molecules without the impregnation of lignin. Therefore, their structure and their physical and mechanical properties differ from that of vegetable cellulose (Iguchi et al., 2000).

During the last three decades especially research and development of the biotechnological production was intensified (Andriani et al., 2020; Esa et al., 2014; Huang et al., 2014; Hussain et al., 2019; Iguchi et al., 2000; Ul-Islam et al., 2020). 30% of the production costs account for the growth media. It has been tried to reduce these costs by the use of agro-by products as melasse, vinasse and others which are available in high quantities (Revin et al., 2018). However, up to now the use of BC prevailed only in the medical field and in some food applications (nata de coco). During biotechnological production growth times are in the range of several days to achieve cellulose concentrations of up to 10 g/L (Andriani et al., 2020; Revin et al., 2018). Therefore, in comparison to plant cellulose the manufacture of bacterial cellulose is not competitive. The interesting technical parameters of bacterial cellulose as long nanofibres with high tensile strength do not balance this economic disadvantage yet. Finally, flexible soft fibrous materials can only be achieved by freeze drying. Convection drying leads to thin but compact films.

4.11 Cultured Meat and Cultured Skin

Latest developments show intensive research to manufacture different animal-like tissues including the skin by cell culture techniques. First products are already marketed as meat substitutes for the food industry. Reasons to look for alternatives to the animal based meat production are manifold. The digestion of cellulose based food to grow ruminants (e.g. cattle, sheep, goat) leads to climate gas emissions though often sheep and cattle are the best option to use arable land with low soil fertility. Other animal feed as e.g. soy and crops are in competency with human food, however. Alternative sources of easily digestible proteins should further be cruelty free, prevent diseases, and religious arguments are discussed as well. A large scale production and the market availability of serious amounts of cell-culture based meat alternatives are still missing (2022) (Siddiqui et al., 2022; Ye et al., 2022). The estimated production costs are in the range of 63 US$/kg meat. The main costs are for medium and the bioreactors (2/3) and labor (almost 1/3) (Garrison et al., 2022), and regarding emissions of climate gases and safety aspects cell culture meat is not necessarily of advantage (Chriki & Hocquette, 2020; Lynch & Pierrehumbert, 2019).

The developed cell-cultured products are fibrous cakes similar to minced meat and finally presented as burger or nuggets (Woll & Böhm, 2018). The resulting protein materials are not tissue like. The complex structure of the skin with the different layers (Fig. 1a) and the intensively woven extracellular collagen fibres has not yet been developed. In vitro skin models, which are used in the pharmaceutical and the cosmetic industry usually consist of a thin collagen layer and especially the living part of the skin of several microns, which is the epidermis with its barrier function and the cells in proliferation mode (Flaten et al., 2015). Therefore, a cell-culture based substitute for skin as raw materials for tannage to manufacture leather like-materials appears not realistic in near future.

4.12 Fibrous Plant-Based Materials

Examples for fibrous plant based materials are manifold and the reasons for their selection are different. Usually, plant based raw materials are used as fabrics supports, as additives and as fillers. Supports are e.g. cotton, linen or hemp, and binders as natural rubber or alginate, and combinations thereof (for examples see section leather board).

The fabrics are manufactured by textile techniques as e.g. weaving, or knitting or by entanglement techniques to get non-wovens. The fabrics can be stabilized by chemical binders and / or coated with polymers. E.g. Pinatex® is offered as leather alternative. It is prepared by separation of the cellulosic fibres of pineapple plants from the leaves mechanically. The resulting long fibres are washed, dried and purified again by carding to remove any impurities. The resulting flock is combined with 20% polylactic acid (PLA) as the binder and compacted into a nonwoven. Finally, a chemical top coat based of polyurethane is applied (Anonymus, 2021). Because PLA can be completely (Singhvi & Gokhale, 2013) and polyurethane mostly manufactured from bio-based raw materials the composite Pinatex® achieves in sum a bio-based material content of ca. 95%. SnapPap® in contrast appears paperlike. It is a cellulose based nonwoven with the fibres bound by an acrylic polymer.

Often, plant based components are just used as design elements, however. Nuo® is a wood coated textile. A veneer is glued on a textile, ground, optionally finished by a lacquer and then cut in small triangles to achieve flexibility (Fig. 1f). Cork, is used as well and also the leaves of Teak (TeakLeaf®) which are glued on a textile support (Fig. 1e). The latter falls as well into the structural category of a coated textile. The textile support on the reverse side is built by two non-woven layers. The middle layer is cellulose-based, the fibres are stuck together with an acrylic acid-based polymeric binder, and the basic support on the reverse side is polyester-based. On the top, the TeakLeaf® material is coated by a transparent waxy polyolefin film. The leaf of teak fulfils desired optical needs.

5 Test Methods for Technical Performance, Comfort, and Longevity

The final application of a material defines the required material properties and their performance is tested by numerous different methods. In particular, for materials used in the transport sector these methods are manifold and comprise physical test, tests for processability, chemical tests, test for requirements on comfort behaviour, longevity, behaviour under various climate conditions, flame retardance, and many more. A reduced set is used in the shoe sector. To get a first impression about the performance of a material and in order to reduce the testing effort, it is useful to measure only selected parameters.

Many of the materials, which emerged the market do not meet all technical requirements, which are evaluated by these tests. Especially, some of the chemical requirements need to be approved only for selected materials. The comparative study (Meyer et al., 2021) showed the technical performance of a selection of materials, which has recently been extended by cork material, leather board, ELeather and a wood veneer coated textile.

5.1 Technical Performance

The values for tensile strength and tear resistance reflect the mechanical stability, parameters which correlate with the processability of the material and with the longevity. Flex resistance is measured to get an impression about the stability of the coating during use. The water vapour permeability and water vapour absorption give hints about the wearing comfort. Figure 2 shows the results of these tests for the different materials at a glance. The values are normalized on the highest value measured for each parameter. The exact values are summarized in Table 1.

5.2 Mechanical Stability—Processability

Tensile strength and tear strength represent one of the most important mechanical properties for materials used for shoes, gloves, apparel goods but as well upholstery applications. They reflect the processability and to some extent the longevity of a material.

The coated textile and their variants show tensile strengths up to 21 N/mm^2. Their mechanical stability is determined by the textile support e.g. of the artificial leather (PUR coated textile) and the similar composed materials Desserto®, Appleskin® and Vegea® but as well of wood-textile, cork material and Teak leaf®. Each of these materials consists of a coated woven textile and some are covered with coatings which appear to aim mainly design effects (Teak leaf, wood textile, cork). The mechanical stability of these types of coated fabrics mainly depends on the properties of the support. Therefore, the choice of the fabric has to be reasonable for the intended use regardless of the use of fibres of natural or artificial origin. Usually nonwovens show lower stabilities than woven fabrics. To increase the bio-based content it is easy to exchange synthetic fabrics against natural fibres. New developments use recycled fibres from polyesters or semi-chemical fibres as Lyocell-fibres.

The properties of naturally grown tissues vary broadly. Leather exhibits the inherent anisotropic fibre orientation of the grown skin tissue and shows a very

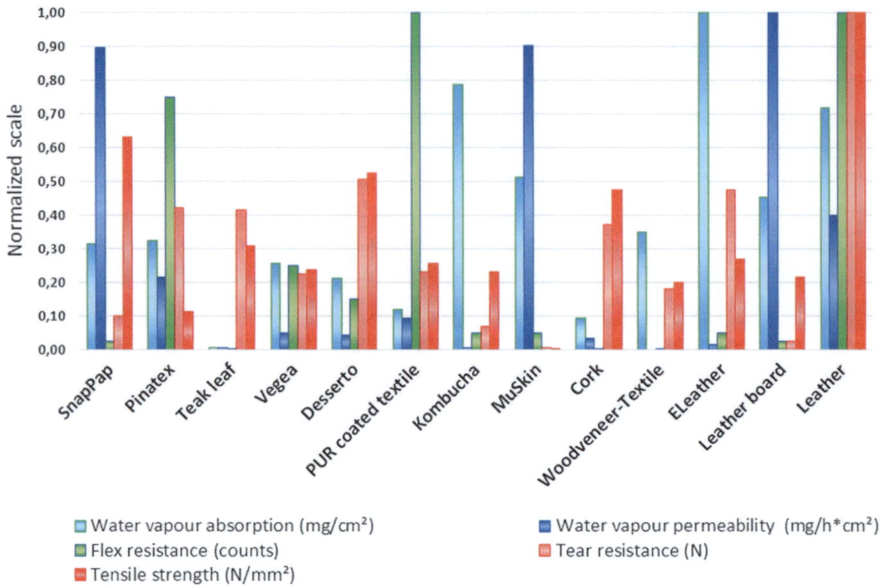

Fig. 2 Comparison of a selection of materials in regard to their technical performance. The values in the legend show the maxima used for the normalization. The value for the water vapour permeability of wood—textile is missing, due to which the test method is not applicable for this material

high mechanical stability (tensile strength, tear strength), representing the highest value for the tensile strength and tear strength within the survey (Table 1). Tensile strength exceeds the specification of >15 N/mm^2 for chrome tanned upper leather for shoes (ISO 20942:2019, n.d.). In contrast, the values of the Muskin® sample are extremely low. The fibres are oriented perpendicular to the loading direction in the tensile test pulling scenario. The tissue does not generate stability. The stability of the fungi fibres itself was not yet measured to get an impression which mechanical values can be potentially achieved.

The tensile strengths of the non-woven materials made of natural plant fibres range from 4 up to 25 N/mm^2. Strength depends on fibre properties and fibre bonding. Despite a satisfying tensile strength, the tear resistance of SnapPap® is low due to the short fibre length, which cannot be leveraged by the polymeric binder. That of Pinatex® is even lower. Bonding of the nonwoven is achieved by a polymeric binder and the density is not as high as that of SnapPap®, though pinapple fibres itself show a high tensile strength (Mukherjee & Satyanarayana, 1986).

Table 1 Values of flex resistance (>80.000 counts), water vapour absorption and water vapour permeability (>0.8 mg cm^{-2} $h^{-}1$), tear resistance and tensile strength (>15 Ncm^{-2}) measured of different materials to compare longevity, comfort behaviour and processability. (limits for shoe upper leather in brackets according to ISO 20942)

Unit	–	mg/cm^2	mg/cm^2*h	N	N/mm^2
Material	Flex resistance	WVA	WVP	Tear resistance	Tensile strength
SnapPap	5,000	3.7	10.3	7.5	24.9
Pinatex	150,000	3.8	2.5	31	4.5
Teak leaf	100	0.1	0.1	30.7	12.2
Vegea	50,000	3.0	0.6	16.6	9.4
AppleSkin	50,000	1.7	0.4	18.4	14
Desserto	30,000	2.5	0.5	37.2	20.8
PUR coated textile	200,000	1.4	1.1	17	10.2
Kombucha	10,000	9.2	0.1	5.1	9.2
MuSkin	10,000	6.0	10.4	0.5	0.2
Cork	1,000	1.1	0.4	27.5	18.8
Woodveneer-textile	1,000	4.1	n.d	13.4	8.0
ELeather	10,000	11.7	0.2	35	10.7
Leather board	5,000	5.3	11.5	2	8.6
Leather	200,000	8.4	4.6	73.6	39.5

5.3 *Physical Parameters—Comfort*

A pleasant wearing comfort of shoes, gloves, or apparel is related to the water vapour permeability (WVP) of the material. Humidity of the body has to be transported through the clothing material to the material surface to prevent sweating and condensation of the humidity on the skin. In addition, the comfort is enhanced by the ability of a material to absorb water vapour (WVA) (Slater, 1977). Comfortable water vapour permeability limits are specified in ISO 20942 to be >>0.8 mg/(cm^2h). Leather, Muskin®, and SnapPap® exceed the water vapour permeability by far, as well as leather board (Fig. 2, Table 1). The latter is surprising, because the binder of the leather board is natural rubber, which is water vapour impermeable. However, the high content of leather fibres markedly enhances the permeability. Pinatex® and PUR-coated textile still fulfil the ISO 20942 requirement. The WVP of all other materials is insufficient. The values of the wood veneer have not been evaluated as the structure with the slits in the wooden coating did not allow a correct sealing in the measurement system. To summarize, the higher the WVP is the more open the structure appears. A coating with polymer films tightens the surface against vapour.

The water vapour absorption (WVA) as well correlates with the comfort behaviour. The materials in contact with the skin should appear dry. Fibres or material components have to absorb water and transport it to the surface (Grabowska-Polanowska et al., 2020). The measured WVA of the investigated materials set shows high values for Kombucha, leather, ELeather and Muskin®, presumably due to the polar nature of the natural polymers. In leather board, the surface of the fibres seems to be coated to some extent, because the values of WVA are lower. The lowest values for water vapour absorption are observed for materials that contain a significant amount of synthetic polymer with less polarity. This observation confirms the results of e.g. (Grabowska-Polanowska et al., 2020), who investigated different shoe materials.

5.4 Longevity

Longevity is a complex parameter, which not only depends on the robustness of a material against mechanical damage or wear as a consequence of use. But, it is also dependent on the usage scenario (e.g. shoe for opera or for trekking), the material as part of a more complex product (e.g. automotive), options to repair or restore the material (expensive accessories or shoes), moral attrition though function is still given, and as well fashion aspects. Here, we will focus on the physical damages, which are characterized by many different technical parameters. Materials for many kind of shoes and especially the upper material must resist intensive bending and convex and concave deformation during usage. Typical use cases are creases of shoes but also of bags, when the volume is covered by a lid. The susceptibility against bending depends on the viscoelastic behaviour of the polymer material, which acts as the finishing material as well as the bending stability of the fibres, the binder and the interaction of the different layers of a material.

The flexometer test is used to assess the long-time resistance against bending. Grade 0 is the best rating indicating that the material itself and the coating layers show no cracks after flexing. A grade of 2 (only very small cracks in top layer of the coating) is usually accepted to pass the test. When a grade of >2 is observed, flexing has to be stopped and the number of flex cycles is noted. Leather, Pinatex®, and PUR-coated textile fulfilled the specification of >80.000 flex cycles according to ISO 20942. The flex resistance of all other materials was found to be insufficient for the proposed applications shoe upper (Fig. 2, Table 1).

6 Outlook

While leather making has a centuries-old history and the manufacture of synthetic alternatives as artificial leather and coated fabrics is known since several decades the market demands for new materials, sometimes named next-gen materials. Designers and scientists not only aim to achieve function and longevity, but these new materials should as well fulfil additional requirements as being fossil-free, degradable, bio-based, climate neutral, and they should be vegan or edible. But, these materials do not achieve the necessary technical requirements, yet.

Leather is a material which integrates many required functions as mechanical stability, flexibility, long life time and breathability already in its raw material, the skin. Tanning and post-processing save this grown structure and the properties which are unique in nature. The biological function of the skin is the best amongst other properties such as mechanical protection and temperature and humidity regulation. Its core structure is the collagen chain, which builds triple-helices, fibrils, fibres and finally tissues (Meyer et al., 2021). This is as well a principle in nature. Natural structural materials are usually basing on a low number of key motifs on nanoscale, which are combined in hierarchical structures to achieve final macroscopic properties. However, currently it is not possible to manufacture such structures synthetically (Wegst et al., 2015). Growing tissues in biotechnological processes is still a challenge and far from being economic on a technical scale.

In artificial leathers the aimed functions as mechanical stability, water vapour permeability, but as well design and haptic, are dispersed in several technically combined layers. The textile support is applied when leather-like mechanical stability is required. Fabrics are not only applied on the backside of artificial leathers but also inside of a material e.g. ELeather. The foamed layer in artificial leather and the fibrous layers in ELeather as well as the fine hyphae of the fungi material account for the softness and the nubuk leather-like haptic.

If a soft haptic is not needed compact films are applied which are composed of synthetic or bio-based polymers. First synthetic leathers were prepared of PVC without the foaming layer, and fabrics are coated with polyurethanes, silicones, polylactide and other synthetic polymers. Alternative bio-based binders are plant based gelling agents (Xanthan, Guar, Alginates). However, these have to be plasticized not to get brittle. Foaming of these bio-based gelling agents leads to very unstable layers which are usually known in the food industry as wine gums or marsh mellows. An exception is natural rubber, which is a natural polysaccharide that is already used as binding agent to manufacture leather board. Top coats, based on synthetic polymers are applied to every material if sufficient fastness against abrasion, sweat, dyes, light, and soiling have to be achieved.

The newly designed materials, which are offered to the market, try to fulfil some of the above mentioned additional requirements. But the synthetic polymer layers in the artificial leather variants (e.g. Desserto®, Vegea® and AppleSkin®) can only be filled partly with natural by-products to increase the bio-based content because the synthetic polymer is needed as the binder. The hierarchical fibrous structure as found in leather is not achieved by these systems.

Pinatex® and SnapPap® do not appear leather-like and one would not associate that these are offered as alternatives to leather. Nevertheless, nonwovens, which are bound with bio-based binders and coated with synthetic polymer films to achieve sufficient wear behaviour, are materials which can be used for similar applications as leather, though design and haptic are completely different.

The analyses showed that neither the artificial leather variants Vegea®, AppleSkin® and Desserto® nor the other materials investigated achieved the same functionality as leather. Furthermore, the additional requirements lead to conflicts of objectives. A readily biodegradable synthetic material cannot be water proof and long-lasting and biopolymer based materials without a synthetic top coat will always show low fastnesses.

Not least, the producibility and the availability of bio-based raw materials have to be discussed. If parts of the technically produced materials are exchanged by bio-based materials this will require sufficient amounts of these raw materials at the industrial scale. Their use for technical or fashion applications should be prohibited, if these agricultural by-products are used as feed for animals. Furthermore, bio-based raw materials are not available in the same amounts and equal quality as naphta, the raw material for synthetic polymers, but only for product niches. Upscaling of handcraft processes necessarily leads to similar challenges as they are known for classical industrial processes. They need higher amounts of energy, waste water treatment, prevention of air pollution and sophisticated logistics.

To evaluate the impact of processes to manufacture different materials for the same use, many life cycle assessments (LCA) have been performed from leather and as well synthetic alternatives (examples are Alvarenga et al., 2013a, 2013b; Bishop et al., 2021; Brugnoli & Král̆, 2012; Catalán et al., 2019; Degieter et al., 2022; Fröhling & Hiete, 2020; Hildebrandt et al., 2021; Hultkrantz, 2018; Joseph & Nithya, 2009; Navarro et al., 2020; Notarnicola et al., 2011). Unfortunately, the results are mostly not comparable. The exact manufacturing techniques of the alternative materials are often not discovered and an intensive collaboration with the manufacturer is necessary to use correct input data. To compare the processes in regard to their environmental impact the knowledge is summarized from websites or logical conclusions from microscopic pictures and analytical results of samples. Often, the materials are not manufactured for exactly the same use, the borders for the assessment are not the same or cannot be the same (e.g. use of synthetic raw materials compared to bio-based), but most important, the useful life is not the same (Hildebrandt et al., 2021). Therefore, the direct comparison of materials for the same use is still an open field for research and standards have to be defined to compare structurally completely different products claimed for similar use.

A recent work compared the LCAs of almost 90 fossil and bio-based polymers. Only 27 could be really compared due to the authors used similar methodologies. It was shown, that the values for seven compared impact categories (energy use, ecotoxicity, acidification, eutrophication, climate change, particulate matter formation, ozone depletion) varied between 200 and 400% and partly more than 10 times (Walker & Rothman, 2020). The authors summarise that a correct comparison between the different LCAs is not possible. But, LCAs have only low relevance if they cannot be compared. It is necessary to homogenize the standards not only to compare LCAs of the final materials but as well to compare the evaluations of the raw materials.

Today, it is widely accepted that the massive carbon dioxide emissions which are caused by burning of fossil resources as oil, coal and gas in high amounts since the beginning of industrialisation lead to the change of the global climate with substantial consequences to global and regional living conditions. More than 80% of these fossil resources are still burnt for energy production (heat, electricity). In the energy sector a decarbonisation and the change to renewable resources is necessary and aimed for the next decades. However, the current material concepts basing on polymers are mainly carbon based.

The bioeconomy approach tries to replace fossil-based polymers by renewable raw materials as part of a circular economy. Figure 3 summarizes cascades of plant based and animal based polymer materials and their degraded intermediates. Each of these intermediates are already used as product. The cascades reflect the recent discussion how nature based raw materials can be used more effectively. The natural synthesis power of plants and animals has to be used and it should be used effectively. Materials should not be degraded from the polymer (cellulose, collagen) to energy or monomers (monosugars, amino acid) in one step. Instead, the intermediates have to be used as materials as effectively as possible. After the end of life of a polymer material it should be processed in further products as mentioned by the cascade use. This will allow to use bio-based raw materials most economically.

The principle to use plant based fibres and wastes of coconut, apple, banana, pineapple, oranges, cactus, wine, fatty acids or corn but as well animal based structural proteins as the skin to manufacture materials or components of them is helpful as long as their use for technical purposes does not compete with the food or feed chain. The terms leather, textiles, coated textiles, artificial leathers and the so called next-gen materials are only umbrella terms for hundreds of materials offered to the market.

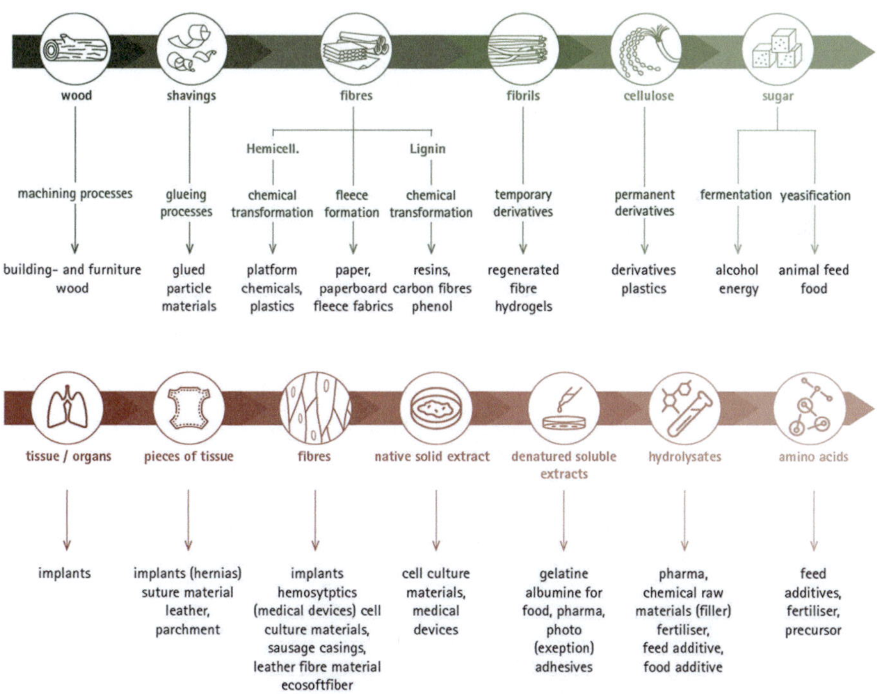

Fig. 3 Cascade use of animal and plant based materials Bioeconomy Cluster of the ZUSE association, Germany (ZUSE-Association, 2022)

This diversity is a result of technological development, creativity of designers, but as well of political discussions and legal frameworks. The cascade concept will only be possible if the intermediates can be recycled or the components can be separated after use. This will as well require intensive additional investigations.

If new materials will be designed the hierarchical principle can as well be used inspired by nature (Wegst et al., 2015). It is an open question and a field of research however, whether this principle can be achieved by technological processes and in economic scale. Biotechnology is not economic yet to produce tissue like multilayer structures for technical purposes and the use of biotechnological processes to manufacture materials is still at the beginning (Haaf et al., 2021; Kircher, 2021; Popp et al., 2021; Schilling & Weiss, 2021). In regard to the fungal materials as well as the cell cultured meat doubts remain, whether biotechnology will get a relevant influence for the manufacture of such materials.

Finally, in regard of sustainability a new economic mindset has to be discussed. A change of the business models away from fast fashion towards slow fashion, renting, leasing, updating, repairing and circular economy instead of liner economy has to be achieved. The increase of longevity, reduction of waste, water use and chemicals at every step of production is necessary and consumers have to understand products more under functional views than as entertainment (Niinimäki et al., 2020).

References

Albano, Y., Habel, K. R., Kohler, J., Morgenstern, B., Hausoul, P. J., & Palkovits, R. (2020). Biomass as source of furan-based plasticizers: an alternative to phthalates. plasticiser.org.

Alvarenga, R. A., Dewulf, J., De Meester, S., Wathelet, A., Villers, J., Thommeret, R., & Hruska, Z. (2013a). Life cycle assessment of bioethanol-based PVC: Part 1: Attributional approach. Biofuels. *Bioprod. Bioref., 7*, 386–395. https://doi.org/10.1002/bbb.1405

Alvarenga, R. A., Dewulf, J., De Meester, S., Wathelet, A., Villers, J., Thommeret, R., & Hruska, Z. (2013b). Life cycle assessment of bioethanol-based PVC: Part 2: Consequential approach. Biofuels. *Bioprod. Bioref., 7*, 396–405. https://doi.org/10.1002/bbb.1398

Andriani, D., Apriyana, A. Y., & Karina, M. (2020). The optimization of bacterial cellulose production and its applications: A review. *Cellulose, 27*, 6747–6766. https://doi.org/10.1007/s10570-020-03273-9

Anonymus, 2021. Ananas anam Impact report 2021; https://www.ananas-anam.com/ananas-anams-2021-impact-report (accessed: 28.01.2023).

Auras, R. A., Lim, L., Selke, S. E. M., & Tsuji, H. (Eds.). (2022). Poly(lactic)acid, in: Poly(Lactic Acid). Wiley. https://doi.org/10.1002/9781119767480.fmatter

Bishop, G., Styles, D., & Lens, P. N. L. (2021). Environmental performance comparison of bioplastics and petrochemical plastics: A review of life cycle assessment (LCA) methodological decisions. *Resources, Conservation and Recycling, 168*, 105451. https://doi.org/10.1016/j.resconrec.2021.105451

Brugnoli, F., & Kráľ, I. (2012). Life cycle assessment, carbon footprint in leather processing. Eighteenth session of the leather and leather products industry panel, UNIDO. https://leatherpanel.org/sites/default/files/publications-attachments/lca_carbonfootprint_lpm2012.pdf

Catalán, E., Komilis, D., & Sánchez, A. (2019). A life cycle assessment on the dehairing of rawhides: Chemical treatment versus enzymatic recovery through solid state fermentation. *Journal of Industrial Ecology, 23*, 361–373. https://doi.org/10.1111/jiec.12753

Cerimi, K., Akkaya, K. C., Pohl, C., Schmidt, B., & Neubauer, P. (2019). Fungi as source for new bio-based materials: A patent review. *Fungal Biology Biotechnology, 6*, 17. https://doi.org/10.1186/s40694-019-0080-y

Chriki, S., & Hocquette, J.-F. (2020). The myth of cultured Meat: A review. *Frontiers in Nutrition, 7*, 7. https://doi.org/10.3389/fnut.2020.00007

Covington, A. D. (2009). *Tanning chemistry: The science of leather*. Royal Society of Chemistry.

Cuthbertson, W. W. (1956). Silicones—Textile and other applications. *Journal of the Textile Institute Proceedings, 47*, P788–P794. https://doi.org/10.1080/19447015608665330

Degieter, M., Gellynck, X., Goyal, S., Ott, D., & De Steur, H. (2022). Life cycle cost analysis of agri-food products: A systematic review. *Science of the Total Environment, 158012.*

Dickson, J. H., Oeggl, K., & Handley, L. L. (2003). The iceman reconsidered. *Scientific American, 288*, 70–79. https://doi.org/10.1038/scientificamerican0503-70

Esa, F., Tasirin, S. M., & Rahman, N. A. (2014). Overview of bacterial cellulose production and application. *Agriculture and Agricultural Science Procedia, 2*, 113–119. https://doi.org/10.1016/j.aaspro.2014.11.017

Flaten, G. E., Palac, Z., Engesland, A., Filipović-Grčić, J., Vanić, Ž, & Škalko-Basnet, N. (2015). In vitro skin models as a tool in optimization of drug formulation. *European Journal of Pharmaceutical Sciences, 75*, 10–24. https://doi.org/10.1016/j.ejps.2015.02.018

Fröhling, M., & Hiete, M. (Eds.). (2020). *Sustainability and life cycle assessment in industrial biotechnology, advances in biochemical engineering/biotechnology*. Springer International Publishing. https://doi.org/10.1007/978-3-030-47066-1

Furtwengler, P., & Avérous, L. (2018). Renewable polyols for advanced polyurethane foams from diverse biomass resources. *Polymer Chemistry, 9*, 4258–4287. https://doi.org/10.1039/C8PY00827B

Gandia, A., van den Brandhof, J. G., Appels, F. V., & Jones, M. P. (2021). Flexible fungal materials: Shaping the future. *Trends in Biotechnology, 39*, 1321–1331.

Garrison, G. L., Biermacher, J. T., & Brorsen, B. W. (2022). How much will large-scale production of cell-cultured meat cost? *Journal of Agriculture and Food Research, 10*, 100358. https://doi.org/10.1016/j.jafr.2022.100358

Grabowska-Polanowska, B., Kwiecień, J., & Gajewski, R. (2020). Measurement of the inside microclimate of footwear constructed from different material sets. *FIBRES & TEXTILES in Eastern Europe, 28*, 79–84. https://doi.org/10.5604/01.3001.0014.3801

Grömer, K. (2016). *The art of prehistoric textile making: The development of craft traditions and clothing in Central Europe*. Naturhistorisches Museum Wien. https://doi.org/10.26530/OAPEN_604250

Haaf, A., Hofmann, S., & Schüler, J. (2021). Measuring the economic footprint of the biotechnology industry in Europe. *Industrial Biotechnology, 17*, 117–124. https://doi.org/10.1089/ind.2021.29249.aha

Haneef, M., Ceseracciu, L., Canale, C., Bayer, I. S., Heredia-Guerrero, J. A., & Athanassiou, A. (2017). Advanced materials from fungal mycelium: Fabrication and tuning of physical properties. *Science and Reports, 7*, 41292. https://doi.org/10.1038/srep41292

Heidemann, E. (1993). *Fundamentals of leather manufacture*. Roether.

Hildebrandt, J., Thrän, D., & Bezama, A. (2021). The circularity of potential bio-textile production routes: Comparing life cycle impacts of bio-based materials used within the manufacturing of selected leather substitutes. *Journal of Cleaner Production, 287*, 125470. https://doi.org/10.1016/j.jclepro.2020.125470

Hosney, H., Nadiem, B., Ashour, I., Mustafa, I., & El-Shibiny, A. (2018). Epoxidized vegetable oil and bio-based materials as PVC plasticizer. *Journal of Applied Polymer Science, 135*, 46270.

Huang, Y., Zhu, C., Yang, J., Nie, Y., Chen, C., & Sun, D. (2014). Recent advances in bacterial cellulose. *Cellulose, 21*, 1–30. https://doi.org/10.1007/s10570-013-0088-z

Hultkrantz, M. (2018). An overview on the environmental impacts of synthetic leather made of hemp fiber with preliminary lifecycle assessment.

Hussain, Z., Sajjad, W., Khan, T., & Wahid, F. (2019). Production of bacterial cellulose from industrial wastes: A review. *Cellulose, 26*, 2895–2911. https://doi.org/10.1007/s10570-019-02307-1

Iguchi, M., Yamanaka, S., & Budhiono, A. (2000). Review bacterial cellulose—A masterpiece of nature's arts. *Journal of Materials Science, 10.*

Illa, M. P., Sharma, C. S., & Khandelwal, M. (2019). Tuning the physiochemical properties of bacterial cellulose: Effect of drying conditions. *Journal of Materials Science, 54*, 12024–12035. https://doi.org/10.1007/s10853-019-03737-9

ISO 20942:2019. (n.d.). Leather—Full chrome upper leather—Specification and test methods.

Jem, K. J., & Tan, B. (2020). The development and challenges of poly (lactic acid) and poly (glycolic acid). *Advanced Industrial and Engineering Polymer Research, 3*, 60–70. https://doi.org/10.1016/j.aiepr.2020.01.002

Jia, P., Xia, H., Tang, K., & Zhou, Y. (2018). Plasticizers derived from biomass resources: A short review. *Polymers, 10*, 1303. https://doi.org/10.3390/polym10121303

Jones, M., Gandia, A., John, S., & Bismarck, A. (2020). Leather-like material biofabrication using fungi. *Nature Sustainability*. https://doi.org/10.1038/s41893-020-00606-1

Joseph, K., & Nithya, N. (2009). Material flows in the life cycle of leather. *Journal of Cleaner Production, 17*, 676–682. https://doi.org/10.1016/j.jclepro.2008.11.018

Kircher, M. (2021). Bioeconomy—Present status and future needs of industrial value chains. *New Biotechnology, 60*, 96–104. https://doi.org/10.1016/j.nbt.2020.09.005

Lynch, J., & Pierrehumbert, R. (2019). Climate impacts of cultured meat and beef cattle. *Frontiers of Sustainable Food System, 3*, 5. https://doi.org/10.3389/fsufs.2019.00005

Maisels, A., Hiller, A., & Simon, F. (2021). Chemisches recycling für Kunststoffe: Status und perspektiven. *Chemie Ingenieur Technik, 93*, 1742–1750. https://doi.org/10.1002/cite.202100115

Martinz, D., & Quadros, J. (2008). Compounding PVC with renewable materials. *Plastics, Rubber and Composites, 37*, 459–464. https://doi.org/10.1179/174328908X362917

Meyer, M., Dietrich, S., Schulz, H., & Mondschein, A. (2021). Comparison of the technical performance of leather, artificial leather, and trendy alternatives. *Coatings, 11*, 226. https://doi.org/10.3390/coatings11020226

Mühlethaler, K. (1949). The structure of bacterial cellulose. *Biochimica et Biophysica Acta*, 527–535.

Mukherjee, P., & Satyanarayana, K. (1986). Structure and properties of some vegetable fibres. *Journal of Materials Science, 21*, 51–56.

Navarro, D., Wu, J., Lin, W., Fullana-i-Palmer, P., & Puig, R. (2020). Life cycle assessment and leather production. *Journal of Leather Science Engineering, 2*, 26. https://doi.org/10.1186/s42825-020-00035-y

Niinimäki, K., Peters, G., Dahlbo, H., Perry, P., Rissanen, T., & Gwilt, A. (2020). The environmental price of fast fashion. *Nature Reviews Earth and Environment, 1*, 189–200. https://doi.org/10.1038/s43017-020-0039-9

Notarnicola, B., Puig, R., Raggi, A., Fullana, P., Tassielli, G., Camillis, C. D., & Rius, A. (2011). Life cycle assessment of Italian and Spanish bovine leather production systems. *Afinidad LXVIII, 553*, 167–180.

Patti, A., & Acierno, D. (2022). Towards the sustainability of the plastic industry through biopolymers: Properties and potential applications to the textiles world. *Polymers, 14*, 692. https://doi.org/10.3390/polym14040692

Popp, J., Kovács, S., Oláh, J., Divéki, Z., & Balázs, E. (2021). Bioeconomy: Biomass and biomass-based energy supply and demand. *New Biotechnology, 60*, 76–84. https://doi.org/10.1016/j.nbt.2020.10.004

Revin, V., Liyaskina, E., Nazarkina, M., Bogatyreva, A., & Shchankin, M. (2018). Cost-effective production of bacterial cellulose using acidic food industry by-products. *Brazilian Journal of Microbiology, 49*, 151–159. https://doi.org/10.1016/j.bjm.2017.12.012

Ruiz-Herrera, J. (2012). *Fungal cell wall: Structure, synthesis, and assembly*. CRC Press.

Schilling, C., & Weiss, S. (2021). A roadmap for industry to harness biotechnology for a more circular economy. *New Biotechnology, 60*, 9–11. https://doi.org/10.1016/j.nbt.2020.08.005

Scionti, G. (2010). *Mechanical properties of bacterial cellulose implants (masterthesis).* Chalmers University of Technology.

Siddiqui, S. A., Khan, S., Murid, M., Asif, Z., Oboturova, N. P., Nagdalian, A. A., Blinov, A. V., Ibrahim, S. A., & Jafari, S. M. (2022). Marketing strategies for cultured meat: A review. *Applied Sciences, 12*, 8795. https://doi.org/10.3390/app12178795

Siegmund, F., Veit, D., & Gries, T. (2009). Biopolymere in textilen Anwendungen. Polylactid Polyhydroxyalkanoate. *Chemie in Unserer Zeit, 43*, 152–158. https://doi.org/10.1002/ciuz.200900469

Singh, I., Samal, S. K., Mohanty, S., & Nayak, S. K. (2020). Recent advancement in plant oil derived polyol-based polyurethane foam for future perspective: A review. *European Journal of Lipid Science and Technology, 122*, 1900225.

Singhvi, M., & Gokhale, D. (2013). Biomass to biodegradable polymer (PLA). *RSC Advances, 3*, 13558. https://doi.org/10.1039/c3ra41592a

Siu, E. (2022). 2021 State of the art industry report: Next gen materials. https://materialinnovation.org/reports. Accessed 28 Jan 2023.

Slater, K. (1977). Comfort properties of textiles. *Textile Progress, 9*, 1–70. https://doi.org/10.1080/00405167.1977.10750095

Smith, W. C. (2010). *Smart textile coatings and laminates.* Woodhead publishing series in textiles.

Stather, F. (1957). *Gerbereichemie und Gerbereitechnologie.* Akademie-Verlag.

Träubel, H. (1999). *New materials permeable to water vapor.* Springer.

Ul-Islam, M., Ullah, M. W., Khan, S., & Park, J. K. (2020). Production of bacterial cellulose from alternative cheap and waste resources: A step for cost reduction with positive environmental aspects. *Korean Journal of Chemical Engineering, 37*, 925–937. https://doi.org/10.1007/s11814-020-0524-3

Vandelook, S., Elsacker, E., Van Wylick, A., De Laet, L., & Peeters, E. (2021). Current state and future prospects of pure mycelium materials. *Fungal Biology Biotechnology, 8*, 20. https://doi.org/10.1186/s40694-021-00128-1

Villarreal-Soto, S. A., Beaufort, S., Bouajila, J., Souchard, J., & Taillandier, P. (2018). Understanding kombucha tea fermentation: A review. *Journal of Food Science, 83*, 580–588.

Walker, S., & Rothman, R. (2020). Life cycle assessment of bio-based and fossil-based plastic: A review. *Journal of Cleaner Production, 261*, 121158. https://doi.org/10.1016/j.jclepro.2020.121158

Wegst, U. G. K., Bai, H., Saiz, E., Tomsia, A. P., & Ritchie, R. O. (2015). Bioinspired structural materials. *Nature Materials, 14*, 23–36. https://doi.org/10.1038/nmat4089

Woll, S., & Böhm, I. (2018). In-vitro meat: A solution for problems of meat production and meat consumption? *Ernährungs Umschau, 12–21.* https://doi.org/10.4455/eu.2018.003

Yang, Y., Zhang, M., Ju, Z., Tam, P. Y., Hua, T., Younas, M. W., Kamrul, H., & Hu, H. (2021). Poly(lactic acid) fibers, yarns and fabrics: Manufacturing, properties and applications. *Textile Research Journal, 91*, 1641–1669. https://doi.org/10.1177/0040517520984101

Ye, Y., Zhou, J., Guan, X., & Sun, X. (2022). Commercialization of cultured meat products: Current status, challenges, and strategic prospects. *Future Foods, 6*, 100177. https://doi.org/10.1016/j.fufo.2022.100177

Zaehnsdorf, J. W. (1890). *The art of bookbinding: A practical treatise.* G. Bell & Sons.

Zhang, C., & Yang, J. (2020). A history of mechanical engineering. *Springer Singapore, Singapore.* https://doi.org/10.1007/978-981-15-0833-2

Zhang, C. J., Wang, L., Zhao, J. C., & Zhu, P. (2011). Effect of drying methods on structure and mechanical properties of bacterial cellulose films. *AMR, 239–242*, 2667–2670. https://doi.org/10.4028/www.scientific.net/AMR.239-242.2667

ZUSE-Association. (2022). https://www.zuse-gemeinschaft.de/forschung/cluster-biooekonomie: download Jan/12/2023.

A Framework of Augmented Design Process: Development of Footwear Concept Designs Using Artificial Intelligence

Kaustav SenGupta

1 Introduction

When Weiskrantz, L. coined the term 'blindsight' as an expression of seeing without really 'seeing through eyes', we never thought there would be a time when a concept design would be realized without intense human intervention (Weiskrantz, 2008)! Nicholas Humphrey, in his new book 'Sentience: The Invention of Consciousness', raised the question, 'What would be wrong—or insufficient for survival—with deaf hearing, scentless smell, feelingless touch or even painless pain?' the same way we do ask today what will be the design without a designer? Or rather, will AI be an enabler for the designers to conceptualize faster, explore better and use their brains for more disruptive designs? We are living in a fantastic time when the machine is taking over the 'conceptualizing' and 'realizing' so that the creative brains of designers can be used for the better pursuit to break human boundaries (Humphrey, 2022).

1.1 Understanding Design and the Science Behind It

Today, design is universal and inherently optimistic. Mostow mentioned that design is a crucial human activity (Mostow, 1985). As a psychological phenomenon, it is an interesting kind of complex problem-solving, as yet incompletely understood. In economic terms, design pervades our high-tech society. It is safe to say that billions of dollars are spent annually on various kinds of design. In his Design Methods book, John Christopher Jones, the founding member of the Design Research Society (1966), stated that design is 'to initiate change in man-made things'. Design can be viewed as a dialectic between the designer and what is possible.

K. SenGupta (✉)
National Institute of Fashion Technology, Rajiv Gandhi Salai, Taramani, Chennai 600113, India
e-mail: kaustav.sengupta@nift.ac.in

S. K. Janardhanan and L. A. Zugno (eds.), *Emerging Trends in Leather Science and Technology*, https://doi.org/10.1007/978-981-99-9754-1_9

Herbert Simon, in his book 'The Science of Artificial' mentioned that the natural sciences focus on how the world is, whereas design sciences are concerned with how things should be. He further stated that the world we live in today is much more a man-made or artificial world than it is a natural world (Simon, 2019). Almost every element in our environment shows evidence of human artifice. Simon defines 'artificial' as 'Produced by art rather than by nature; not genuine or natural; affected; not pertaining to the essence of the matter'. Hence, most of the 'artificial' is designed. The process of design is useful not only for building useful products, services, or solutions but also for pursuing a more creative and open-minded approach to life. A good design allows one to look good and make a positive first impression on potential customers, but it can take a lot longer than this to alter that perception once a first impression has been made. Hence, it is essential to ensure that the design is able to make a mark in the first instance. A good design multiplies business by demanding attention and increasing exposure because the users are always attracted to things that appears good and function well. The usability is crucial in context to product design.

According to Paul Rodgers and Alex Milton in their book Product design, the lines between specialized fields like lighting, furniture, graphics, fashion, interaction, and industrial design are frequently blurred (Rodgers & Milton, 2011). Products like eyeglasses, scissors, cameras, fly swatters, trash cans, vases, fruit bowls, telephones, door mats, clothes hangers, razors, bottle stoppers, kettles, cigarette lighters, fire extinguishers, cutlery, salt and pepper shakers, shelving systems, MP3 players, and computers can all fall under this category. Whether in the home, workplace, or public space, product design seeks to improve quality of life through everything from chairs and lights to consumer and environmental items. Product design is another economic activity that can benefit organizations by ensuring they produce and market goods that intrigue, please, or challenge customers. They also emphasized that a product designer's job involves a variety of academic fields, including art, science, and commerce, in order to create tangible artefacts. These fields include marketing, management, design and engineering. Product design is rapidly being recognized as a crucial strategic instrument for influencing consumer desire and underlying emotional values. Products that are more usable, appealing, dependable and cost-effective and create stronger emotional relationships are some of its benefits for the consumer. The consumer's loyalty to the product may increase due to these advantages. Product design is rapidly being recognized as a crucial strategic instrument for influencing consumer desire and underlying emotional values. Products that are more usable, appealing, dependable and cost-effective and create stronger emotional relationships are some of their benefits for the consumer. Due to these advantages, the consumer may become more devoted to the product in question.

1.2 AI and Bigdata in the Context of Design

In a generic sense, Artificial Intelligence (AI) is a collection of incredibly powerful, fast and effective computer systems or models and algorithms that can cohesively work and deliver outputs or reacts like human beings and imparts an appearance of human-like natural intelligence. Most AI-enabled systems are task-driven and address the mechanization of complex, knowledge-intensive tasks. Design is certainly such a task. Thus, AI is a natural discipline for the study of design, and in fact, a number of AI researchers have recently been studying design, especially hardware and software design. Developing a better model through design is the key to the research problem is AI-based designs today. AI is also becoming an enabler for better product design. With big data, efficient deep learning models (ANN) and visualization capabilities, generative AI has become the crucial design enabler. UX designer Miklos Phillips referred IBM CEO Ginni Rometty wrote that AI had become an 'augmented intelligence' to design. In his article, 'the present and Future of AI', he emphasized that AI primarily focuses on efficiency and speed. Due to AI's increased speed and efficiency, designers can produce designs more quickly and affordably. The speed with which AI can analyse massive volumes of data and make design recommendations will be its greatest strength. On the basis of that data, a designer can select and approve alterations. Multiple prototype versions can be A/B tested with users to see which designs are the most successful. An AI design tool might be used for quick design prototyping where simple designs are scanned in, a few parameters are supplied, and a library of pre-existing UI components comes to life to produce a prototype that aligns with a company's needs. Additionally, designers must perform time-consuming daily activities like translating the same visuals into several languages and localizing products. The localization of show banners and the customization of artwork are already being done by Netflix employing augmented intelligence technologies. The technology 'reads' the master version and then quickly generates the customized and localized images. A huge time saver, all a designer needs to do is review the visuals, accept or reject them and manually change them if necessary. The conventional position of a designer has already been replaced by an AI system that has created millions of different packaging designs for Nutella. The graphic identity for Nutella was generated by an AI programme using data from a database of thousands of patterns and colours to produce seven million distinct iterations that have been sprayed across the front of jars in Italy. In less than a month, all seven million jars had been sold.

In this chapter, we will explore the domain of AI, and it's cohesive usage in concept design.

1.3 Understanding 'Footwear' and Its Social Impact

The shoes' history predates civilization's settings, making them one of the 'useful objects' in general. We may claim that old models were created by primitive societies solely for defensive functions. Still, in the most advanced societies, footwear has begun to be conceptualized with a growing complexity between aesthetics, usefulness and innovation in technology and materials (DiRoma, 2017).

The use of footwear in India can be traced way back to 200 BC. We can see features of the kings wearing boots in various coins and paintings of the Kushan period (130 BC to 185 AD) and the Gupta period (320–550 AD). Footwear has been part of ancient Indian society for thousands of years. As per the great epic Ramayana, Bharata returned from Chitrakuta carrying Rama's paduka (toe-knob sandal or khadau) and ruled the Ayodhya after placing the paduka on a throne in Nandigram as Rama's proxy. While leaving the princely life in the 5th century BC, Lord Buddha is mentioned in scriptures as having taken off his sandals. In the ancient book, 'Laws of Manu' (800 BC), leather shoes are mentioned as a suitable gift for a Guru or teacher (Ancient Footwear of Bharata Dr. Uday Dokras 2020, Indo Nordic Author's Collective). In ancient Indian texts Rigveda, Yajurveda Samhita, Atharvaveda, Brahmans and Panini Grammer, footwear has been mentioned with the Sanskrit term 'Upanah' or 'Upanat'. This footwear (sandals and shoes) was made from grass, wood and leather. The sculptures in the Konark Temple, dedicated to Sun God (750 AD) have sculptures of women wearing High heeled shoes. The Ramalingeswara temple near Warangal has 850-years-old sculptures with high-heeled shoes. In India, footwear has gone beyond its basic requirement of covering feet and evolved into an identity of self or imagery of an individual. At the archaeological site at Chandraketugarh, about 35 km northeast of Kolkata, footwear with a raised heel and floral motifs used around 200 BC were found. The term paduka (Sanskrit of footwear or foot marks) is a compound word made up of two Sanskrit words, namely, 'pada' (foot) and 'ka', a diminutive ending literally meaning 'small'. The word pada or 'foot' is cited in the ancient Hindu scripture, Rigveda representing the universe, namely the Prithvi (earth), Vayu (Air), Akash (Sky) and the element of the realm beyond the sky. It also means the footprints of divine figures and gods. Hence, the design process of ancient Indian footwear indeed was laced with spiritual philosophy and practices beyond physical need.

Egypt is rich with archaeological evidence, some of which indicate ancient Egyptians' footwear. Skin appears to have been the primary component used in the manufacture of footwear during the predynastic period of Egyptian history, which Egyptologists call the Predynastic period. The sandals typically consist of a single sole layer kept in place by leather straps that are integrally cut out of the same piece of leather as the sole (shoes were not yet commonly worn during this time). The Old Kingdom and First Intermediate Period saw a minor improvement in this scenario due to the society's increased emphasis on vegetable fibres. Most of the footwear was made with vegetable fibres, and the technique in which a horizontal bundle of material is fastened to another pile by wrapping with a strip of vegetable material is referred to as 'sewing'.

Most of the footwear discovered in Egypt during excavations dates to the New Kingdom. This is partly attributable to the archaeological record, which is far better preserved than earlier times, the traces of which have either been removed or are still buried beneath the New Kingdom ruins and are thus inaccessible. Egyptologists typically devote more time than others to studying this period. But this rise is also a result of an increase in the number of people wearing shoes. People may have even owned multiple pairs; based on the more than 80 pairs of shoes and sandals interred with King Tutankhamun, this was undoubtedly the case for royalty. The variety of sandals grew, and in the later New Kingdom, completely closed shoes were first observed. There are only a few specimens from the Late Period. Pharaonic footwear, especially leather shoes, was extraordinarily vibrant and most frequently featured red, green, or a combination of the two colours.

1.4 Contemporary Footwear Design Process

Researchers Lei and PengWen-li presented a flow chart per the information flow in the footwear design process while proposing a DIMS (Design Information Management System) (Lei & Wen, 2012).

The process is as shown (Fig. 1).

The above process flowchart appears linear and lacks steps such as concept design, modelling, etc., which are crucial in the footwear design process.

Hernandez et al. exhibited a general design process of footwear production in SMEs in Colombia, which indicated steps including concept development, cutting and modelling, assembling, etc., but still needs more crucial details, especially in the concept development stage (Hernández et al., 2012) (Fig. 2).

As per Rana Cheikha, shoe expert and advisor, shoe production is lengthy and detailed. The steps involved can vary between 70 and 300 or even more. Essentially, the process involves Research and experimentation, defining the concepts, defining the style, selection of shoe shapes and materials, developing the design and collection, selection of shoe last, pattern making and cutting, pattern cutting, assembling the shoe with the closer, creation and testing of prototype or sample shoe, assessing and adjusting the prototype, re-making the prototype based on agreed adjustments, assessing the look and fit of the revised prototype and placing the shoe into production.

In general, footwear design is a complex process involving the collaboration of various experts, including users, sample makers/artisans, footwear engineers, material scientists and even anthropologists. However, one must understand the differences between footwear design and fabrication. Footwear design is a set of complex tasks primarily focusing on finding the most versatile specifications that can cover a

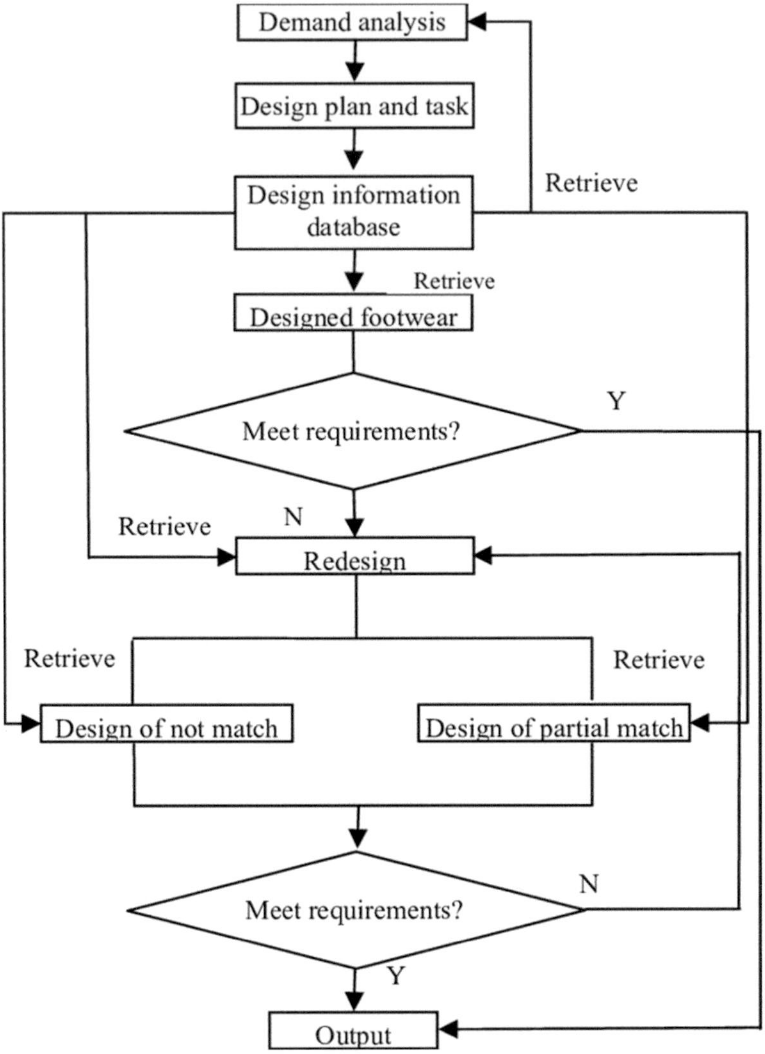

Fig. 1 The application flow of footwear design information management (Lei & Wen, 2012)

broad spectrum of the user's biomechanical (body), psychological (mind) and kinesthetic requirements. In contrast, putting a shoe together once the specifications have been set is a significantly streamlined technical process.

The modern footwear industry is most interested in implementing new manufacturing technologies, particularly those involving additive manufacturing, 3D printing, robots, the Internet of Things, machine-to-machine communication, 3D scanning and remote use. In the context of Industry 4.0, for instance, which the German government supports, Adidas has returned to domestic production this year following several

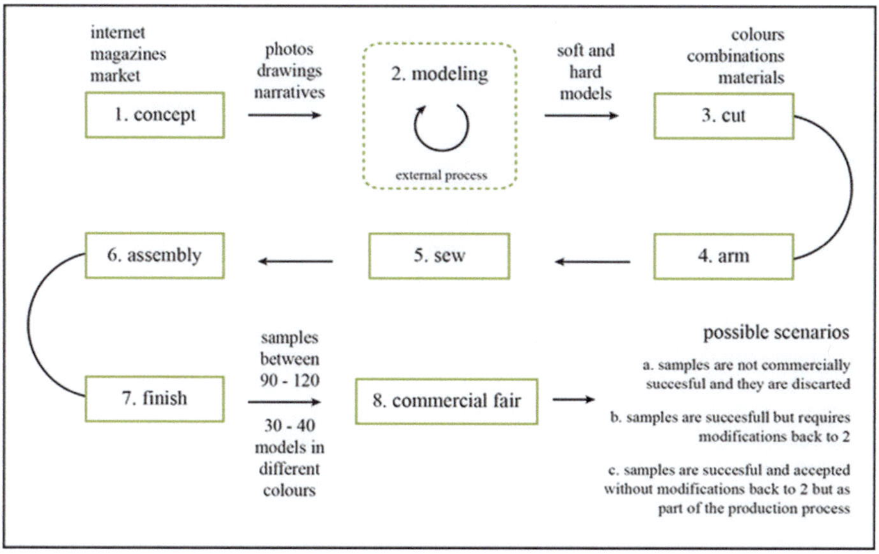

Fig. 2 General design process of footwear production in SMEs in Colombia (Hernández et al., 2012)

years of productivity-enhancing delocalization techniques. 'The three-dimensional modelling technologies allow to produce an elaborate exact project, detailed, and increasingly similar, even if virtual, to the final product'. The gap between the design process and manufacturing are minimized by sharing a single 3D file across various professions, including engineers, marketers, communicators and industrial designers (DiRoma, 2017).

The contemporary footwear design process starts with analysing the user's requirement, the feet measurement using the 3D scanner, consideration of ergonomics is crucial while the development of last using mechanized process, using 3D printing technology to print the customized lasts, masking and developing the flat patterns, designing the upper, outsoles, insoles and insteps, etc. In many cases, the entire shoe itself is 3D printed.

Lien Ya and Chien Hsu proposed an innovative method of footwear development using the 3D scanner and 3D printing (Lin & Chen, 2015), as presented (Fig. 3).

The above method reduces the time in the prototype manufacturing process. Still, it may only apply to some materials used to develop footwear.

According to the study by Rahman et al., Fit and Comfort, followed by style, appear to be the essential evaluative clues for footwear users (Rahman, 2018) (Fig. 4).

However, according to J. H. Thornton, the design of shoes, like all articles of apparel, is considerably more vulnerable to the whims of fashion than the design of any other produced items. This is especially true for women's clothing. However, given that shoes are among the oldest products humans have ever manufactured, it is conceivable to trace back some of their basic qualities, which we might call 'basic

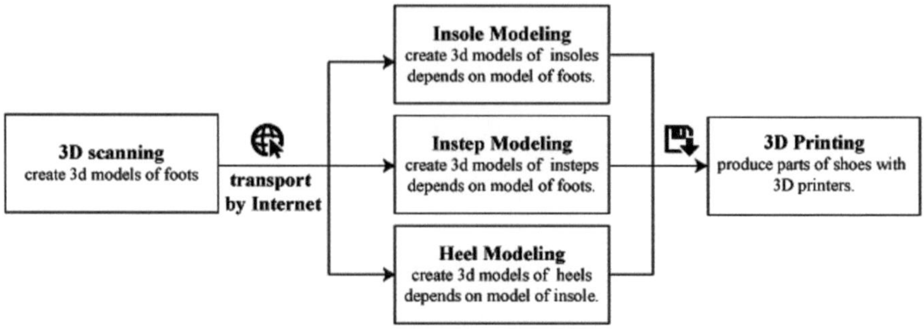

Fig. 3 Innovative producing process (Lin & Chen, 2015)

Footwear	N	Mean	S.D.
Product Cue			
Fit	21	4.95	.218
Comfort	21	4.81	.402
Style	21	4.52	.602
Price	21	4.43	.598
Color	21	4.43	.598
Quality – workmanship	21	4.38	.865
Durability	21	4.33	.913
Wardrobe coordination	21	4.19	.680
Fabric	21	4.19	.928
Ease of care	21	3.57	.978
Brand name	21	2.86	1.276
Country of origin	21	2.62	1.117

Fig. 4 Importance of footwear evaluative cues (Rahman, 2018)

design'. In the author's opinion, a well-designed object, whether a shoe or a rabbit hutch, may be compared to a three-legged stool, where the seat serves as the design and is supported by the three legs of purpose, material and method. The designer must have a clear understanding of the aim and always keep it in mind. Every other factor should be secondary to this; none may prevent the design from being realized when the item is ultimately employed for its intended use. The purpose for which the design is intended, followed by the material's availability, determines the material, or what the item is constructed of, in the main. After being selected, it substantially controls the third leg's construction technique. Thornton finally addressed the procedure or means of producing the piece. Therefore, the designer must thoroughly understand the actual manufacturing process that his design will go through to become a reality. As described, the three-legged stool of design would stand securely if it weren't for the fourth interchangeable leg of fashion has to be inserted so frequently, especially when it comes to apparel, with the stool often tipping over. The four legs of purpose, material, method and fashion serve as the stable base for the ideal design, just as the four legs of the ideal stool.

Jane Schaffer & Sue Saunders emphasized that becoming familiar with the qualities of material and hardware is a valuable exercise for any footwear designer (Schaffer & Saunders, 2012). They stated that the footwear designer must know about the last, the heel, proportion and scale, material, colour and texture, fittings and trims, as well as soles (and insoles) to design a pair of good shoes. They also mentioned that a designer must respond to the client brief and develop their way of approaching the process, adapting the method that suits one as an individual and discovering fast-track solutions to meet deadlines.

However, the above processes focused on something other than the initial conceptualization stage to crystallize the new ideas into products but more on realizing the product through sampling or prototyping.

1.5 AI-Enabled Design Tools, Software and Platforms

Miklos Phillips in 2020 states AI will shape design tools, designer roles, workflows and procedures. It emphasized the need to change our perspective from 'human vs machine' to 'human plus machine' in light of the passing of time. The theoretical underpinnings of artificial intelligence and machine learning have remained relatively unchanged over the past few decades. The statistical methods are unchanged. In the 1990s, theories were developed on the underlying technology of neural networks, a machine learning technique based on how physical brains work. The compilation and analysis of countless millions of tagged photos (from the source), or pictures with some pre-attached description of their contents, is what the AI generation of images depends on. These descriptions and photos are then processed by neural networks, which train to link specific and intricately detailed aspects of the image—such as its compositions, colours and shapes—with particular words and phrases. The billions of connections with various weights that are generated by a straightforward stimulus are then placed on top of one another to create novel arrangements of shape, colour and composition. Tagging and training the dataset takes a lot of time and manual labour.

There are numerous generative AI tools available right now. Dall E is one among them and is a flexible instrument. A software programme called Dall-E with restricted access was launched by OpenAI (an AI research centre) in January 2021. This software allowed users to enter a straightforward text description of an image they had in mind, and in a short amount of time, the programme would interpret their suggestion almost uncannily well, on par with an illustrator or an Adobe-trained designer, but much more quickly and without cost (The Guardian, 2023). DALLE utilized a Transformer, a deep learning architecture that first appeared in 2017 and has since come to be the standard for processing sequential input and text encoding, as well as a variational autoencoder (VAE), a model trained to encode an image into a low-dimensional probability distribution and then decode it off of it. VAE can also generate new images by sampling from the intermediate distribution and feeding that sample through the decoder.

The Transformer has trained autoregressively on concatenating a sequence of text tokens and image tokens, which was acquired with the VAE trained beforehand. This involved training the Transformer to attempt to predict future values based on past ones. The entire text prompt is provided to the Transformer during sampling, which then generates each token, in turn, to be later decoded by the VAE to produce the entire output image. When it was initially released, DALL-E displayed genuinely astounding results—it also had certain drawbacks, the most obvious of which was the difficulty it encountered when attempting to produce photorealistic images instead of comical or artistic-looking ones. The way text material is produced has altered thanks to a product made by OpenAI called ChatGPT. According to The Guardian, if you ask ChatGPT to summarise the Book of Job in the poet Allen Ginsberg's manner, it will make a passable effort in seconds. You could even ask it to convert Ginsberg's poem Howl into a presentation from a management consultant's slide deck. The public was fascinated by these programmes' abilities to create bizarre new worlds in both words and images. The urge to experiment on one's own led to the growth of literature on the ins and outs of using these tools to their fullest potential, particularly how to structure inputs to get the most interesting results. The open-source Stable Diffusion and separate commercial efforts like Midjourney identify and produce photos differently. The way creatives are produced is evolving thanks to AI-driven automated technologies like Dall-E, Midjourney and ChatGPT.

Let's take the CALA product as an example. CALA is a fashion platform established in 2016 for designers looking for an approachable way to translate concepts into physical items. The service is accessible via a mobile app and its website. It has been adjusted to work with CALA's platform. Still, it shares much in common with AI art generators like DALL-E 2 and Stable Diffusion. Designers are instructed to choose a basic style from a list of 25 alternatives, such as a jumper, blouse or tote, rather than submitting a text prompt in one long string of text. Designers then alter the style using generative AI in response to two text cues. The first uses adjectives and materials to describe the design—the second talks about desired trims and features such as cuffs or zippers.

The CALA founders think this alternate user interface will aid designers in focusing on key features. 'We've put a user interface on top of the prompt engineering. We want individuals to arrive at a significant conclusion as soon as feasible. CALA displays six results at a time, and any of them can be added to the design platform for more iterations. It is important not to misinterpret CALA's implementation as a one-click design tool. Designers still have to use the CALA platform and provide their talents. The founders anticipate that AI will dramatically lower the entry barrier for aspiring designers and provide seasoned designers with a means of overcoming creative obstacles. Shutterstock (a service offering stock photos, images and videos) launched an API that helps generate prompt-based images and designs easily.

Midjourney is yet another modern tool. A San Francisco-based independent research lab called Midjourney, Inc. developed and hosts the AI-based programme and service known as Midjourney. According to the website, Midjourney, like OpenAI's DALL-E and Stable Diffusion, creates visuals using natural language inputs.

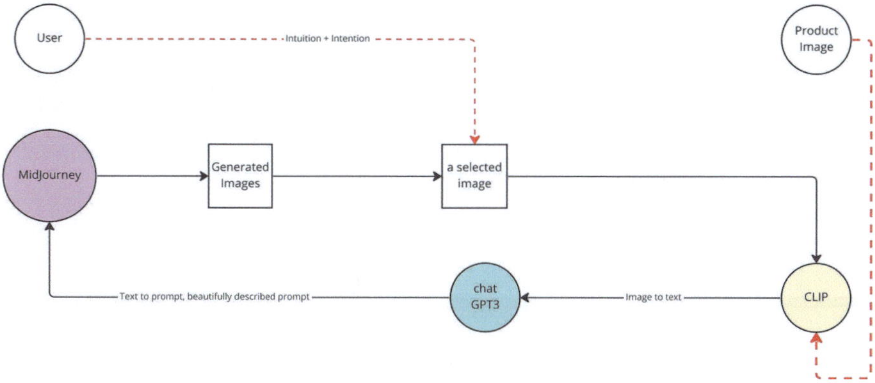

Fig. 5 Process diagram (Jang, 2023)

The nature and function of all the technologies mentioned above (Text-to-image generators) are remarkably similar. Researchers compared three of these most often utilized design tools such as Stable Diffusion, Midjourney and DALL-E-2. The second-generation DALL-E model (DALL.E 2) is smaller than its predecessor but, in the author's opinion, is superior. While DALL-E 2 can create almost anything, it employs a technique called unCLIP that is sufficiently advanced to produce visuals that were once challenging for us humans to convey. It still has some restrictions. An artificial intelligence model called CLIP (Contrastive Language-Image Pre-Training) was created by OpenAI and is trained to recognize the connection between verbal and visual information. CLIP can read and understand natural language descriptions of images and videos and classify and recognize objects and scenes within visual content.

Here is a flowchart of usage of CLIP in Midjourney versus ChatGPT (Fig. 5).

The ChatGPT-3 algorithm can be used to alter the text extracted from a picture by the CLIP model in a way that results in a new prompt. The generative AI image model can take this prompt as input to produce a new image, which can then be processed again by the CLIP model to extract new text so that the cycle can continue.

DALL-E 2 produces far more complex output and is ideally suited for corporate use because it is trained on millions of stock photos. Emad Mostaque, the author of Stable Diffusion, believes that DALL-E 2's most robust feature that sets it apart from other picture generators is inpainting. DALL-E 2 creates much better pictures when it has more than two characters, compared to Midjourney or Stable Diffusion (Fig. 6).

Midjourney is a tool that is primarily renowned for its beautiful design, Mr Chowdhary continued. According to him, the image it produces virtually never resembles a photograph but rather a painting. Some artists consider it to be a student of art. In his opinion, Midjourney is a budding artist with a distinctive flair. And when you use my name to construct an image, it's like asking a student of art to create something that was inspired by my work, an artist once said. Midjourney uses a discord bot to make

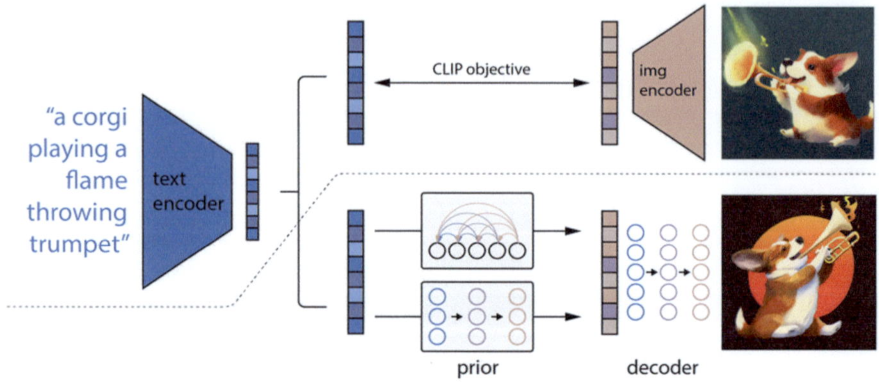

Fig. 6 Process diagram for DALL-E 2 (AI Network, 2022)

and receive calls to AI servers, and most activity occurs there. A lively community of more than a million individuals exists on Midjourney, where everyone contributes magnificent works of art. Midjourney and DALL.E 2 refuse to become fully open-source. However, Stable Diffusion claims to be an open-source model everyone can access. Additionally, Stable Diffusion has a good grasp of contemporary visual illustration and can create incredibly intricate artworks. It cannot, however, decipher intricate original cues. A simple picture generator like Cryon (formerly DALL.E mini) can provide those cues, whereas Stable Diffusion cannot. Stable Diffusion excels in producing intricate creative graphics but fails to produce generic visuals like logos. The limitations of the tools are another distinction. Unlike Midjourney or DALL.E2, Stable Diffusion has no restrictions.

1.6 AI-Enabled Footwear Concept Design—A Case Study

Gen Next designers are already using AI-enabled tools and platforms. Here I am furnishing an example by a product designer (Bhurke, 2023) as published in his article 'Artificial Intelligence in industrial design—using AI in your design process'. He used one of his sketches, added it in Midjourney and added the prompt 'a detailed sketch of a sneaker, futuristic, 4 k, studio lighting, pen and paper' (Fig. 7).

The outcome of the process is shown (Fig. 8).

Further, he selected one of the generated sketches and added the prompt 'shoe design, leather, DSLR, studio lighting, sneaker'. The result is shown (Fig. 9).

According to the designer, the results are not too crazy, but they look great.

Next, he used the Stable Diffusion as Dreamstudio. The designer stated that Dreamstudio works fairly similarly to Midjourney, except that it has a bit more control over how much quality and quantity one is looking for (Fig. 10).

Fig. 7 The original sketch (Bhurke, 2023)

The designer can import an image reference as well in the Stable Difffusion and can type in a prompt for what one is looking for more contextual detailing. The user then controls exactly how much of the reference image influences the output. Lower image strengths will make the sketch more loosely based on reference image, and higher percentages make it closer to the image.

Here is an output with 60% Image Strength (Fig. 11).

Stable Diffusion also has another feature where one can take a reference image and erase parts of the image to let the software interpret and complete the image based on the given prompt. So if the designer is looking for variations of more specific areas in an image, it can be done using this feature.

The demo screen is shown (Figs. 12 and 13).

Finally, the designer opts for rendering with 80%. The result is shown (Fig. 14).

Next, the designer used DALL-E which has inpainting, but the result is not as crisp as what Stable Diffusion provided. The work screen is shown (Fig. 15).

Fig. 8 Outcome from original sketch in Midjourney (Bhurke, 2023)

Finally, he used Vizcom which has a Sketchbook-like interface. The designer wrote that this platform allows one to draw something, provide a prompt for it, dial in the sketch's influence on the output, and then also let one draw over the generated image, allowing the designer to fix it up and regenerate something that's closer to what is imagined (Fig. 16).

The designer mentioned that it took less than 10 min from the first step to the final outcome. According to him, this tool could help visualize early concepts better or even provide a helpful underlay for a more refined concept sketch.

However, the designer and their imagination get reflected in the machine in the end. As Georgie Gilbert put it in her article, 'Our imagination is a powerful tool to kick start the process. Some people believe it's just all the Artificial Intelligence doing the work, but it is not; it's a relationship, a collaboration between man and machine'.

Here is a concept design using ai (Fig. 17).

Fig. 9 Concept designs generated by Midjourney (Bhurke, 2023)

Although the AI-enabled designs, as shown above, are quick, elaborate and unique, many challenges need to be addressed with human intervention, especially in the context of functionality and ergonomics. There is no checkpoint to confirm that the footwear produced by AI algorithms is safe and comfortable for wearers. It requires careful human intervention for iteration, testing and validation of the designs to ensure that they meet the highest quality and performance standards.

Hence, an 'augmented concept design process' involving human and machine intelligence has been proposed.

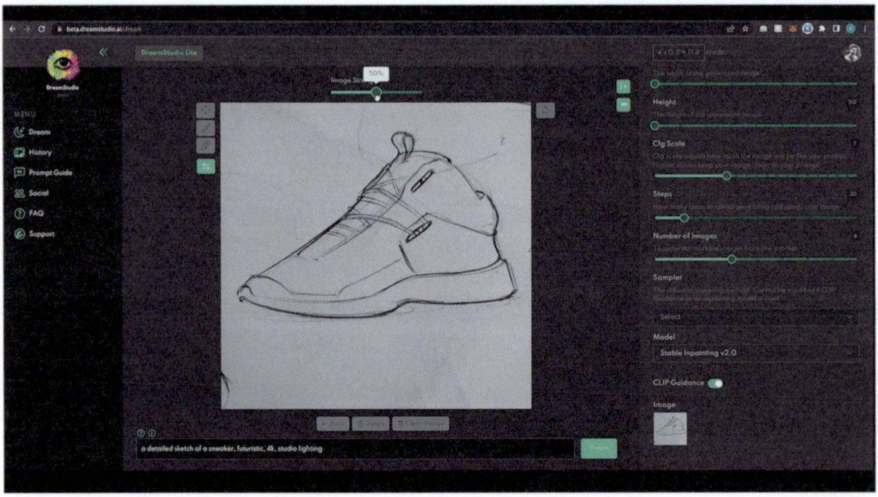

Fig. 10 Dreamstudio work-screen (Bhurke, 2023)

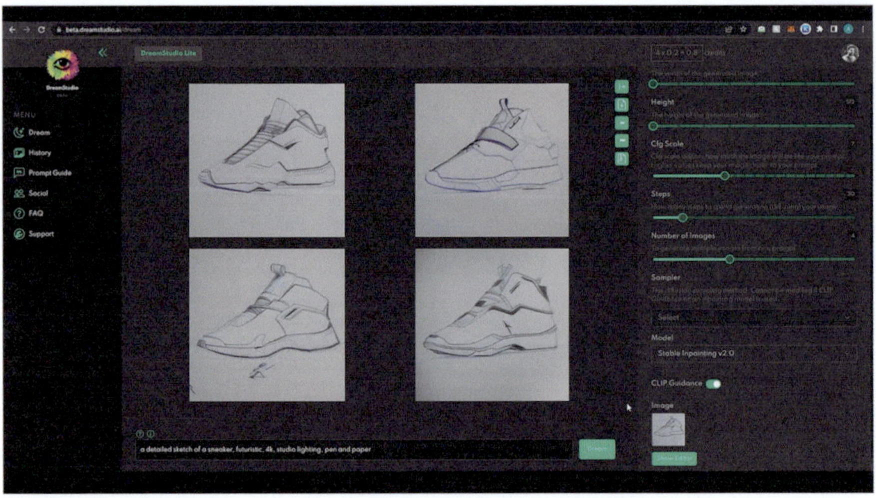

Fig. 11 Outcome from dreamstudio work-screen (Bhurke, 2023)

1.7 'Augmented Concept Design Process': Collaboration Between Human and Machine

I propose a five-step-augmented method of creating footwear concept design. The steps largely integrate the initial concept generation process, imprinting it to modify and manually enhancing the functional detailing to convert the concept into a 3D design finally. The detailed process is described below.

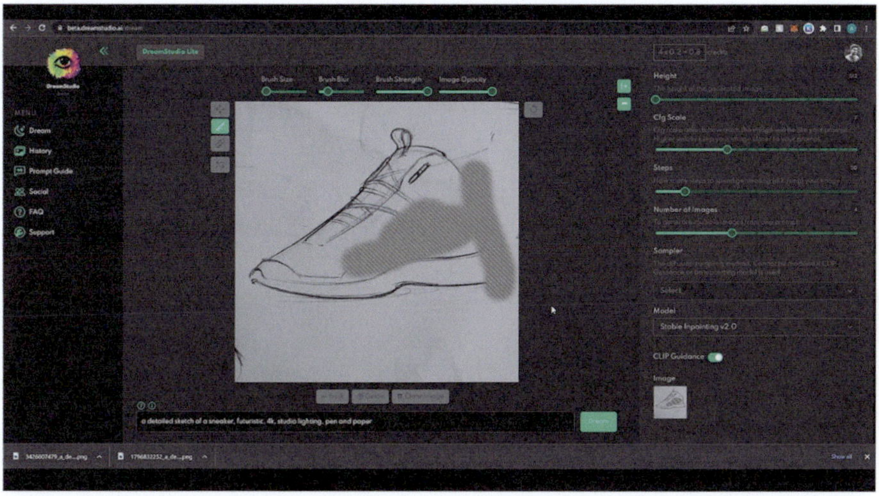

Fig. 12 Inpainting—Erasing part of image to be reinterpreted in dreamstudio (Bhurke, 2023)

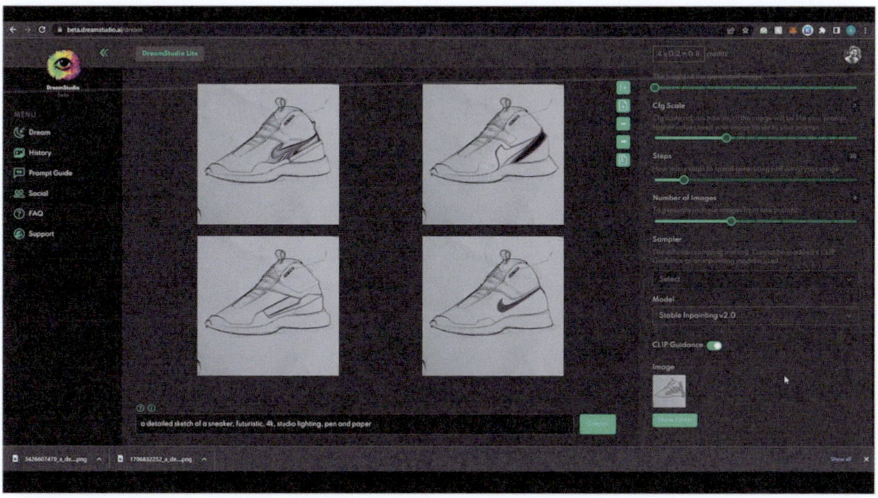

Fig. 13 Inpainting results on sketch reference in dreamstudio (Bhurke, 2023)

Once the designer adds input (Concept sketch of footwear and a Text prompt) in any AI-enabled platform (i.e. Dreamstudio, Midjourney, Dall E, etc.), the variations of sketch concepts get generated in the AI-enabled platform. From the generated concepts, the designer selects one sketch concept and in paints the same in the platform to modify the detailing in the selected concept. The result generates rendered concepts from which the designer selects one rendered concept and transfers it into Procreate or any similar software for manual enhancement of functional detailing.

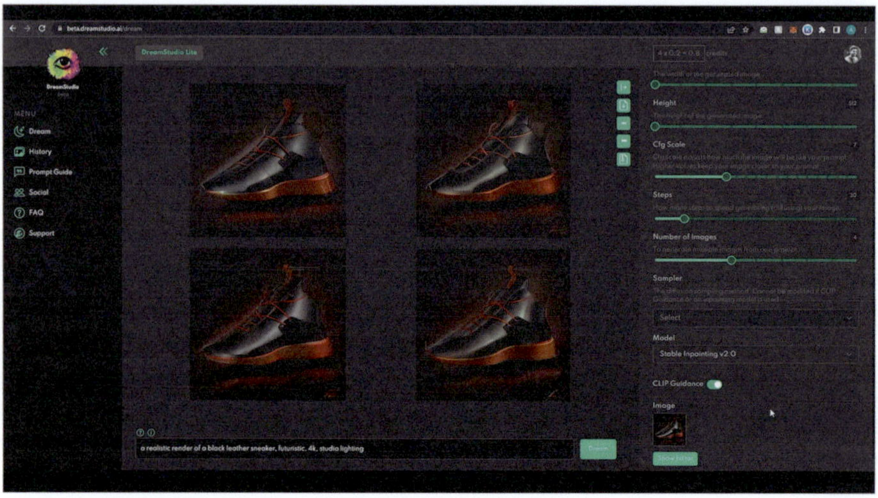

Fig. 14 Dreamstudio work screen with 80% rendering (Bhurke, 2023)

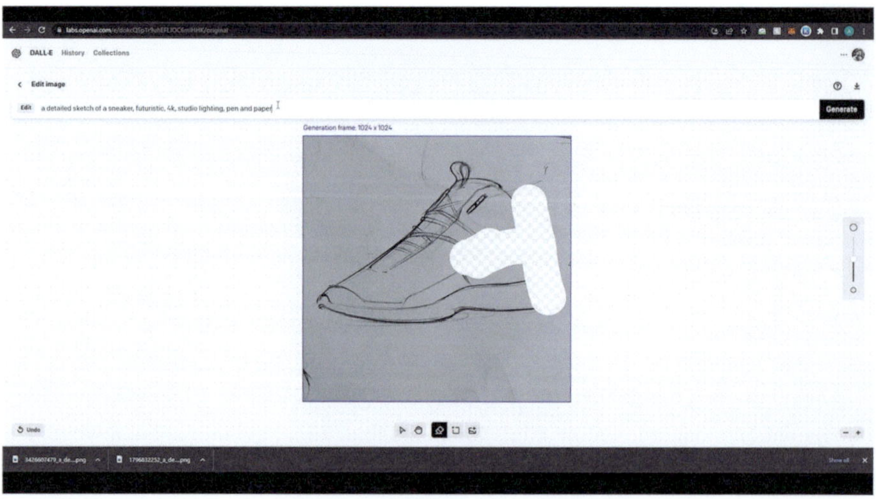

Fig. 15 The work screens and outcome in DALL-E (Bhurke, 2023)

The designer then takes the final design to Gravity Sketch in VR software for 3D design development. The 3D design is further rendered in the Keyshot, and the best output is exported as the final result.

The whole process is shown in the below flowchart (Fig. 18).

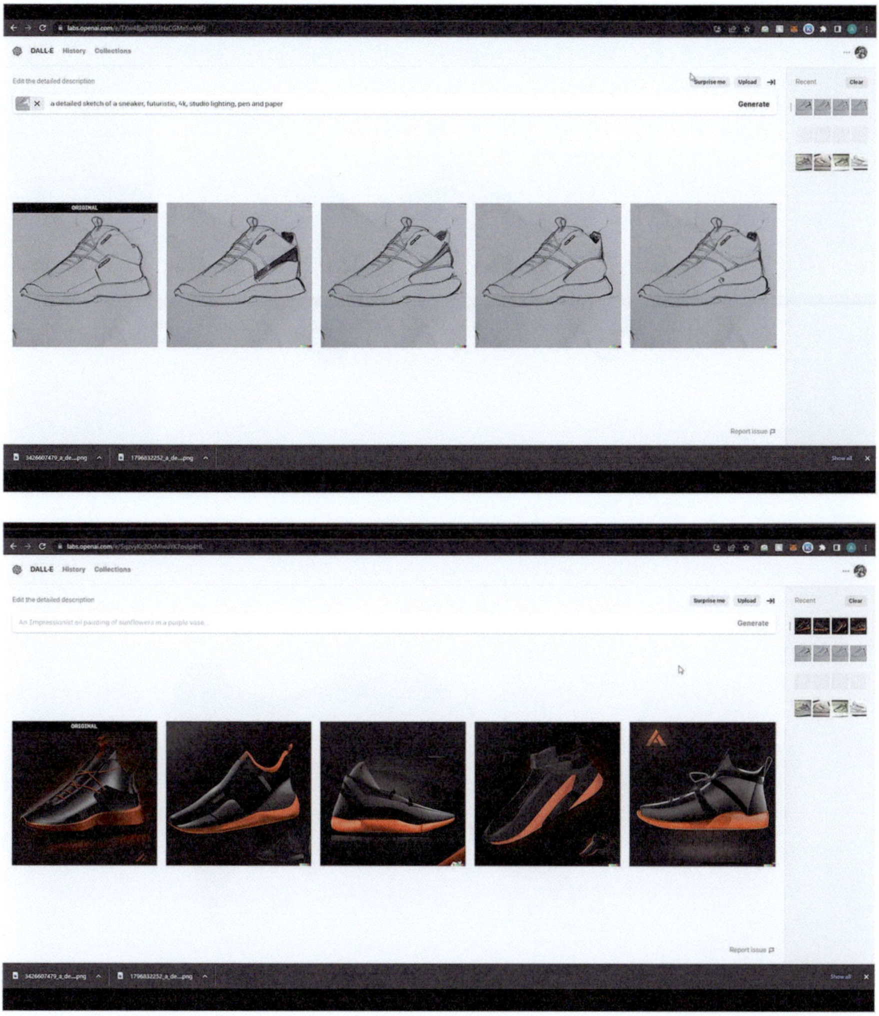

Fig. 15 (continued)

2 Conclusion

AI-generated footwear designs reflect a significant innovation in the domain of design. This process provides designers with tools and techniques to create customized and sustainable footwear with a much faster timeline. While challenges are there to be addressed, the potential benefits of this technology are significant.

They wrote that what is happening right now is a predictable situation. It's simply how the business operates. A company will choose a more practical strategy if it can

Fig. 16 The concept design outcome using Vizcom (Bhurke, 2023)

save money. Because they feared being replaced by machines during the nineteenth-century industrial revolution, some English textile workers deliberately damaged textile machines. Of course, some jobs were replaced by machines (usually those requiring a lot of heavy lifting or repetitive labour), but not all of them. AI tools operate similarly. AI will not entirely replace human ingenuity; instead, it will enhance it. When you consider the main driver behind tool invention, it becomes clear that increasing productivity at the workplace was the main driver—In AI, the same holds. AI will make our work more productive. In any era of product design, even AI design, the quality of your ideas and your capacity to comprehend user issues and develop solutions that benefit people are crucial.

Fig. 17 Concept footwear design using AI by @ai_clothingdaily

The following degree of expertise required for footwear designers is 'prompt engineering', which is the ability to formulate instructions in a way that the system can understand to produce outcomes that are most closely aligned with expectations. According to tech gurus, prompt engineering would become a sought-after and well-paid job description in a 'no code' or 'low code' future, when human language would be the most effective manner of dealing with intelligent computers. The boundaries of our ideas would be the limits of AI's inventions.

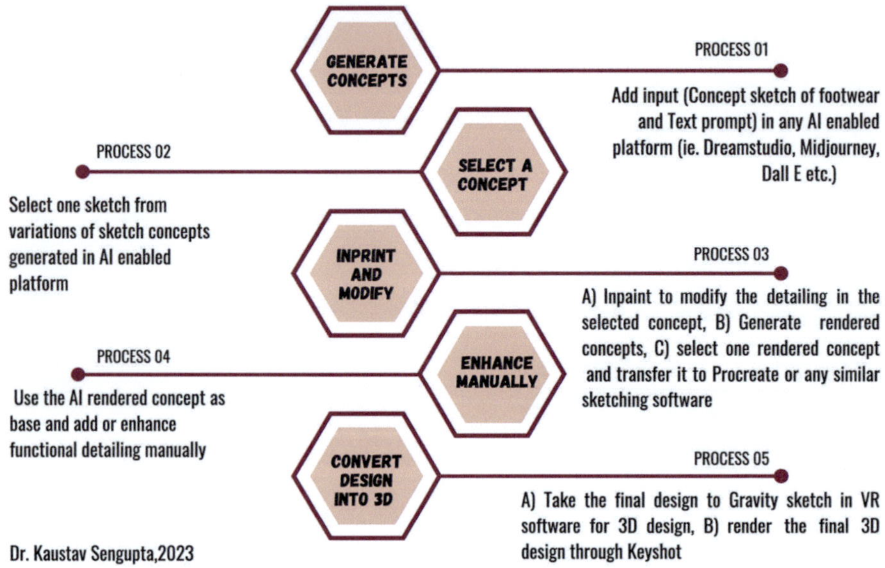

Fig. 18 'Augmented concept design process' for footwear design with AI and human intelligence

References

AI Network. (2022). Dall·e 2: Meaning, limitations, and solutions. *Medium.* Retrieved April 15, 16, 2023, from https://medium.com/ai-network/dall-e-2-meaning-limitations-and-solutions-a988c87ddeae

Bhurke, A. (2023). Artificial Intelligence in industrial design - using AI in your design process. Retrieved April 16, 2023, from https://medium.com/AKBhurke/artificial-intelligence-in-industrial-design-using-ai-in-your-design-process-midjourney-stable-f44353ff8331

DiRoma, A. (2017). Footwear design. The paradox of "Tailored shoe" in the contemporary digital manufacturing systems. *The Design Journal, 20*(sup1). https://doi.org/10.1080/14606925.2017.1352780

Hernández, P., Bhamra, T., & Bhamra, R. (2012). Sustainable product service systems in small and medium enterprises (smes): Opportunities in the leather manufacturing industry. *Sustainability, 4*(2), 175–192. https://doi.org/10.3390/su4020175

Humphrey, N. (2022). *Sentience: The invention of consciousness.* Oxford University Press.

Jang, J. (2023). CHATGPT-*3 x midjourney.* INDEED. Retrieved April 16, 2023, from https://www.indeed-innovation.com/the-mensch/chatgpt-midjourney-design/

Jones, C. (1966). Design Methods - John Wiley & Sons Inc.

Lei, L & Wen, P. (2012). *Research on design information management system for leather goods.* Elsevier publication.

Lin, L. Y., & Chen, C. H. (2015). Innovation and ergonomics consideration for female footwear design. *Procedia Manufacturing, 3*, 5867–5873. https://doi.org/10.1016/j.promfg.2015.07.891

Mostow, J. (1985). Toward better model of design process. *AI Magazine, 6*(1).

Rahman, O. (2018). Online consumer choice: Footwear design and visual presentation. *Journal of Textile and Apparel, Technology and Management, 10*(4).

Rodgers, P., & Milton, A. (2011). *Product design.* Laurence King Publishing Ltd.

Schaffer, J., & Saunders, S. (2012). *Fashion design course. Accessories—Design practice and process for creating hats, bags, shoes and more*. Thames & Hudson.
Simon, H. A. (2019). *The sciences of the artificial*. The MIT Press.
The Guardian. (2023). The stupidity of AI. Retrieved April 15, 2023, from https://www.theguardian.com/technology/2023/mar/16/the-stupidity-of-ai-artificialintelligence-dall-e-chatgpt?utm_source=pocket-newtab-intl-en
Weiskrantz, L. (2008). *Blindsight: A case study and implications*. Clarendon Press.

Policies for Promoting Leather Trade

Capacity Building of Global Leather Sector for Inclusivity and Sustainable Development

Swarna Vinodh Kanth and Habib Hussain

1 Introduction

1.1 Skill Development Panorama of the Global Leather Sector

The pandemic, divided the world with a stark contrast between the massive job losses in hard-hit sectors like leather sector and the positive job growth in industries like information technology and finance. In 2020, the pandemic triggered losses equivalent to 255 million full-time jobs and $3.7 trillion of income, making it the most severe recession since the year 1930 (Global Skills Report, 2021). The pandemic economy has hence created a complex landscape that threatened and left millions of leather sector workers unwary for the new post-pandemic future. Many of these jobs have not come back. The scale of this challenge means that, perhaps more than ever, policymakers worldwide in the leather sector need to work towards an intimate understanding of the skills related to leather/non-leather/vegan leather-like industries that drive sustained employment and economic growth. The entire leather/non-leather/vegan leather-like and associated industry worldwide is based on the export/import market and the supply chain, which went divested and faced a major hard-hit by layoffs of employees. The shifts in the economy globally are expected to stick and imply a steep uphill climb for the displaced workers who could have previously shifted from leather to another growing service sector job. A recent study

S. V. Kanth (✉)
CSIR-Central Leather Research Institute, Adyar, Chennai 600 020, India
e-mail: swarna@clri.res.in

H. Hussain
A.V. Thomas Leather and Allied Products Private Limited, Chennai 600 032, India
e-mail: habib@avtleather.com

S. K. Janardhanan and L. A. Zugno (eds.), *Emerging Trends in Leather Science and Technology*, https://doi.org/10.1007/978-981-99-9754-1_10

of eight major economies of the world showed that over 100 million workers—about one in 16 workers—that includes the leather industry will need to find a different occupation by 2030 post-COVID (ILO Monitor Report, 2021a). This is 12% more than was estimated before the pandemic, and up to 25% more than was estimated after the pandemic for advanced economies. The trend is in line with the global leather/non-leather/vegan leather-like sector. Hence, jobs and skills in leather sector are pressing challenges for developing countries (Sedik & Yoo, 2021). Paradoxically, there is also a growing concern of leather/non-leather/vegan leather-like skills mismatch with the new age digital era post pandemic. The sector worldwide should have Global inclusive skills training ecosystems to support new age skills acquisition and continued upgrading to digital systems (Manyika et al., 2021). Figure 1 describes the global leather sector landscape. This emphasises the potential growth of the leather industry from 2020 (@ USD 600 Billion) to 2030 (@ USD 1200 billion) at a calculated annual growth rate (CAGR) of 6.4%. The revenue distribution among each continent is shown. Asia occupies the top position followed by Europe in the revenue generation from the leather sector (Gul, 2014).

The bar graph also shows the difference in the revenue distribution of different products from the leather sector. This is highly achievable by the mentioned key players of the leather industry worldwide. The newer education systems should interface and interact with other sectors like mechanical, electronic and information technology along with labour markets and the entire world leather industries currently need to coordinate skills development with changing leather/non-leather/vegan leather-like industry needs (Maclean et al., 2012). The evolved sector should

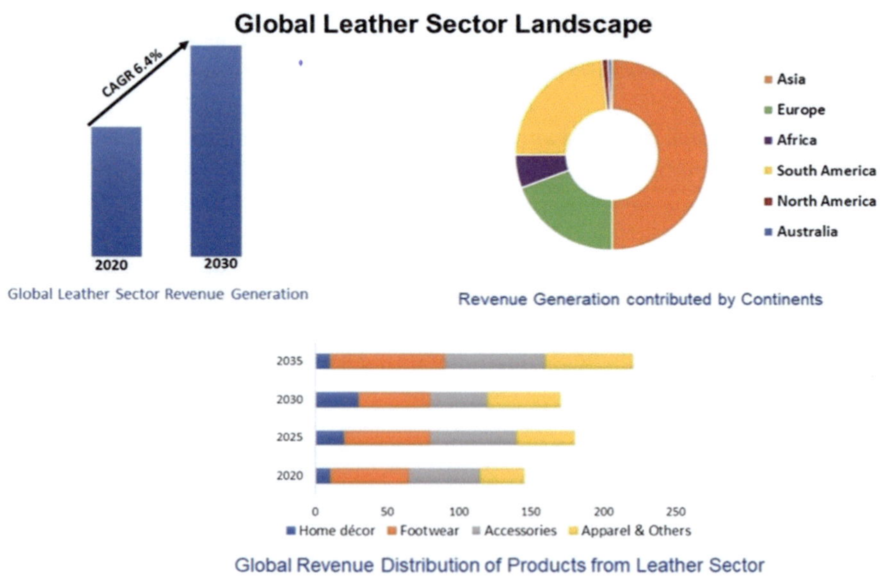

Fig. 1 Global leather sector landscape

globally generate good practices and lessons from different countries and regions that can better inform skills development policy making native to their requirements as a futuristic model.

1.2 Skill Gap and Institutional Capacities for Skill Development

At the global level, the leather/non-leather/vegan leather-like sector trade is estimated to be at $ 1200 billion USD in the year 2030 (Market analysis Report, 2022). Leather trade across the world is growing more rapidly potentially leading to a deeper level of regional integration. Asia-Pacific countries like China, Vietnam and India are already world leaders in the production and export of leather/non-leather/vegan leather-like products with smart applications and services at small and medium sized enterprises (SMEs) as they play a crucial role in creating employment. Asia's population, long considered a concern, has turned into a desirable demographic dividend with skilled, educated and employable people (Wilmoth, 2011). With the increase in the share of value-added products in leather and its products production globally, it is expected that the footwear and leather product segments will have a significant share of this projected growth. Within the footwear and products segments the need for skilled man power is the largest at the operational level, supervisory, manufacturing, designer, merchandiser and brand production (World footwear business condition survey Report, 2022). As per study, based on the current growth trends, it is estimated that the footwear segment alone will need about 20 million trained youth out of the 50 million manpower for the entire global leather sector that include leather, footwear, accessories, upholstery, bags, saddlery and other leather/non-leather/vegan leather-like products during the next 10 years (Fig. 2). The availability of contemplable and growing work force in developing nations of Asia, South America and African countries, which is often referred to as country's demographic dividend, has to cater to the increasing demand of the leather industry and enhance their institutional capacity building and quality improvement of more highly qualified workforce and dominance of leather/non-leather/vegan leather-like and its associated products knowledge and skill intensive jobs before quantity expansion of manpower. Although the socio-economic challenges in every continent are not the same and the demand for changing leather sectorial skill needs is somewhat different, overall directions for institutional capacity building towards regional cooperation within inter- and intra-countries globally could provide a long-term goal.

Fig. 2 Global leather sector demography and manpower requirement

1.3 Education, Employability, Employment, and Entrepreneurship of Skill Development

The Global leather/non-leather/vegan leather-like industries are delivering the desired employment outcomes for the 4 E's of skill development as given in Fig. 3. The 4 E's of the leather sector worldwide have five labour market transitions: unorganised to organised, farm to nonfarm, rural to urban, subsistence self-employment to decent wage employment and school to work are part and parcel of the current global leather sector (Sabharwal, 2012).

1.3.1 Education

Education is one of the 4 E's of skill development. Primary education is necessary for the proper understanding and to develop the behavioural aspects of the individual. With education, the skill development of the person in the later stages will be easy and smooth. The new-age requirements post pandemic need global standards that align occupational codes with national specific framework and infrastructure for leather and associated skills development, entry gate assessment and exit gate certification (Sabharwal, 2012). The future requirements of the educated skill workforce demands better regulations of quality, better product knowledge and incentives aligned for the creation of green ecosystem of the global leather sector in the future. Real-time interfacing of the demand and supply side of the growing leather sector, permitting the change of job descriptions, curriculum, job posting and assessments will be the futuristic framework of the new age leather sector skills.

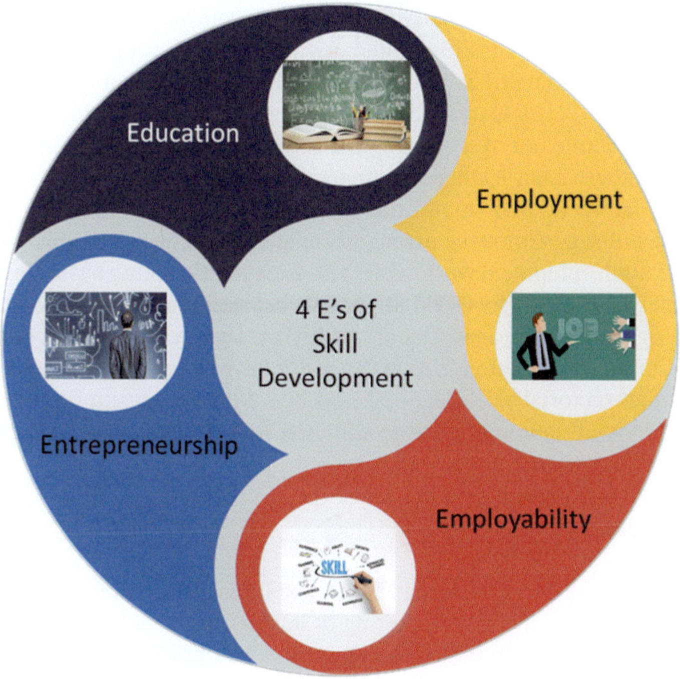

Fig. 3 4 E's of skill development

1.3.2 Employment

Generating employment has been a major focus of most of the governments. Promoting employment to the workforce will result in the propagation of the nation in terms of GDP. The real time establishment of the exchanges of job seekers and job providers need to be brought in action to provide employer connection, provide employee service and turn themselves into potential workforce. The "learning by doing" and "learning while earning" by the modern vibrant global apprenticeship regime for the leather sector will help in improving the employment outcomes globally (Sabharwal, 2012). Apprenticeship programmes ensures direct employer involvement in relevance of the leather and its associated skills obtained.

1.3.3 Employability

Employability skills relate to a host of skill sets which are desirable for an employee to ensure s/he continues his/her employment in the long run and help the company meets its objective. The retaining of skilled employees in the industry has been a key aspect to be considered in human resource (HR). Human capabilities need to be improved to bridge the gap of mismatch problems that arises from the divergence in what employers are seeking globally in the leather sectorial candidates in addition to qualifications and the actual skills candidates have. Skilled people for employability is more than just qualifications for improving the supply of educated manpower to leather/non-leather/vegan leather like product industries. Addressing the challenges for the post-pandemic era of technological challenges associated with industries of leather/non-leather/vegan leather like materials and products, gradual novice pathways within the vocational stream has to be given adequate attention for gradual upgrading of skills, performance management and competencies (Sabharwal, 2012). Currently, skills development worldwide with vocational training has wide open entry and exit gates and has to focus on alignment between various parts of the leather skills, training and the futuristic global employment ecosystem that includes assessment, curriculum, certification, soft skills and jobs. Market demands and trends for new age leather sectorial skills shall take on board training institutions and include certification for well-defined futuristic skill sets in their curriculum and courses.

1.3.4 Entrepreneurial

The leather sector has huge potential for the development of entrepreneurship. Many small-scale business in the manufacturing of leather products have contributed a lot in terms of the export potential of the respective countries. The shift of culture in developing entrepreneurship minds in leather/non-leather/vegan leather like products need to be taken up vibrantly worldwide with regulatory and policy framework across countries. The lack of pragmatism in public policy for the manufacture of innovative leather products for new age applications in not recognising entrepreneurial motivations and structures for what they are costing the futuristic leathers and products skills education dearly for the global leather sector. Global entrepreneurs in the leather sector can be increased by a more conductive regulatory framework of the individual governments of countries across the globe. Leather and products production entrepreneurial skills centre around attitudes (soft skills), such as persistence, networking and self-confidence on the one hand and enabling skills (hard skills) on the other hand, including basic start- up knowledge, fashion forecasting, branding initiatives, business planning, financial literacy and managerial skills (Sabharwal, 2012). A lot of efforts are being practised for the effective entrepreneurship education in the leather sector worldwide with policies and framework focusing on developing the entrepreneurial competencies and skills, which are transferable and beneficial in many work contexts of leather and products manufacture. The entrepreneurship curricula in embedding entrepreneurship in both formal and informal products manufacture, shall include coaching teachers and trainers, strengthen the institutional framework and capacity building. The future leather sector shall focus more on not just strengthening the capacity of the individuals but also involve in bringing up entrepreneurial culture in society.

2 Methods and Discussions

2.1 Global Investment in Leather Sector Skill Development

Investments in the skilling of the workforce have the potential to drive competitiveness, innovation and equity. Economic growth doesn't follow the "one size fits all model" but demands a growth of a specific skill set in the workforce globally. The promoting of leather sector worldwide and to enhance the social, economic and ecological development, large investments in the leather sector is essential through proper policy development and planning. Investments in technology and HR of the global leather sector will create comprehensive skills development and training policies that meet the demands of the labour market. Skill development will be an integral part in keeping the workforce accelerated with the digital transformation post pandemic. The world economic forum seeks to provide 10 billion people with better education, skills and jobs by 2030 that includes leather sector by investing in some of the largest reskilling programmes ever undertaken. Regional cooperation with success models on skill development from Singapore, Hong Kong, China, Republic of Korea, India, Australia and the European countries have overcome issues, concerns and challenges related to specific pathways of economic development in each country as the rapid expansion of knowledge-led leather sector. Many countries have adopted the private-public partnerships models to develop the livelihoods of workforce and to have a sustainable future of the leather industries worldwide. The past four decades have hit workers who perform routine work hardest, and made manual service jobs that pay lower wages an increasingly central job category[11]. With the COVID 19 recession, the labour markets were polarized as companies increased their automation during the downturns in leather and leather product manufacture[12]. Many top performing countries showed higher overall skills and these skills help in productivity of sectors like leather sector. The leather sector is also stepping up investment in automation and Artificial Intelligence (AI) either somewhat or significantly. The result is reduced economic opportunity and social mobility for the less educated (Autor et al., 2021).

The demand requirements of product segment wise of employees in the global leather sector is given in Fig. 4. Most of the leather sector is dominated by manufacture of footwear and required skilled employees with design capability and creative outlook for product proposition. With the demand increasing the productivity of the leather sector, a shift in the upskilling with digital needs like AI and machine learning is forecasted that necessitates for more graduates and higher learning degrees. The current development of the skills includes nurturing an in-house skill work force at all levels using unique training model scattering to core belief systems of the global leather industry demands. Best practices demand pristine bubble of total focus with tailor-made training programs, with interdisciplinary computability and pedagogic innovations. Building and nurturing finesse and design skills of specialization and standardization with constant skill upgradation to meet global dynamics with cross industry and country partnership skill archetypes will be the new normal for the

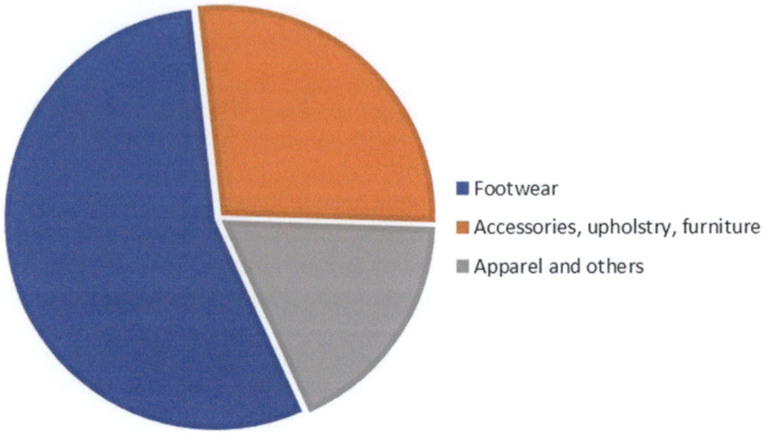

Fig. 4 Employment and entrepreneurship contributed by products

new-age global leather sector. The sector shall focus on the intrapreneurial capability skill sets with quality conscious and professional capable skills for elite performance producing skill workforce.

2.2 *Corridors to Opportunity of Skill Development for a New Global Leather Sector*

The onset of investments in leather and products, skills development across nations of the world offers the future a unique opportunity to build more inclusive, modern, and scalable education systems. Figure 5 shows the distribution of employees in the global leather sector. Most of the leather sector is dominated by semi-skilled employees with no proper graduation. Wages are comparatively low for unskilled workers as compared to skilled and semi-skilled. With a shift in the upskilling of the industrial sector, it is forecasted that the leather industry will have more graduates doing technical analysis and increasing the productivity of the industry. Reskilling the millions of leather and products manufacture workers affected by the impact of the pandemic and the growth of automation have realised that the world leather sector requires innovative skills. The future of work demands not only continuous upskilling and reskilling, but an individual's ability to succeed in leather and product manufacture education and career starts with their upbringing and environment. As the workforce emerge, research shows that the top skills needed for entry level jobs of the future leather sector are more accessible by the growth of digitalization. The digital era skills are attainable with hours, not years, of online learning. In less than 60 h a worker can gain technical knowledge on the skills required to start his career. This has been brought by the digital learning platform of the leather global sector and

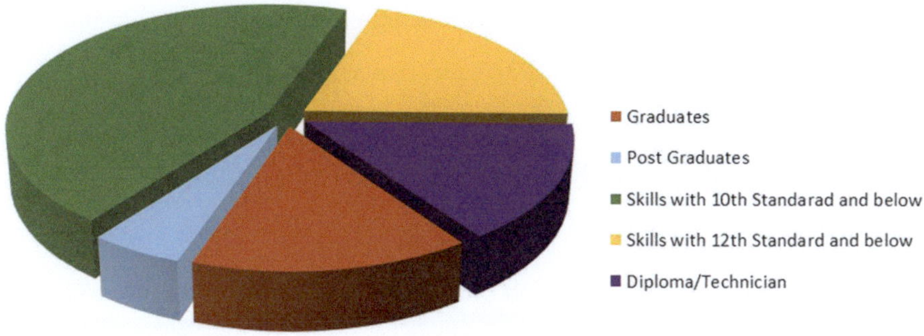

Fig. 5 Profile of registered employees of the global leather sector

adapting to quickly changing technology and digital economy. The lifelong learning concept is a very important aspect to the workforce, where they can keep themselves updated with the currently growing trends in the sector. The digital online learning has facilitated learners even if they are a parent as a small amount of dedicated work in upskilling can result in better career results. The global leather sector has already paved its path with digital skills linking and online training as a manufacturing sector.

2.3 *The Role of Learning in Upskilling Pathways*

The demand for workers evolve as they move through their education and career. Many institutions world-wide are taking part in facilitating the learners in all possible ways[6]. For students, online learning can be a valuable complement to traditional education. For adult learners, digital online learning can make lifelong learning more accessible. Whether an individual is a working parent that needs to spread their coursework across nights, weekends, or the daily commute, or a recently unemployed professional that can dedicate concentrated hours over the span of a few weeks, the self-paced nature of learning in today's context has become digital and online which is a is flexible, affordable, and designed to suit the busy working adult. The same is applicable for the leather manufacture industries as well. Approximately, 75% of active learners spend less than three hours per week on their coursework. Almost 20% spend less than an hour a week, 35% spend between one and two hours, and another 20% spend about three hours. The top 25% dedicate significantly more time. While some learners make rapid progress in a matter of a couple of weeks, reskilling global leather sector requires diligence and commitment (Maclean et al., 2012). The role of these emerging skills for upskilling and reskilling is fueling future economic growth for the leather sector. By designing training programs with this framing in mind, leather sector workforce leaders are working with employers to bridge gaps and ease the transition for workers into high-demand jobs in tune with the global demands. Despite the path chosen, learners acquire skills in ways that fit into their schedules.

2.4 Cross-Cutting and Foundational Skills Curriculum Development

Strengthening foundational skills and direct measurement of skills are needed for the global leather sector because when countries move up the value-added chain of production, employment shifts away from jobs involving routine cognitive and manual tasks toward jobs requiring tasks such as critical thinking, fast production and complex communication. However, measurement of skills largely focuses on proxies such as years of education or qualifications. More years of education do not necessarily mean higher skills. Qualification-based measures also ignore the fact that skills can be acquired informally and outside the education and training system through work experience which is predominant in the leather sector. A recently developed instrument by the Organisation for Economic Co-operation and Development (OECD), the programme for international assessment of adult competencies, aims to provide internationally comparable direct assessment of foundational skills of the adult population (ages 16–65) in literacy, numeracy, and problem solving. There is a growing demand in the leather sector for cognitive skills involving understanding, interpretation, analysis, and communication. There could be a steep decline in the demand for routine cognitive skills and a sharp increase in the demand for nonroutine interactive skills with changing demands of the future in the leather/non-leather/vegan leather-like industries. While skills training systems in this sector need a base of strong foundational skills, they also must adapt rapidly to changing market requirements. This has implications for the design of curriculum for leather/non-leather/vegan leather-like industries and its delivery, the type of courses on offer, and their regular renewal. While routine skills of the leather sector are easiest to teach and test, they are also the ones that are easiest to digitize, automate, and outsource for the new normal working of newer leather/non-leather/vegan leather-like Industries. The digital economy requires foundational and cross-cutting skills that cut across professions and industries. Digital literacy is enabling workers in leather sector to participate in increasingly tech-heavy work environments across continents; human skills that are expected to stand the test of time; and career management skills that facilitate job transitions in the newer evolving leather sector. JOB READINESS does matter to the new leather sector business world: About a quarter of all adults in the OECD do not know how to use a computer. Adults who are not digitally literate have a lower rate of labour force participation and tend to work in lower skilled jobs. The curriculum development for such foundation and cross cutting skills, need pathways that involve preparation, basic research on field study forecasting industry trends and workplace analysis with core operations curriculum mapping and measurement of related degrees or certificates as given in Fig. 6. Cognitive, wellness, social and

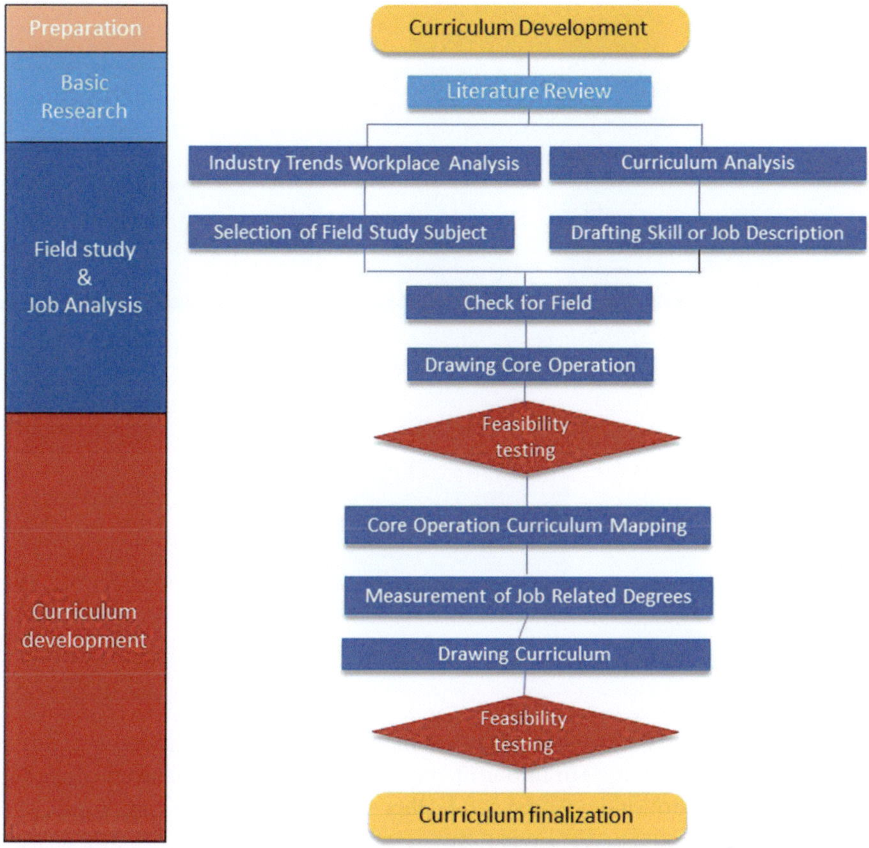

Fig. 6 Curriculum development for foundation and cross cutting skills

emotional skills and cross industry training models shall be are part of the new leather sector skill development.

2.5 *New Skills for New Times*

The future of 21st century leather sector skills are problem solving, critical thinking, innovation, communication, team work etc. also there has been an emphasis on new "green skills" for this sector. The green skills contribute to the green economy by the manufacturing sector that includes leather and allied sector and associated by using eco-friendly processes and techniques to produce green goods and accessories. The leather sector workforce learning green skills shall contribute to sustainability of its industries in a very practical manner. In the coming years many educational and skill training institutions and the skilling sector of the global leather sector shall focus

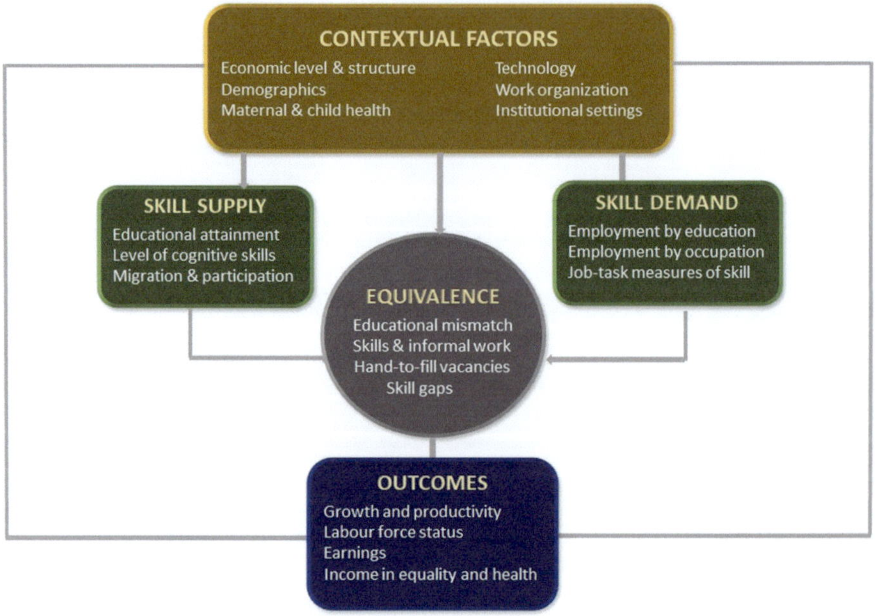

Fig. 7 Measurement of skills in conceptual framework

on the green skills to contribute to the sustainability of its sector. Initiatives such as certificate or elective courses in carbon management or any new technological/digital/green skill in leather/non-leather/vegan leather-like sector shall be considered, thus giving a vocational subject the same status as an academic subject. The way forward can be that schools themselves can become the means by which students are trained to learn new skills that can help them with their careers and contribute to sustainability in a very practical way. Such new education and skill requirements will call for measurement of skills in conceptual framework as explained in Fig. 7. In the new global leather sector business world it's no longer the big established organizations that dominate the small ones, it's the fast, technologically advanced, environmentally sustainable with innovative skill workforce and adaptive organizations will dominate the future leather sector. The global leather sector is embracing the change taking place and evolving new skills as big is no longer advantage, speed is essential.

2.6 Leather Sector's Green Growth and Green Skills

Green growth and conditions for a green economy are high on the agenda of the G20 and the United Nations. The green economy is defined as an economy that aims at reducing environmental risks and ecological scarcities for sustainable development without degrading the environment and needed for building green future. The leather sector globally is implementing strategies and policy directions to be among the top green growth industries and shall focus to be completely green by 2050. 24 million new jobs are likely to be created for the global leather sector with the right policies to promote a greener economy. Core technologies in improving the efficiency of green technologies, End-of-pipe green technologies for leather and product manufacture, expanding use of clean energy technologies, Non-CO_2 processing and also CO_2 capture in the entire spectrum of the leather sectorial processes, green supply chain, inventory and storage processes for leather and products and processing, water quality and management in greener ways with an alternative to water resources for leather processing, waste recycling through greener pathways, enhancement R&D in monitoring and processing for hazardous substances in leather and products during and after manufacture, new age R&D requirements for virtual reality in leather products manufacture, developing and confirming green growth infrastructure for leather and products industries, promote and increasing green industrial exports and creation of green jobs for the above will be the focus for the entire global leather sector. The zero-liquid and solid discharge implementation, carbon footprints, energy self-reliance, development of future growth engine of green technology, making and fostering the green industry, sophistication of the industry structure, establishment of bases for green economy, formation of green transportation, green marketers, green revolution of lives, and realization of the world's renowned green growth industrial leather sector. The strategies are being developed for greening of existing industries and the promotion of new age green leather and product manufacture, advancement of the leather products manufacture, and bringing this sector green infrastructure and working environment to daily life of the workforce and becoming a role model for the international community as a green leader with cleaner processes in the manufacturing of leather and its associated products.

2.7 *Green Skills and Curriculum Development*

Green skills in leather sector mean the ability required to execute green jobs and to produce new green products necessary for jobs relating to manufacturing, services, and marketing of the products produced. Pertaining to the green growth worldwide, the traditional jobs can be incorporated into the green industry for making innovations in low-carbon technologies in addition to technical ability. The standard set for this sector will focus on creating an increase in the demand for green occupations and creates green skills for the leather and product manufacture as given in Fig. 8.

The curriculum redesigning with mapping of occupational profiles, strengthening the measurement and monitoring of learning outcomes by change of vocational education and training for green skills under the green growth policy of the world leather sector will be to improve teachers training for developing young leather technicians and scientists who are concerned about the sustainability of the green world.

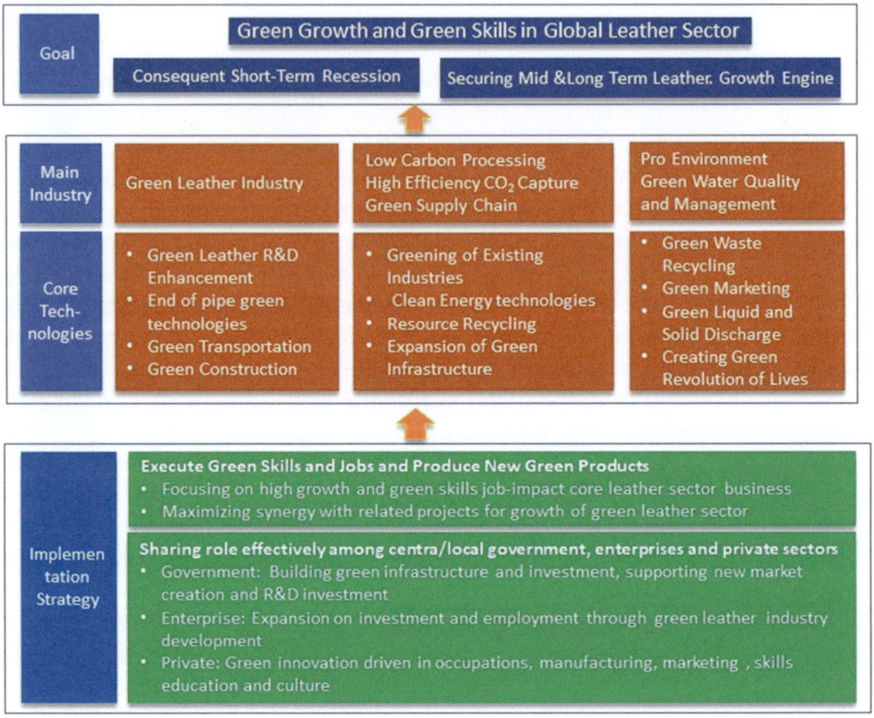

Fig. 8 Sustainability of global leather sector through green growth and green skills

3 Global Leather Sector Skill Landscape

3.1 North American Leather Sector Skill Development

North America (facing social challenges) has a whopping 134 fortune 500 companies and has the world's 70 largest digital companies (Gammarano, 2020; Gauthier, 2021; Rossel, 2018). The growth of the region will be hindered if the social disparities are not addressed as in Europe and Asia on the number of skills that include leather and products manufacture. Despite accounting for only 8.2% of the GDP (Addati et al., 2018), the digital economy of this continent has contributed to about 86% of US labour productivity growth in recent years. The skill gaps hindered the workers to shift to an online mode of work after the pandemic. In the US, 1 in 3 workers lack the foundational digital skills to perform their jobs, and these gaps resulted in over 17 million high school graduates deciding to cancel their college plans in the US after the pandemic due to an affordability crisis due to both cost and structure that necessitates them to look for new 't pools and diversify their workforce (DeCarlo, 2020; Shamika et al., 2019). The statistics show that in April 2020 alone about 1.7 million private-sector jobs were lost in New York State. Efforts were taken to

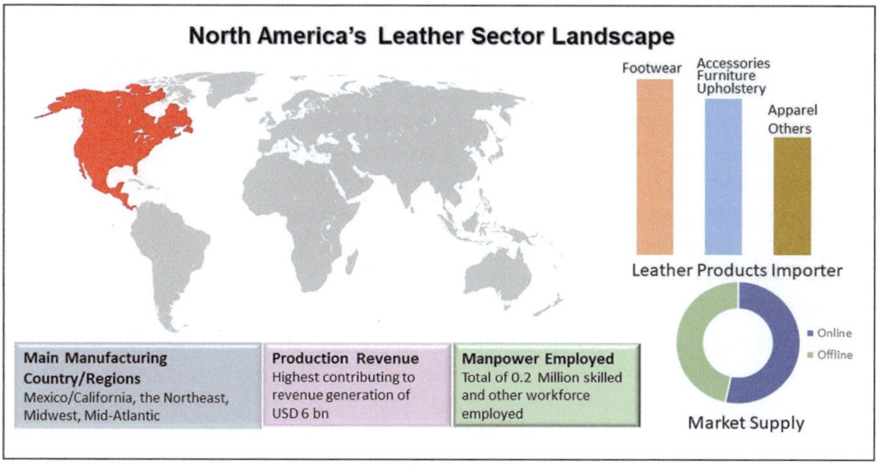

Fig. 9 North America's leather sector landscape

help the unemployed workers to reskill and upskill and the employers to bring in cross-sector partnerships to connect workers to real, local jobs. Even though the West Coast remains the leader in Technology skills, driven by global leadership of technology hubs like San Francisco and Seattle, they are mere importers of leather and its products (Erumban & Ark, 2019; Ronan et al., 2021). Despite ninety percent of the nation's innovation sector employment growth in the last 15 years was generated in just five major coastal cities: Seattle, Boston, San Francisco, San Diego, and San Jose, California, they have no major innovations coming from the leather sector as they are mere traders of leather products (Eligon, 2012). Canada is leading in security engineering and ranks among the most secure countries in terms of the number of cyberattacks (Atkinson et al., 2019; Lichfield, 2019). This digitization could support the global leather sector as employees but will to some extent benefit the leather products sector through imports of leather products are more into their country. In North America, most of the countries are major importers of leather. Their highest imports are footwear followed by upholstery and apparel respectively as given in Fig. 9. Even in North America, the online and offline channels go on par with each other. The production revenue comes to a meager around $6 bn with a total of 0.2 million manpower employed. Major manufacturing regions and countries are California, the Northeast, Midwest, Mid-Atlantic and Mexico.

3.2 South American Leather Sector Skill Development

Major investments in the leather sector of the South American continent will boost the economy in the coming years, where emerging strengths are hindered by skill gaps and inequality in this continent. The labour market inclusion and manufacturing skills promotion in this continent will be enhanced by strategic investments and has the potential to increase shared prosperity. The leather sector in South American countries is helping women, who were among the hardest hit by the pandemic, reintegrate into the labour force. Manufacturing sector is estimated to result in an average loss of 14% of GDP with gender gaps in labour force participation and skills education (Devadas & Kim, 2020). Recently the government of this continent are understanding that producing many graduates with new skills and training will be the global currency of modern economies (Whaites, 2019). Before the pandemic, two out of every five young people were neither studying nor working and 55% of workers in the region worked in the informal economy (Leary, 2019). Workers often do not have the skills that companies need, and companies are having a hard time finding the talent they require to grow their businesses. Around 50% of South American firms claim they cannot find candidates with the skills they need (Pages, 2017). These nations are focused on transitioning from Business Process Outsourcing (BPO) to the emerging Knowledge Process Outsourcing (KPO) industry, which is expected to be less susceptible to automation in the coming years. Brazil, Argentina and other countries in the South American continent show strength in leather and leather products manufacturing. Since the leather sector is transitioning into a green economy the momentum in the leather industry has gained a lot of opportunities. The challenge faced in one of the aspects of green revolution is the digitization of the monetary funds, which is a crucial part in the leather sector. About 5% of this region's population does not have a bank account and more than 85% of the transactions in the leather sector here are handled by cash. Leather and product manufacturing businesses that embrace digital payments realize a number of benefits, such as the expansion of markets, reduction in fees, and the simplification of bookkeeping and automation of manual work. Among the countries of this continent, Brazil is making drastic changes in the opening bank regulations and instant payment system. This extension of digital transformation will employ people will technical knowledge and will also be beneficial in financial management to be transparent in the industry (Cepal, 2020). Another possible intervention to promote the development of leather business skills could be to help workers in small businesses gain better access to larger firms that operate in the formal leather sector of the economy (Maxwell et al., 2019). Countries like Barbados, Chile, and Brazil have traditionally had some of the strongest economies in leather and product manufacturing technology. Barbados has made moon-shot innovations and is extending its arms by investing in digital upskilling nationwide to transform its traditional sectors like the leather sector and diversify into new sectors. The start-up Chile innovation by the government of Chile has gained prominence and has been replicated in more than 50 countries (Terry et al., 2020). Venezuela, Uruguay, Argentina, and Costa Rica score the highest in Leather sectorial skills in the region. Each country in the continent is making forward towards digitization and a green economy through innovations and upskilling of the most important resource—HR. South America or Latin America has a market value of $140bn in the leather sector (Tan, 2021). The manpower employed comes to around 7.2 million including the both skilled and semi-skilled workforce. In this continent, Upholstery and apparel contribute to a similar value whereas footwear is ranked

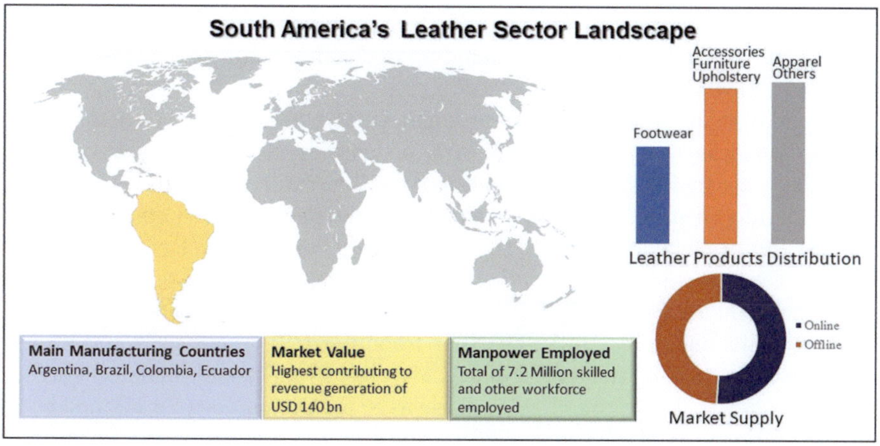

Fig. 10 South America's leather sector landscape

lower as given in Fig. 10. The online and offline channels are equally distributed. Major countries producing leather are Argentina, Brazil, Colombia and Ecuador.

3.3 *European Leather Sector Skill Development*

After the pandemic, the European continent is recovering its position in terms of skills being a global skills leader and facing talent shortages (Metsola, 2021). Most of the top countries of the world in skills fall under the European continent and this was achievable by having high technology compared to other regions. Nevertheless, Europe faces skill shortages in key leather and digital talent pools compared to other regions (The Global AI Agenda Report, 2021). Over 70% of businesses in the region claim that the lack of staff with adequate digital skills is an obstacle to investment. The data science talent pool in Europe is about half of that in the US (IFR Report, 2019). There is an imbalance in the talent pool emerging in the European union as most of the people end up in the UK which has the advantage of adopting of technology and enhancing the lives of the people in terms of living standards (EU Commission Report, 2021a). To emerge strongly in the economic recovery, there are a number of priorities for the region. In Western Europe, policymakers are investing in digital transformation skills related to Internet of Things (IoT) and manufacturing innovation which is also applied to leather sector. By 2023, Europe is expected to account for 25% of worldwide IoT spending (EU Commission Report, 2020). The manufacturing sector in the EU plays a key role in research and innovation, being responsible for 64% of private-sector research development expenditure and 49%

of innovation expenditure in Europe. Investing in streamlining the regulatory environment for sustainability projects including the leather sector presents yet another opportunity (IMF Report, 2020a, 2020b).

Outside of its traditional strength in Finance, the UK exhibits poor performance in many business skills. Today, only 66% of working-age graduates in business-related subjects are in high-skilled employment (International Bank for Reconstruction & Development Report, 2020). To address the gaps between employer needs and the skills of the workforce, the UK Government launched the Build Back Better Business Council—a new group which will see business leaders work directly with the government to fuel the COVID-19 economic recovery (Pioneering Reforms to Boost Skills and Jobs, 2021; The EIU Report, 2020). However, some Eastern European countries—like Russia, Belarus, Ukraine, Estonia, and Lithuania—are top performers in HR. These countries continue to be top destinations for HR and back-office outsourcing (UNDP Report, 2020). The region is a global leader in Machine Learning, with Finland leading the way. Finland has been leading the way in AI education with a recent push by its government and the University of Helsinki to teach 1% of the world basic AI skills. Europe as mentioned is recovering its position by skilling and digital upgradation of the continent (EU Commission Report, 2021b). This will create a positive impact in the leather sector as well due to the demand for skilled employees with knowledge of digital and automation techniques. European leather qualifies for the highest standards in terms of production quality. Apparel occupies the largest domain followed by upholstery and footwear in this sector as shown in Fig. 11. The European leather sector production revenue grosses to about $116 USD. The total manpower of both skilled and other workforce is around 1.3 million. Compared to other continents, the online segment occupies a major part compared to the offline. Major countries involved in leather and product manufacture include Italy, Spain, France, Portugal, Türkiye, Germany, Poland, Romania and Finland.

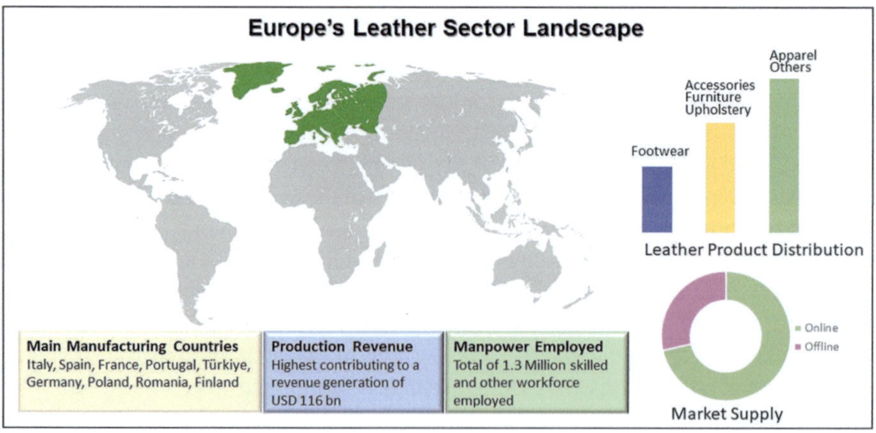

Fig. 11 Europe's leather sector landscape

3.4 Asia and Middle East Leather Sector Skill Development

Levels of economic development vary widely in Asian continent, with some of the richest countries (such as Japan) and some of the poorest countries on earth (such as Bangladesh) (ILO Report, 2020). Asia contains the largest number of poor of any region in the world, with the incidence of those living below the poverty line remaining as high as 40% in some countries in Asia (UNESCAP 2011) (Asia-Pacific Trade and Investment Report, 2011). A contrast lies between the developed and developing countries of the largest continent. Countries such as India, China, Japan, Hong Kong, Taiwan, and Singapore demonstrate high proficiency across the skill domains, whereas South and Central Asian countries rank near the bottom of the global rankings. These countries with low skills performance were the most hit during COVID-19 and were struggling to cope up with the pandemic. The difference in performance within the Asian continent is due to Urban-Rural digital divides (ILO Monitor Report, 2021b). The statistics show that 70% of the urban households in the region have access to the internet, compared to only 36% of rural homes. The manufacturing sector including the leather/non-leather/vegan leather-like of the Asian subcontinent has both formal and informal employment. Currently, 1.3 billion of the world's2 billion informal workforces& leather sector resides in Asia Pacific, and rural workers are twice as likely to work in informal arrangements and more predominant in the leather sector. The governments of the countries in Asia are investing in computer literacy and other foundational skills training via mobile devices, which tend to be more widely adopted than computers, which are a promising area for new skills development and the leather sector takes precedence. The adoption of learning operations and controlling through remote devices can be a very effective technology in the leather sector, particularly where the machines are controlled by manpower, which requires a lot of time and the production efficacy also gets reduced.

Beyond addressing these labour market challenges, the Asian region has an opportunity to invest in skills that support several globally leading sectors, such as e-commerce, logistics, and manufacturing. Supporting the core industries like leather, in the five years before the pandemic, Asia Pacific generated about three-quarters of global retail growth and two-thirds of online growth. Retail is leapfrogging development stages that the US and Europe once toiled through. Engineering skills particularly those that relate to improving supply chains could support the vast logistics Infrastructure investments taking place across the region that compliment many industries like leather. By 2030, 12 of the top 15 global trade lanes will involve Asian countries, and many of the top trade lanes already involve major regional players like China, Japan, India, and Vietnam. Transforming the manufacturing sector by accelerating the adoption of digital skills presents yet another opportunity. Most Asian countries have been slow to adopt Industry 4.0 technologies but some countries including Singapore, Thailand, and Vietnam have drafted plans for encouraging manufacturers to adopt Industry 4.0 technologies. Among these countries, Vietnam occupies a prominent position in footwear manufacturing, which involves a lot of automation and contributes to industry 4.0. China and India also draw the benefits of Industry 4.0 (Arbulu et al., 2018). The Chinese government included implementation guidelines in its "Made in China 2025" strategy, and the Indian government is progressing to "National Policy for Advanced Manufacturing".

Vietnam is the overall leader in Business including the leather sector (Lee, 2021). This result is driven by strong scores in communication, entrepreneurship, leadership management, strategy and operations. Vietnam was among the most resilient economies in the region in 2020 and one of the few to not face a single quarter of contraction. Vietnam has also been a major beneficiary of the supply chain relocation trend out of China over the past several years. China, Hong Kong, and Singapore maintain their position as regional finance hubs. The trio tops the region in finance skill proficiency. All three also continue to score high on other measures of finance-sector leadership, such as business environment, reputation, and availability of talent. Japan is the world leader in technology skills. Japan has long been a leader in industrial automation and robotics and has recently made significant investments in accelerating the commercialization of quantum computing with initiatives like the Quantum Innovation Initiative Consortium.

The Middle East is home to several global skill leaders seeking greater diversification and global integration, as well as some emerging players. Israel and UAE lead the region in skills performance, but they follow different formulas. Israel is strong in technology and data science, whereas the UAE is the regional business leader, but lags in technology and data science skills. These challenges can be tackled by digital skills development. Technologies supported by automation and skilled employees in this region will shorten the duration of trade as well as facilitate the issues as and when occurred. The era ahead offers much opportunity for the young, fast-developing nations of the region like India. By 2050 more than 300 million young people will be coming onto the job market a potential addition to the workforce not much smaller than the region's total population today. This means that, between now and then, the continent needs to create 30 million jobs a year to keep pace with the number of new labour market entrants like the leather sector. The skill development of the global leather sector systems will need to be in a constant state of innovation to keep up and ensure that the region's future economy is more global, inclusive, and resilient. On comparison between the two countries of the continent like China& India, Saudi Arabia despite its much strength in data science and technology is trying to invest heavily in innovation through programs like vision 2030. They are doing well by giving a lot of tax exemptions to corporates that move in their region, whereas Israel being strong in the development of technology and devising formulations for a different chemical process, ranks lower that their regional peers in terms of business skills. These gaps may be explained by a lack of on-the-job training from local employers. In these two countries leather industry is meagre. In recent years, Israeli employers have voiced concern about the inadequacy of vocational skills available locally. There is generally less vocational and work-based training available in Israel than in many other countries. The leather sector specifically in the Middle East countries has potential in the raw material i.e. the hides and skins. Many tanneries in this region have showed strength in the superior tanned wet blue leather. The Riyadh

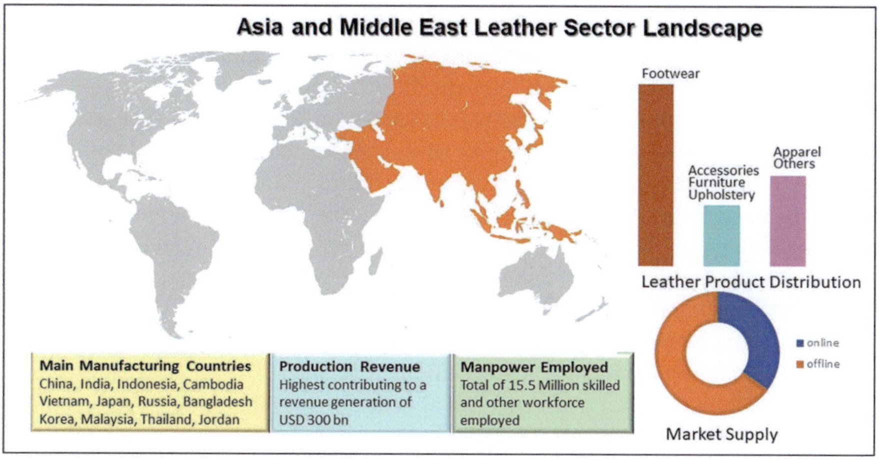

Fig. 12 Asia and Middle East leather sector landscape

modern factory and other tanneries have shown potential and the inclusivity of technology and a skilled work force will enhance the productivity of this region in the Asian continent (IMF Report, 2020a, 2020b).

The VUUCA model of skill development has become the new normal by the Asian countries, as the leather business world-wide is Volatile (major changes frequent and sudden) Unpredictable (change is not following a predictable pattern) Uncertain (there were many surprises) Complex (problems and opportunities are complex, with many different elements) Ambiguous (confusion from contradictory information). The frequent changes and demands from the global industries has created challenges to the extent that the industry will stay in a place or shift to another location; however the future of the global leather sector will be to become an adaptive industry with skilled manpower to coexist (Fig. 12).

3.5 Africa's Leather Sector Skill Development

In Africa, skill investments have the potential to promote labour market inclusion and support a quicker recovery of global trade relations, which is promoting labour market with inclusion of skills. There are three major hinderances of the African economy in the leather sector. First, the large amount of export of cheaper goods from China to all other countries, second the slump in the demand of goods in the Europe due to lockdowns and the poor supply chain management of the continent. Although the African continent is rich in resources like hides & skins, coffee, cocoa and rubber, many skilled and educated workforce is not being employed to explore the resources extensively. The same applies for the state of skills in leather sector in Africa (Youth Empowerment Report, 2020). Gradually countries like Ethiopia are partnering with global institutions for skill development in leather and allied industries to develop the skill of their workforce by the means of exchange programmes and promoting investment into the country by giving subsidies for new industries.

As it looks to diversify and move towards the digital economy, sub-Saharan Africa can take advantage of young workforce and growing middle class (Goedde et al., 2019). Africa is forecasted to host 42% of the world's youth (15–24-year-olds) by 2030, and more than double its youth population by 2055. Africa could be two to three times more productive if it intensified agricultural and manufacturing productivity that includes the leather industry and leather tanning industry. Today, more than 60% of the population of sub-Saharan Africa works as small-scale farmers, and about 23% of sub-Saharan Africa's GDP comes from agriculture and small manufacturing industry which include leather (Signe & Johnson, 2018). Second, fostering financial inclusion. Digital platforms can also boost agricultural and manufacturing productivity through prompt payments for produce, information sharing, and agro-industrial activities. Another strategy is to remove restrictive barriers to especially intra-African trade as well as trade with other regions. Finally, supporting Africa in becoming the next great manufacturing hub. Business-to-business spending in Africa's manufacturing is projected to reach USD 700 billion by 2030, USD 250 billion more than it did in 2015. It has the potential to be "the world's next great manufacturing centre," potentially capturing part of the 100 million labour-intensive manufacturing jobs that will leave China and India by 2030. The percentage share from the leather business and jobs will continue the same trends as other businesses. The African leather industry is propagating gradually in terms of both production and employment. Accessories and upholstery occupy a prominent position now followed by footwear and apparel respectively. There is very less online channel distribution compared to offline. The major manufacturing countries in this region are Ethiopia, Nigeria, Uganda, Kenya, Tanzania, South Africa, Egypt, Tunisia and Mauritius. The revenue generation is around $34bn with a total manpower of 2.1 million from the entire continent (Fig. 13).

4 Recommendations for Non-aligned and Other Developingcountries (NAM) Countries

Leather industry for NAM countries across the globe, is increasingly corroborated principally on skills as it provides a strong base to foreign policy of its members. The transition of traditional leather and its products processes to sustainable leather world ensures that the NAM countries could learn skills to adapt and rebuild the industries to meet the strategic autonomy, which is the need of the hour of today's world. Workforce leaders of the leather sector should choose NAM as a platform to play a critical role to provide an active leadership and by baring a torchbearer for smaller countries at multilateral programs. NAM platform should present opportunities to build more inclusive, modern, and scalable leather and its products skills education

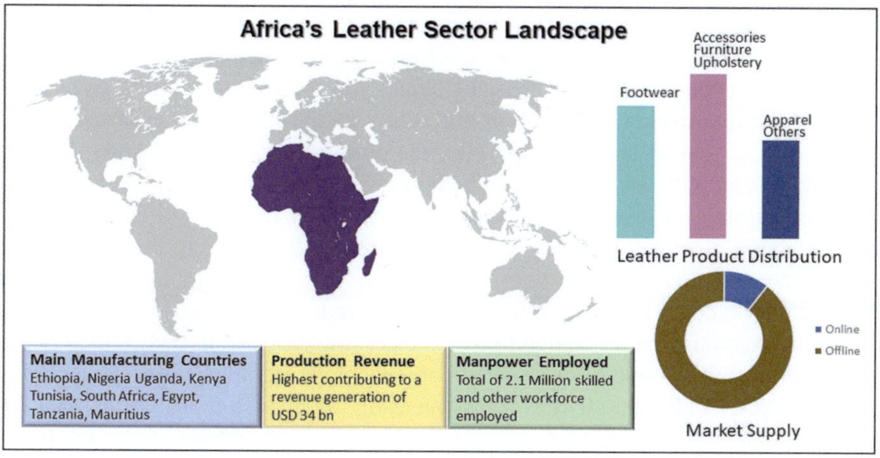

Fig. 13 Africa's leather sector landscape

with scalable workforce and sustained moving forward to garner support overcoming many disputes. The training investments of the leather sector have the prospective to support the individual nations competitiveness, innovation and equity. Each country can help themselves with skills and jobs with new technologies more accessible. NAM can provide a platform for reskilling the millions of leather and its product workers and coming up with innovative public-private solutions and a strong position for economic development without compromising the sovereignty of their land.

5 Conclusions

Education and skills development are crucial to achieving inclusive growth and sustainable leather sector for green growth. For workers, entrepreneurs, governments, and economies across the globe, growth in this sector is increasingly predicated on skills. Governments worldwide are considering policy options to strengthen the human resource base required to serve the leather sectorial growth, to increase employment and entrepreneurship. While COVID-19 disrupted leather industry entrepreneurial systems and jobs worldwide, the transition to rebuilding their enterprises, ensured people could relearn skills to adapt and rebuild their career pathways. Leather and its products workforce development leaders worldwide are playing a critical role in ensuring skills for growth and competitiveness that can be extended and sustained moving forward. Reskilling the millions of leather workers that have been impacted by the double disruption of the pandemic and automation means that now more than ever, the world needs innovative private-public solutions and every country have their skills policy measures in place. Addressing skills mismatch calls for not just upskilling but also "Right Skilling". Relearning skills education, post pandemic moment presents an opportunity to build leather sector into a more inclusive, modern, and scalable workforce. These skills development investments align skills training policies with economic and industrial policies that have the potential to support increased competitiveness, innovation, and equity for the global leather sector around the world. The jobs in leather sector after the pandemic and of the future are open to everyone and with the help of new technologies, they are ensuring Industry-led, more accessible every day with a conducive regulatory framework. Emerging and modernizing economies are very much concerned with upskilling their workforce to meet the demands of knowledge-intensive and high-technology industries of the future leather sector. However, in the entire world today, it can take just a few weeks or months to acquire the specialized skills in emerging professions contributed by the leather sector skills councils and research organisations globally in close association with the leather industries. The world is working together supporting qualification frameworks and quality assurance strengthening the skills and workforce by developing systems increasing the prestige of skills training for improved productivity, integrating employment and entrepreneurial services to build a world where anyone, anywhere has the power to transform their life through learning in this modern era. The global leather skills sector has definitely set a benchmark on how to be "useful and relevant" to the society concerned through capacity building. This sector with its dominance as a global player will continue to utilize its skill expertise for new challenges and solutions adhering to its sectorial growth and development, and majorly building lives from the bottom.

Acknowledgements We would like to express our special gratitude to Dr. T. Ramasami for his guidance and support. Our sincere thanks and appreciation to Mr. G. Sathiamoorthy, Ms. Kanimozhi Balaji, Mr. Z. Shoaib Khan, Dr. (Mrs) K. Saranya, Ms. K. Devi, Ms. M. Yuvashree, Mr. M. Suriya, Ms. M. Kokelavanthana, Ms. D. Ruby Shekinah and Ms. R. Monica for the timely help and inputs.

References

Addati, L., Cattaneo, U., Esquivel, V., & Valarino, I. (2018). ILO report—Care work and care jobs for the future of decent work. In *International Labor Organization* (pp. 1–526). United Nations.

Arbulu, M., Lath, V., Mancini, M., Patel, A., & Tonby, O. (2018). Industry 4.0 report—Reinvigorating ASEAN manufacturing for the future (pp. 1–26). Digital Capability Centre, McKinsey.

Atkinson, R. D., Muro, M., & Whiton, J. (2019). The case for growth centres report. In *How to Spread Innovation Across America* (pp 1–94). Brookings Institute.

Autor, D., Mindell, D., & Reynolds, E. (2021). The work of the future report—Building better jobs in an age of intelligent machines. In *Future of work initiative* (pp. 1–100). Massachusetts Institute of Technology.

Cepal N. U. (2020). OCED report—Latin America economic outlook 2020: Digital transformation for building back better. In *The organization for economic cooperation and development* (pp. 1–3).

DeCarlo, S. (2020). *Fortune report—Fortune global 500* (pp. 1–3).

Devadas, S., & Kim, Y. E. (2020). World bank report—Exploring the potential of gender parity to promote economic growth. In *Research and policy briefs* (1–5). The World Bank.
Eligon, J. (2012). *NY times report—Silicon prairie takes root in the great plains* (pp. 1–3). The New York Times.
Erumban, A. A., & Ark, B. V. (2019). EU commission report—productivity & innovation competencies in the midst of the digital transformation age. *An EU-US European Commission.* Europe (pp. 1–4).
EU Commission Report. (2020). *A green COVID-19 recovery and resilience plan for Europe* (pp. 1–52).
EU Commission Report. (2021). *Advanced manufacturing* (p. 3). European Commission.
EU Commission Report. (2021). *Digital skills and jobs* (pp. 1–3). European Commission.
Gammarano, R. (2020). ILO report—Education pays off, but you have to be patient. In *Spotlight on work statistics* (pp. 1–4). International Labour Organization.
Gauthier, J. F. (2021). *GSER report global startup ecosystem* (pp. 1–2).
Global Skills Report. (2021). *The world's top trending skills benchmarked across over 100 countries* (pp. 1–54).
Goedde, L., Ooko-Ombaka, A., & Pais, G. (2019). *McKinsey report—Winning in Africa's agricultural market* (pp. 1–13). McKinsey.
Gul, E. (2014). Skill development report—Efficacy of skill development techniques: Empirical evidence. *Journal of Education and Educational Development*, 134–144.
IFR Report. (2019). Brexit: *UK falling back in global automation race—Robot sales down 3%* (pp. 1–4). International Federation of Robotics.
ILO Monitor Report. (2021). *COVID-19 and the world of work*, 7th ed. (pp. 1–35). International Labour Organization.
ILO Monitor Report. (2021). *COVID-19 and the world of work*, 7th ed. (pp. 1–3). International Labour Organization.
ILO Report. (2020). *Asia–Pacific employment and social outlook: International labour organization* (pp. 1–94).
IMF Report. (2020a). *Central, Eastern, and South Eastern Europe after COVID-19: Securing the recovery through wise public investment* (pp. 1–3). International Monetary Fund.
IMF Report. (2020b). *Navigating the pandemic: A multispeed recovery in Asia* (pp. 1–2). International Monetary Fund.
International Bank for Reconstruction and Development Report. (2020). COVID-19 and human capital, Europe and Central Asia economic update. *The World Bank. Fall, 2020*, 1–8.
Leary, N. (2019). *Five Latin American nations growing as outsourcing hot spots* (p. 1). Nearshore Americas. https://nearshoreamericas.com/five-latin-american-places-growing-as-outsourcing-hot-spots/. Accessed 17 Feb
Lee, Y. N. (2021). CNBC report—This is Asia's top-performing economy in the Covid pandemic—it's not China. In *Consumer news and business channel, Asia economy* (pp. 1–7).
Lichfield, G. (2019). Wired magazine report—The future is Canada—and it's open for business. *Wired Magazine*, 1–6.
Maclean, R., Jagannathan, S., & Sarvi, J. (2012). Growth report—Skills development for inclusive and sustainable growth in developing Asia-Pacific. In *Technical, and vocational education and training: Issues, concerns, and prospects* (vol. 19, pp. 1–387). Springer.
Manyika, J., Smit, S., Ellingrud, K., Meaney, M., & Robinson, O. (2021). *COVID-19 report—The future of work after COVID-19* (pp. 1–8). McKinsey Global Institute.
Market analysis Report.,. (2022). Leather goods market size. *Share and Trends Analysis and Segment Forecast Report, 2023–2030*, 217.
Maxwell, J., Pavon, F., & Epstein, Y. (2019). *IADB report—Supporting technology adoption through sector skills councils* (pp. 1–4). Interamerican Development Bank.
Metsola, R. (2021). *EU commission report—Digital skills and jobs, shaping Europe's digital future* (pp. 1–3). European Commission.

Pages, C. (2017). *WEF report—In Latin America, companies still can't find the skilled workers they need* (pp. 1–5). World Economic Forum.
Pioneering Reforms to Boost Skills and Jobs. (2021). Department for Education, Government of the United Kingdom (p. 4). https://www.gov.uk/government/news/pioneering-reforms-to-boost-skills-and-jobs. Accessed 17 Feb
Ronan, S. E., Herrmann, A., & Rosenbaum, D. J. (2021). *World business report—Business and professional services* (pp. 1–6). World Business Chicago.
Rossel, C. (2018). *World bank report—Human capital index* (pp. 1–9). World Bank Open Data.
Sabharwal, M. (2012). *Sustainable growth report—Education, employability, employment, and entrepreneurship: Meeting the challenge of the 4Es, skills development for inclusive and sustainable growth in developing Asia-Pacific* (p. 82).
Sedik, T. S., & Yoo, J. (2021). Monetary report—Pandemics and automation; will the lost jobs come back? In *International monetary fund working paper* (pp. 1–26).
Shamika, N. S., Fredriksson, T., & Garces, P. F. (2019). UNCTAD report—The digital economy report. In *United Nations conference on trade and development* (pp. 1–194).
Signe, L., & Johnson, C. (2018). Brooking report—The potential of manufacturing and industrialization in Africa. In *Africa growth initiative* (pp. 1–36). Brooking Institute.
Tan, F. (2021). *From a crazy experiment to economic thrive!* (pp. 1–4). Startup Chile. https://startupchile.org/en/our-impact/. Accessed 17 Feb
Terry, C. A., Graham, B. G. L. (2020). *IADB report—Time to pivot: The Caribbean as a world leader* (pp. 1–2). Interamerican Development Bank.
The EIU Report. (2020). *Growth in slow motion: Western Europe to lag behind global recovery average* (pp. 1–3). The Economist Intelligence Unit.
The Global AI Agenda Report. (2021). MIT Technology Review (p. 1).
Trade, A.-P., & Report, I. (2011). *Post-crisis trade and investment opportunities* (p. 3). Economic and Social Commission for Asia and the Pacific.
UNDP Report. (2020). Ministry of labour and social protection of the population. In *UNDP and Coursera join forces to support 50000 Azerbaijani citizens who lost their jobs during COVID-19 gain new skills and knowledge* (pp. 1–2). United Nations Development Program.
Whaites, A. (2019). *OCED report—Supporting SME development in Latin America and the Caribbean* (pp. 1–4). The Organization for Economic Cooperation and Development (OCED).
Wilmoth, J. (2011). World fertility report—United Nations Department of Economic and Social Affairs (UNDESA), Population Division (p. 1).
World Footwear Business Condition Survey Report. (2022). Issue: 7, *Portuguese shoes* (pp. 1–4).
Youth Empowerment Report. (2020). Office of the special advisor on Africa (pp. 1–3).

Indian Leather and Footwear Industry—Government Policies Which Shaped Growth

R. Selvam

Graphical Abstract

This paper analyses the major policies and schemes which have led to the transformation of Indian leather, leather products and footwear industry from a mere exporter of raw materials to a leading exporter of value-added products and also the future growth prospects of the industry.

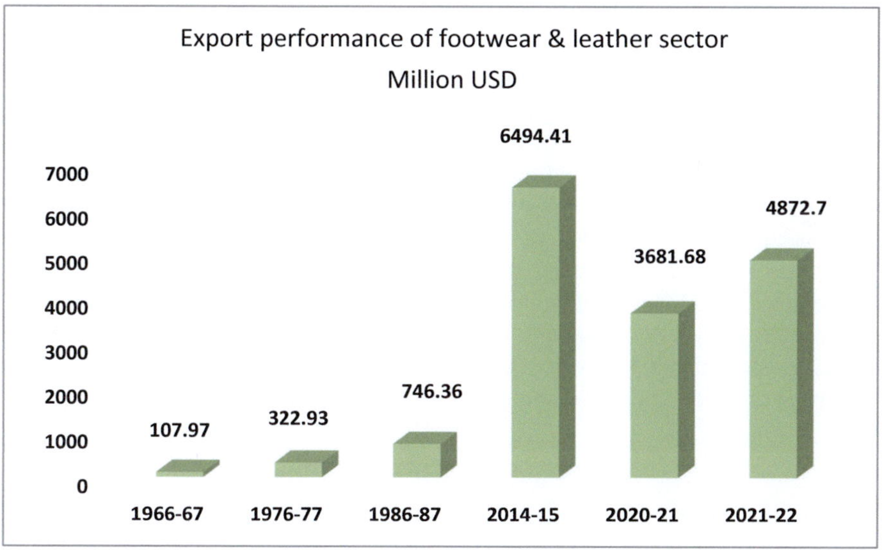

R. Selvam (✉)
Council for Leather Exports, No. 1, Sivaganga Road, Nungambakkam, Chennai 600034, India
e-mail: cle@cleindia.com

S. K. Janardhanan and L. A. Zugno (eds.), *Emerging Trends in Leather Science and Technology*, https://doi.org/10.1007/978-981-99-9754-1_11

1 Introduction

Leather making is an ancient art that has been practiced for more than 7,000 years[1] (https://www.britannica.com). Fresh skins were dried in the sun, softened by pounding in animal fats, and preserved by salting and smoking. Beginning with simple drying and curing techniques, the process of vegetable tanning was developed by the Egyptians and Hebrews about 400 BCE. During the Middle Ages, the Arabs preserved the art of leather making and so improved it that morocco and cordovan (from Córdoba, Spain) became highly prized leathers.

By the 15th century, leather tanning was once more widespread in Europe, and, by the mid-19th century, power-driven machines that performed such operations as splitting, fleshing, and dehairing were introduced. Towards the end of the 19th century, more advanced tanning methods with a wider range of chemicals were introduced.

Today, the footwear and leather industry is a modernized industry which has spread worldwide, supplying high quality products to popular brands all over the world.

1.1 History of Leather Industry in India (Council of Leather Exports (CLE) Records)

While India was a traditional producer of leather, export trade in raw hides and skins and leather began in the 1830s. It began to be recognized that India, with the largest cattle population in the world, could become a potential supplier on the world market. Initially, India exported only raw and cured hides and skins, but by 1850, began exporting tanned hides and skins as well. This was due to a significant technical improvement introduced during the British period.

The major technical improvement that took place in the first decade of the twentieth century, i.e. the development of chrome tanning, led to the major growth of the leather industry in India. Although chrome tanning was introduced in the USA and Europe as early as the 1890s, it was started in India on an experimental basis in a factory set up by the then Government of Madras only in 1904.

1.2 Initial years of Leather Industry in India, After Independence—1947 to 1972

India continued to remain an exporter of raw hides, skins and semi-processed leathers in the first two decades since independence. In the period from 1950 to 1970, Indian leather trade was mostly confined only to the export of raw hides and skins, semi-processed leathers and finished leathers. The export of value-added products was very insignificant. It may be mentioned here that until 1972, exports of all kinds of raw hides and skins as well as semi-processed leathers and finished leathers were allowed freely without levy of any export duty.

As may be seen from the statistics below, leather was the predominant item of export during the period 1966–67 to 1971–72 having a share of 91% in 1996–67 and 89% in 1971–72 as shown in Table 1.

In 1972, Government of India appointed a committee to look into the export potential of the leather industry (known as the Seetharamiah Committee) and made recommendations for large scale exports of value-added products from leather industry. The primary aim was to increase the export potential and to increase foreign exchange earnings for the country. The Committee recommended, among other things, a ban on the exports of raw hides and skins, quota restrictions on export of semi-finished leather, a simultaneous increase in finished leather production capacity, and incentives for increasing finished leather exports.

In addition to the quota restrictions, semi-finished leather was subject to an export duty of 25%. Soon after, in December 1972, semi-finished leather became a canalized item to be exported through the State Trading Corporation (STC). Exports were dechannelized much later, in 1988–89. A large number of incentives were given to exports, following the recommendations of the committee. Cash Compensatory Support for exports was also extended to leather exports in 1973, and Duty Drawback was also provided.

The export of semi-processed leathers like East India (E.I) tanned leather (vegetable tanned leather) wet blue tanned leathers (chrome tanned leather) and crust leathers (ready to finish leather) were restricted through a quota-mechanism with effect from 1st April 1973 (ETC Circular No. 14/73) and a duty of 25% was levied for these items in the same year. This shift in policy resulted in evolution and growth of export of value-added products like leather footwear, footwear components, leather goods and leather garments as shown in Table 2.

1972 to 2000

Since 1972, this sector witnessed significant changes in the policy regime with an emphasis on value addition. This was driven by the market changes at the international

Table 1 Exports from 1966–67 to 1971–72

Year	Leather	Leather products	Total export	% share of leather in export (%)
1966–67	98.49	9.48	107.97	91
1967–68	81.06	10.66	91.72	88
1968–69	103.14	10.91	114.04	90
1969–70	118.95	11.67	130.61	91
1970–71	100.51	13.66	114.17	88
1971–72	121.11	14.73	135.84	89

(value in million USD)

Source CLE records

Table 2 Export of finished leathers and leather products from 1972 to 1995

Year	Semi-finished leather	Finished leather	Leather footwear	Footwear components	Leather garments	Leather goods	Total exports
1972–73	198.92	22.40	12.62	0.63	0.15	5.16	239.89
1973–74	189.54	21.61	13.37	1.03	0.39	6.25	232.19
1974–75	127.97	38.49	20.19	1.62	1.11	8.97	198.37
1975–76	154.78	63.17	22.10	1.09	0.96	10.98	253.08
1976–77	153.13	117.31	28.71	6.76	3.40	13.63	322.93
1977–78	124.15	125.51	22.56	17.59	2.91	21.26	313.98
1978–79	134.48	233.60	27.89	21.70	7.08	31.39	456.13
1979–80	130.66	396.90	36.37	46.19	7.34	46.20	663.66
1980–81	63.71	287.07	42.72	57.05	6.54	48.91	506.01
1981–82	59.42	252.68	33.54	86.18	8.14	45.43	485.38
1982–83	56.33	206.94	20.27	82.10	7.69	40.11	413.45
1983–84	51.58	189.15	22.71	109.83	6.73	41.90	421.91
1984–85	41.48	260.36	22.59	112.05	7.98	48.15	492.62
1985–86	40.89	240.16	27.53	158.63	13.94	70.95	552.10
1986–87	42.00	320.70	64.30	192.54	49.81	76.98	746.34
1987–88	50.65	339.11	89.34	225.97	73.77	89.82	868.67
1988–89	28.48	411.30	82.39	269.37	105.16	121.24	1017.93
1989–90	12.64	416.27	102.90	311.07	199.80	175.79	1218.47
1990–91	6.86	438.85	155.82	318.33	308.22	190.70	1418.78
1991–92	–	294.59	174.76	269.09	298.91	210.53	1247.88
1992–93	–	280.63	164.23	246.00	325.57	261.05	1290.10
1993–94	—	270.25	204.25	253.39	342.28	229.18	1339.55
1994–95	–	382.96	302.49	247.49	387.12	208.83	1626.16

Source CLE records (value in Million USD)
Note
(1) Export during 1995–96 also includes export of Saddlery and Harnesss items (USD 18.53 million and leather gloves (USD 64.68 million)
(2) Export during 1992-93, 1993-94 and 1994-95 also includes export of non-leather footwear valued at USD 12.61 million, USD 40.20 million and USD 14.06 million respectively

level, with the production base of the leather industry shifting from Europe to Asia, due to factors namely high cost of labour, pollution etc.

The second major policy thrust came in the form of the adoption of recommendations made by a second committee (known as the Kaul Committee) in 1979. Its major emphasis was on making available the capital goods needed in the production of leather and leather manufacturers through imports. Accordingly, the import duty on tanning, finishing, footwear and leather goods machinery was reduced to a uniform rate of 25%.

A more integrated view of the problems faced by the industry was adopted in the mid-eighties when the need for producing value-added leather products were recognized and given utmost priority. A third Committee (known as the Pande Committee) that published its report in 1985 concentrated on evolving measures to augment raw material availability, further take forward the modernization process, and promote footwear as the most important item of export. This committee also recommended that imports of finished leather be permitted to compensate for the shortage of raw material and imports of raw hides and skins, wet blue leather, and crust leather were also put on Open General Licence (OGL).

From 1981 onwards, export duty on all sorts of semi-finished leather, raw hides and skins, tanned and un-tanned were levied @ 15% (Notified vide Customs Notification No. 006 dated 6.1.1981). In line with the policy of the Government of India to encourage the export of value-added finished products, the Government withdrew Cash Compensatory Support on finished leather exports from April 1987 and also restricted the All Industry Rates of Duty Drawback for the export of finished leather on sq.ft. basis.

The Ministry of Commerce, Government of India constituted a group in the year 1990 under the Chairmanship of Shri T. S. Vijayaraghavan, the then Joint Secretary to consider the question of phasing out export of finished leather so as to promote and increase higher value-added export products. The Group considered various measures such as immediate ban on certain categories of finished leather, levy of export duty, introduction of regulatory export quotas, notification of minimum export price etc. Among various alternatives available for enforcing phasing out of leather exports, the Group recommended for quantitative restrictions on export as an optimal measure for achieving the desired results. Accordingly, the Government of India issued a Public Notice (Public Notice No. 47/ETC(PN)/90 dt. 31st August 1990), notifying the intention of the Government to bring the export of finished leather under quota and authorizing Council for Leather Exports to allocate and administer quota for export of finished leather. However, on a review of the situation, the Government withdrew the quota notification. But, as a policy, the Government tightened the manufacturing norms of finished leather so as to eliminate possibilities of exporting semi-finished leather under the guise of finished leather. Accordingly, the norms for export of finished leather were notified through an Export Trade Control Order in August 1991, which was further amended in May 1992. Further, as a Policy measure to discourage export of finished leather, the Government imposed export duty at the rate of 10% ad-valorem in the Union Budget 1992–93. It was later reduced to 5% with effect from 1st May 1992 and the same was continued till Union Budget 1995–96, when the export duty on finished leather was abolished. Accordingly, finished leather of goat, sheep and bovine animals and their young ones were granted exemption from levy of export duty (Customs Notification No.68/95 dated 16th March 1995).

Also, the export of finished leather, although being a non-negative item of export was allowed on fulfillment of certain terms and conditions (Public Notice No.15/ETC(PN)/92–97 dated 31st March 1993), stipulating that the exports will be subject to fulfillment of definition of finished leather notified by Directorate General of Foreign Trade (DGFT) from time to time.

In the meantime, considering the resource crunch faced by the tanning industry for modernization, the Committee of Administration (COA) of the Council for Leather Exports (CLE) requested the Government to consider abolition of the 5% export duty on finished leather but instead levy a 5% "Development Fee" on export of finished leather in lieu of the export duty so as to form a Corpus Fund for modernization of the tanning industry. For various reasons, eventually, that measure was not implemented. However, from 1991 till January 2000, the export of EI/vegetable tanned leather, crust leather and a few other leather items was prohibited by putting them on the restricted list of exports under the Export and Import policy (EXIM policy).

1.3 Export Policy on Leathers Since 2000

In order to make the export policy in line with World Trade Organization (WTO) requirements, the same was amended in 2000, by removing any sort of restriction/ prohibition on the export of hides, skins and leathers but instead levying export duties as detailed below.

- 60% export duty on the export of raw hides and skins and semi-processed leathers like wet blue leather, crust leather etc.,
- Exemption on export duty on export of finished leathers which conform to the norms notified by the Government.

In Union Budget 2019–20, the export duty on raw hides and skins and semi–finished leathers was reduced from 60 to 40% while the export duty on EI tanned leather was reduced from 15% to "Nil".

In Union Budget 2022–23, the export duty on raw buffalo hides was reduced from 40 to 30%.

2 Summary of the Export Policy on Leathers

- Raw Hides and skins, semi-processed leathers and finished leather were allowed for export freely without any duty till 1972.
- Export of raw hides and skins was banned in 1973. However, the export of semi-processed leathers like E.I tanned leather (vegetable tanned leather), Wet Blue tanned leathers (chrome tanned leather) and Crust leathers were restricted through a quota-mechanism with effect from 1st April 1973 and a duty of 25% was levied on these leathers.
- From January 1981 onwards, duty of 15% was levied on export of all sorts of tanned and un-tanned hides, skins and leathers.
- Export of finished leather was further regulated by notification of norms through an Export Trade Control Order in August 1991, which was further amended in May 1992. This means that the finished leather which fulfilled the norms alone

could be exported. Further, as a Policy measure to discourage export of finished leather, the Government imposed export duty at the rate of 10% ad-valorem in the Union Budget 1992–93. It was later reduced to 5% with effect from 1st May 1992 and the same was continued till Union Budget 1995–96, when the export duty on finished leather was abolished. Accordingly, the export duty exemption was granted for export of finished leather of goat, sheep and bovine animals and their young ones.

- However, from 1991 till 2000, the export of EI leather and crust leather, and a few other leather items was prohibited by putting them on the restricted list of exports under the EXIM policy (Export Import Policy).
- To make the export policy in line with WTO (World Trade Organisation) requirements, the export policy was amended in 2000, by removing any sort of restriction/prohibition on export of hides, skins and leathers but instead levying export duties on certain types of hides, skins and leathers, as detailed below. Accordingly, 60% export duty was levied on the export of raw hides and skins and semi-processed leathers like wet blue leather and crust leather while there is no export duty on finished leathers which conform to the norms notified by the Government.
- In Union Budget 2019–20, the export duty on raw hides and skins and semi–finished leathers was reduced from 60 to 40% while the export duty on EI tanned leather was reduced from 15% to "Nil".
- In Union Budget 2022–23, the export duty on raw buffalo hides was reduced from 40 to 30%.

2.1 De-Reservation of Leather Industry & Micro, Small and Medium Enterprises (MSME) Development Act

Leather, Leather Products and footwear and host of the items used as inputs/raw materials were reserved for production by Small Scale Industries (SSI) units. The de-reservation and de-licensing of leather, leather products and footwear sector from small scale sector initiated in various stages since 2001 and the implementation of the policy of allowing 100% FDI in the sector since 2002 and also the focused market schemes introduced by the Department of Commerce created the ground work for the expansion of this sector.

The enactment of MSME Development Act in 2006 enabled more industries in the sector to get Government schemes, as the MSME Act defined a new sector called "Medium Enterprises". As a result, about 90% of the enterprises in the leather, leather products and footwear sector were estimated to be MSME units.

The MSME Development Act 2006 defined MSME as follows.

Enterprises engaged in the manufacture or production, processing or preservation of goods as specified below:

- A micro enterprise is an enterprise where investment in plant and machinery does not exceed Rs. 2.5 million;

Table 3 The revised classification of MSME

Classification	Micro	Small	Medium
Manufacturing & services	Investment* <Rs. 1 Cr & Turnover <Rs. 5 Cr	Investment <Rs. 10 Cr & Turnover <Rs. 50 Cr	Investment <Rs. 50 Cr & Turnover < Rs. 250 Cr

Note Rs. 1 Cr = Rs. 10 million
* Investment in plant and machinery or equipment

- A small enterprise is an enterprise where the investment in plant and machinery is more than Rs. 2.5 million but does not exceed Rs. 50 million;
- A medium enterprise is an enterprise where the investment in plant and machinery is more than Rs.50 million but does not exceed Rs.100 million.

2.2 Revised MSME Definition

The Government of India revised the criteria for the classification of Micro, Small and Medium Enterprises (MSME) with effect from July 1, 2020, For the first time, MSMEs were classified on the basis of both investment in plant and machinery or equipment and also turnover. The revised classification of MSMEs is shown in Table 3.

Enterprises have to satisfy both Investment and Turnover Limits to be in a Category. If either one condition is above limit, they will be in the next higher category. However, exports of goods or services or both have been excluded while calculating the turnover of any enterprises whether micro, small or medium for the purpose of classification. With regard to the calculation of investment in plant and machinery or equipment, the Government Notification states that it shall include all tangible assets (other than land and building, furniture and fittings.

The aforesaid revised MSME definition brought in more enterprises in the leather, leather products and footwear sector under MSME ambit. It is estimated that about 98% of the manufacturing enterprises in the sector fall under MSME category, after notification of the revised MSME definition.

2.3 Support Measures for MSMEs

The Ministry of MSME, Government of India is implementing several schemes for the benefit of MSME enterprises, including the MSME units in the leather, leather products and footwear sector. Some of the schemes are highlighted below.

2.4 National Scheduled Caste & Scheduled Tribe Hub (NSSH) Scheme

The NSSH Scheme has four components.

2.4.1 Special Credit Linked Capital Subsidy Scheme for Technology Enablement of Scheduled Caste (SC)/Scheduled Tribes (ST) MSEs

Provides 25% subsidy to SC/ST MSEs on institutional finance up to Rs.1 Crore for procurement of Plant & Machinery (i.e. a subsidy cap of Rs. 25 Lakhs i.e. Rs. 2.5 million) without any sector-specific restrictions on technology upgradation.

2.4.2 Capacity Building and Training Programs

Provides assistance for conducting various entrepreneurship and skill development training programs.

2.4.3 Support for Enhancing Competitiveness Through Various Reimbursement Sub-Schemes/Interventions

Provides reimbursements such as Bank Loan Processing Fee, Bank Charges for Performance Bank Guarantee, Testing Fee, membership/subscription/entry fee of Export Promotion Council(s) etc.

2.4.4 Reimbursement of Membership/Subscription/Entry Fee of Export Promotion Council(s)

The assistance will be limited to 80% or Rs. 20,000 (excluding GST (Goods and Services Tax) and all other applicable taxes), whichever is less, per financial year, to an SC/ST MSE on annual membership subscription fee/one-time subscription charges/entry fee charges by various Export Promotion Councils (EPCs). The details of EPCs are provided on the website of NSSH. The membership of multiple EPCs is allowed within the overall capped amount of Rs. 20,000/- per enterprise in a financial year.

These schemes are envisioned to be implemented through online provision developed on NSSH website: https://www.scsthub.in/nssh-schemes/ and also in the bank portal.

For the Single Point Registration Scheme, please refer to NSIC (National Small Industries Corporation) website http://www.nsicspronline.com/.

3 Special Marketing Assistance Scheme (SMAS) for SC/ST Entrepreneurs

Under the above scheme, marketing support is provided to the SC/ST enterprises for the enhancement of competitiveness and marketability of their products through National SC-ST Hub by way of events namely organizing visits to international exhibitions/trade fairs/seminars abroad, participation in international exhibitions/ trade Fairs abroad, etc.,

3.1 International Cooperation Scheme (IC)

The Ministry of MSME is implementing International Cooperation Scheme (IC) with the objective of providing international Marketing opportunities to MSMEs as well as integrating them into global value chains. Financial assistance is provided to eligible/State/Central Government Organization, registered industries associations and societies/trusts involved in the promotion and development of MSME Sectors.

I. **Market Development Assistance (MDA)**

This scheme provides assistance for participation in International Exhibitions/Trade Fairs abroad and also for organizing international conferences/seminars/workshops/ summits in India and abroad.

II. **Participation of MSME delegations in international exhibitions, trade fairs and buyer-seller meets in foreign countries (PHYSICAL MODE)**

There should be a minimum of 19 participants. The assistance provided includes reimbursement of space rent for MSMEs, economy class airfare and freight charges subject to ceilings. The application process under the MDA sub-scheme has been fully digitized and applications have to be filed online for events to be held in FY 2022–23 from the list of approved events.

III. **Capacity Building of First Time MSE Exporters (FTE)**

Under this component, reimbursement is provided for Registration-cum-Membership Certificate (RCMC) charges of Export Promotion Councils, Export Insurance Premium paid to Export Credit Guarantee Corporation (ECGC) under the ECGC's Small Exporter Policy, testing, and quality certification fee with subject to ceilings.

Support Measures under Foreign Trade Policy

Schemes under Foreign Trade Policy (FTP) have played a vital role in the promotion of exports of leather, leather products and footwear from the country, as enumerated below.

Period: 2004–09

Two major schemes namely Focus Product Scheme (FPS) and Focus Market Scheme (FMS) were introduced. As the name suggests, the Focus Product Scheme provided incentives for notified products while Focus Market Scheme provided incentives for notified markets. Under FPS, 1.25% of FOB was provided as an incentive while under FMS 2.5% was provided. The FPS and FMS were provided as Duty Credit Scrips, which could be used by the exporter to pay the import duties or can be transferred/sold to other exporters.

Period: 2009–14

In this policy, various support measures were enhanced. Duty Credit Scrip under Focus Product Scheme (FPS) enhanced from 1.25 to 2% of FOB (Free on Board). Many new products included under FPS namely travel goods of leather, other articles with outer surface of leather, of composition of leather/of patent leather and other footwear (of leather) with uppers of textile materials were added. Gloves, Mittens and Mits, specially designed for use in Sports were made eligible for higher duty credit scrip of 5% under FPS.

Duty Credit Scrip under Focus Market Scheme (FMS) enhanced from 2.5 to 3% of FOB. 26 new markets were added under FMS, including 16 new markets in Latin America and 10 in Asia-Oceania.

Zero Duty Scheme introduced under Export Promotion Capital Goods Scheme (EPCG) for certain sectors including leather and leather products introduced. Kanpur, Dewas and Ambur recognized as "Towns of Export Excellence" for leather products. For upgradation of export sector infrastructure, 'Towns of Export Excellence' and units located therein would be granted additional focused support and incentives.

To accelerate exports and encourage technological upgradation, additional Duty Credit Scrips shall be given to Status Holders @ 1% of the FOB value of past exports. The duty credit scrips can be used for procurement of capital goods with Actual User condition. This facility was available for sectors of leather (excluding finished leather), textiles and jute, handicrafts, engineering (excluding Iron & steel & non-ferrous metals) etc.

3.2 *The Annual Supplement 2010–11 to the Foreign Trade Policy 2009–14*

In the Annual Supplement, Duty Credit Scrip benefit under Focus Product Scheme (FPS) was further enhanced from 2 to 4% for exports made from 1.4.2010 onwards. Finished leather exports were entitled for Duty Credit Scrip @ 2% under Focus Product Scheme (FPS) for exports made from 1.4.2010 onwards.

The incremental Exports Incentivisation Scheme was announced in Dec. 2012. Under this scheme, the exporters were eligible for 2% duty credit scrip on the FOB value of exports to USA, Europe and Asia (excluding Singapore, UAE and

Hong Kong) on the incremental growth during the period 01.01.2013 to 31.3.2013 compared to the period from 01.01.2012 to 31.3.2012. This scheme was extended for the year 2013–14 also on an annual basis. Also, exports to 53 notified countries in Latin America and Africa were added under this scheme vide DGFT (Directorate General of Foreign Trade) Public Notice No. 3 (RE 2013/2009–14 dated 18th April 2013. The 2% incremental scrip was however applicable only for incremental exports achieved during 2013–14 as compared to the year 2012–13 for notified countries. This scrip was transferable and could be used for domestic sourcing and for payment of Service Tax.

Period: 2015–20

Under FTP 2015–20, the Zero Duty EPCG Scheme was continued. In place of Focus Product/Focus Market Schemes, the Merchandise Exports from India Scheme (MEIS) was introduced. The MEIS Scrip value was initially 2% for finished leather and 3% for leather products and footwear. Additional 2% Scrip was provided for exports made from 1.11.2017 onwards. The MEIS scheme was however stopped w.e.f. 1.1.2021.

This policy is still in force and the new FTP is expected to be notified with effect from 1.10.22.

Fiscal support measures

The Government of India has been providing many fiscal incentives and also duty exemptions/concessions for the overall growth of the leather industry. Besides, the duties incurred on export products are also reimbursed.

Import duty is exempted on the import of raw hides and skins. As mentioned above, the import duty on semi-finished leathers like wet blue and crust leather and also finished leathers were exempted since 2000, but Union Budget 2021–22 removed this exemption, thereby subjecting them to levy of 10% import duty with effect from Feb. 2, 2021.

Interest Equalisation Scheme on Pre & Post Shipment Rupee Export Credit was implemented with effect from 1st April, 2015 till Sept. 30, 2021. This scheme provided a reduced rate of interest for the rupee export credit (packing credit) availed by exporters (pre-shipment and post-shipment credit). The rate of interest equalisation was 5% for MSME units and 3% for other units falling under 416 tariff lines (including leather products under HS Codes 4201, 4205, and also all footwear under 6401–6405) notified and also merchant exporters. This scheme is now continuing with 3% equalization for MSMEs and 2% for the said tariff lines, for the period October 1, 2021, and ends on March 31, 2024.

Goods and Services Tax (GST) rates have been reduced for many leather industry-related items. These include reductions from 28 to 18% for leather products under Chapter 42 and for CETP services from 18 to 12% w.e.f. Nov. 15, 2017.

All Industry Rate of Duty Drawback Under All Industry Rates of Duty Drawback (AIR), the incidence of Customs duties i.e. Basic Customs duty and Social Welfare

Surcharge, on inputs and Central Excise Duty on fuel used in the production of export products (e.g., Diesel for generators) is reimbursed.

Remission of Duties and Taxes on Exported Products (RODTEP) Under RODTEP, unrefunded taxes/duties in export products (like Value-Added Tax (VAT) and Central excise duty incidence on the transport of inputs/raw materials used in the manufacture of export products, VAT, and Central excise duty incidence on the transport of export goods from factory to port; Embedded GST in purchases from unregistered dealers, electricity duty, stamp duty for registering export documents etc.,) are reimbursed.

3.3 Infrastructure Development Projects

The Government of India has provided financial support for implementing several infrastructure projects, as a result of which many common facilities have come up across the country, as detailed in Table 4.

Table 4 List of various infrastructure development projects established by the Government of India for the Indian leather sector

S. No	Name of the project	Scheme/sanction year
1.	Design Studio in Footwear Component Park, Chennai	ILDP—(X Plan)
2.	Design Studio in Leather Goods Park, Kolkata	ILDP—(X Plan)
3.	Upgradation of Design Studio, Kolkata	Central ASIDE
4.	Additional facilities at Ambur Trade Centre	Central ASIDE
5.	Testing Laboratory in Ranipet	Central ASIDE
6.	Common Facility Centre in Jalandhar	Central ASIDE
7.	Multi-Skill Development Centre in Kanpur	State ASIDE
8.	Testing Laboratory in Kanpur	State ASIDE
9.	Trade Centre at Agra	Central ASIDE
10.	Testing Lab. & Design Studio, Agra	State ASIDE
11.	CFC in Melvisharam (on going)	Central ASIDE
12.	Capacity Expansion of Ranitec CETP	Central ASIDE
13.	Capacity Expansion of Madhavaram CETP	Central ASIDE
14.	Capacity Expansion of Vishtec CETP (on going)	Central ASIDE

ILDP-Indian Leather Development Programme, ASIDE—Assistance to States for Developing Export Infrastructure and Allied Activities

3.4 Focus Schemes

The Government has implemented many focus schemes for the leather industry. These include the Tannery Modernization Scheme, the Indian Leather Development Programme (ILDP: 2012–17) the Indian Footwear, Leather & Accessories Development Programme (IFLDP: 2017–21) and the Indian Footwear and Leather Development Programme (IFLDP: 2021–26).

The IFLDP has a total outlay of Rs. 1700 crore (Rs. 17,000 million) and includes sub-schemes namely Mega Leather Footwear and Accessories Cluster Development (MLFACD—Rs.300 crore outlay); Integrated Development of Leather Sector (IDLS) (Rs. 500 crore outlay); Sustainable Technology and Environmental Promotion (STEP) (Rs. 500 crore outlay); Promotion of Indian Brands in Leather and Footwear Sector (Rs. 100 crore outlay); Development of Design Studios (Rs. 100 crore outlay) and Establishment of Institutional Facilities (Rs. 200 crore outlay).

The brief of various sub-schemes is given below.

- IDLS: 30% support on the cost of plant and machinery (40% for North East States) for Micro, Small & Medium units and 20% for other units (30% for North East States)—Ceiling of Rs. 15 crore. 5% additional assistance for procurement of domestically produced machinery.
- Design Studio: 50% of project cost subject to max. Rs. 10 crore per Studio.
- Institutional Facilities: Grant-in-aid for establishment/upgradation of the institutional infrastructure of FDDI (Footwear Design and Development Institute).
- Branding: 50% of total cost (subject to ceiling of Rs. 3 crore per year for 3 years, for developing Indian brands. This will be applicable for companies having exports or domestic sales of Rs. 50 crore
- Mega Clusters: 50% of eligible project cost (70% for North East States) with ceiling Rs. 125 crore for establishment of Mega Clusters.
- Sustainable Technology and Environmental Promotion (STEP): 70% assistance (80% for North East States), with max. assistance of Rs. 200 crore towards establishment, expansion or up gradation of CETPs, Upgradation of constituent Primary Effluent Treatment Plants (PETPs) connected to concerned CETP, Developing secure landfills, Conversion of waste into some by-products and any other techniques for hazardous waste management

3.5 Marketing Support

The Marketing support is provided by the Government in the form of Market Access Initiative Scheme (MAIS) for the various export promotion activities like International Fairs/Buyer Seller Meets/Reverse Buyer Seller Meets etc., conducted by Council for Leather Exports. This scheme has helped many exporters particularly the MSMEs to gain exposure to the global market.

3.6 Ease of Doing Business Measures

The Government of India has taken a number of measures as part of Ease of Doing Business. These include introduction of e-BRC (Bank Realisation Certificate) introduced (realization of export proceeds directly transferred from bank to DGFT service), implementation of Authorised Economic Operator Programme (AEO) of Customs facilitates faster clearance for import and export consignments and 24 × 7 Customs Clearance facility and reduction of Mandatory documents for export & import brought down to 3 *(Bill of Lading/Airway Bill, Invoice-cum-Packing List, Shipping Bill/Bill of Entry).*

3.7 Schemes of State Governments

In addition to the incentive schemes and support measures implemented by Government of Indi, various are also implementing support schemes, which can be availed by the footwear and leather industry. The schemes of some State Governments are given below.

Tamil Nadu

The Tamil Nadu footwear and Leather Products Policy 2022 released on August 24, 2022, provides many additional incentives for setting up manufacturing units of footwear and leather products, in addition to the incentives notified in Industrial Policy 2021.

Large and above projects in "B" and "C" category districts (₹300 crore+) will be eligible for Special Packages, while 'Mega' projects in "A" category districts (₹500 crore+) will be eligible for a 'Structured Package of Assistance" as per the Tamil Nadu Industrial Policy (TNIP) 2021. Considering the labour-intensive nature of Footwear and Leather products (FLP) manufacturing, the eligibility criteria for firms applying for incentives will be relaxed under the "Special Package for FLP Manufacturing". For leather FLP and standalone non-leather footwear and leather footwear and leather products, the eligibility is a minimum investment of Rs. 50 crore and minimum employment of 2,000 people; for component cluster ₹300 crore and 3,000 persons and for allied footwear industries (outside the cluster) ₹150 crore and 100 persons, the policy said. An investor can avail either turnover-based subsidy or fixed capital subsidy as detailed below.

3.7.1 Turnover-Based Subsidy

Considering the higher turnover to asset ratio in FLP manufacturing, a turnover-based subsidy will be applicable as a percentage of the turnover in each financial year. For leather FLP in B category districts, it will be 1.5%, and for C category districts, it will be 1.75%; for standalone non-leather footwear manufacturing 1.75% and 2%, and for component clusters, it will be 1.5% and 1.75%, in B and C category districts respectively. The companies can choose to avail of turnover-based subsidy from the date of commercial production or upon achieving the minimum eligible investment for the structured package in the given industry category type, whichever is later, up to a cap of 4% of cumulative investment in eligible fixed assets per annum for a period of 10 years, the policy said.

3.7.2 Fixed Capital Subsidy

Industries allied to footwear manufacturing or component industries located outside the clusters will be eligible for a Fixed Capital Subsidy of 10% of investment in Eligible Fixed Assets to be disbursed in 10 equal annual instalments. The company can choose to avail of the Fixed Capital Subsidy from the date of commercial production or upon achieving the minimum eligible investment threshold of ₹150 crore and minimum employment threshold of 100 jobs, whichever is later, the policy said.

The package also includes other incentives namely training subsidy, land cost subsidy, stamp duty incentive, enhanced incentive for quality certificate, enhanced incentive for intellectual property creation, interest subvention, transport subsidy, marketing assistance, traceability incentive etc. Investments made from April 1, 2022, will be considered eligible for availing of incentives.

Uttar Pradesh

Stamp duty exemption of 100% in Bundelkhand & Poorvanchal, 75% in Madhyanchal & Paschimanchal (except Gautam Buddh Nagar & Ghaziabad districts) region of the state and 50% in Gautambuddhnagar & Ghaziabad districts.

Employees' Provident Fund (EPF) reimbursement facility to the extent of 50% of employer's contribution to all such new Industrial units providing direct employment to 100 or more unskilled workers.

Reimbursement of net Value-Added Tax (VAT) and Central Sales Tax (CST) or the net amount deposited in the State's account vis-a-vis share of the state under GST, subject to ceilings.

Capital Interest Subsidy to the extent of 5% per annum for 5 years in the form of reimbursement on loan taken for procurement of plant & machinery, subject to an annual ceiling of Rs. 50 lacs.

Infrastructure Interest Subsidy to the extent of 5% per annum for 5 years in the form of reimbursement on loan taken for the development of infrastructural amenities for self-use like roads, sewer, water drainage, erection of power line, transformer, and power feeder, subject to an overall ceiling of Rs. 1 Crore.

Interest subsidy to the extent of 5% per annum for 5 years in the form of reimbursement on loan taken for industrial research, quality improvement, and development of products by incurring expenditure on the procurement of plant, machinery & equipment for setting up testing labs, quality certification labs and tool rooms, subject to an overall ceiling of Rs. 1 Crore.

Exemption from electricity duty to all new industrial units set up in the state for 10 years.

Exemption from electricity duty for 10 years to all new industrial units producing electricity from captive power plants for self-use.

Exemption from Mandi fee for all new food processing units on purchase of raw material for 5 years. The industries which are disallowed for input tax credit under the GST regime, will be provided reimbursement of that amount of VAT/CST/GST paid on purchase of plant and machinery, building material and other capital goods during construction and commissioning period and raw materials and other inputs in respect of which input tax credit has not been allowed. In addition to the above, the facility of interest free loan equivalent to the sum of VAT and CST/state share of GST deposited by industrial units admissible to eligible industrial units under the previous policy will be continued.

3.7.3 Incentivising Employment Generation Units

Generating minimum employment of 200 direct workers including skilled and unskilled will be provided 10% additional EPF reimbursement facility on employer's contribution.

All incentives in the form of reimbursement, subsidies, exemptions etc., will be subject to a maximum of 100% of fixed capital investment (FCI) made in Poorvanchal and Bundelkhand, 90% of fixed capital investment made in Madhyanchal & Paschimanchal (except Gautambuddh Nagar & Ghaziabad districts) and 80% of fixed capital investment made in Gautambuddh Nagar & Ghaziabad districts.

Uttar Pradesh also provides incentives for Mega Projects.

West Bengal

The implementation of a new Incentive Scheme for MSME in West Bengal has been approved with an eye to promoting growth and development of such enterprises in the State. The Scheme is called Banglashree for MSMEs. Banglashree for MSMEs shall generally be applicable to all MSME units in the manufacturing sector which have started production on or after 01.04.2019. The enterprises may be in the private sector, cooperative sector and joint sector undertaking as also companies/undertakings owned and managed by the State Government and the Industrial SHGs.

Any MSME unit commencing its commercial production from 01.04.2019 to 31.03.2025 is eligible.

State Capital Investment Subsidy- An eligible micro or small enterprise located in Zone C, D & E area will be entitled to State Capital Investment Subsidy as detailed below in Table 5.

Table 5 West Bengal state capital investment subsidy for different zones

	Micro enterprise	Small enterprise
Zone C	25% of the fixed capital investment	15% of the fixed capital investment
Zone D	40% of the FCI	30% of the FCI
Zone E	60% of the FCI	40% of the FCI

Note
Zone A: Kolkata Municipal Corporation Area, All municipal areas of North 24-Parganas, All municipal areas of South 24-Parganas, All municipal Corporation and municipal areas of Howrah
Zone B: Districts of Hooghly, North 24 Parganas (excluding municipal areas and Sundarban areas), South 24 Parganas (excluding municipal areas and Sundarban areas), Howrah (excluding corporation and municipal areas), Siliguri Municipal Corporation, Municipal Corporations/Municipal Areas of Paschim Medinipur, Purba Medinipur, Purba Bardhaman, Paschim Bardhaman and Nadia
Zone C: Districts of Purba Bardhaman, Paschim Bardhaman (excluding municipal corporation/municipal areas), Purba Medinipur (excluding municipal corporation/municipal areas), Nadia (excluding municipal corporation/municipal areas), Malda, Jalpaiguri, Murshidabad and Darjeeling (excluding Siliguri Municipal Corporation), Kalimpong, Alipurduar
Zone D: Districts of Birbhum, Purulia, Bankura, Paschim Medinipur (excluding municipal corporation/municipal areas), Uttar Dinajpur, Dakshin Dinajpur, Sundarban Areas of South and North 24 Parganas and Jhargram
Zone E: Cooch Behar District

An additional subsidy of 20% on State Capital Investment Subsidy normally admissible will be provided to all eligible micro and small enterprises wholly owned (100%) by women, SC/ST, and minority common temporary entrepreneurs.

An eligible micro or small enterprise wholly owned by women, SC/ST, and minority entrepreneurs and set up in Zone A and B area will be entitled to State Capital Investment Subsidy as follows:

Micro Enterprise: 15% of the Fixed Capital Investment
Small Enterprise: 10% of the Fixed Capital Investment.

The subsidy normally admissible and additional subsidy taken together will not exceed Rs. 50 Lakh for a Small Enterprise located in Zone C & D area and will not exceed Rs. 75 Lakh for a Small Enterprise located in Zone E area.

3.7.4 Interest Subsidy on Term Loan

An eligible micro, small or medium enterprise will be entitled to Interest Subsidy on annual interest liability on the Term Loan borrowed from a Commercial Bank/Cooperative Bank/Scheduled Banks approved by Reserve Bank of India (RBI)/Regional Rural Banks (RRBs)/Financial Institution (Central and State) for implementation of the approved project as follows.

Micro & Small Enterprise: Zone A & B—55% of interest liability for 5 years. Zone C, D & E—70% of interest liability for 5 years Medium Enterprise: Zone B, C, D & E: The interest subsidy will be 25% of total Term Loan interest paid by the enterprise for 5 years subject to a ceiling of Rs. 175 lakh per year.

In addition, there are several other incentives provided to eligible MSME units namely Waiver of Electricity Duty, Power Subsidy, Subsidy for Energy Efficiency, Subsidy on Stamp Duty and Registration Fee, Subsidy for State Goods and Services Tax (SGST) paid to the Govt of West Bengal., Subsidy for Water Conservation/ Environment Compliance., Subsidy for Standard quality compliance, Work force Welfare Assistance, Subsidy for Patent Registration and Incentive for approved expansion project of an existing enterprise.

Maharashtra

Government of Maharashtra provides a basket of incentives with ceilings ranging from 30 to 80% according to district classification, while 100% assistance is provided for industries in no industry district, naxalism-affected areas and aspirational districts namely Osmanabad, Gadchiroli, Washim and Nandurbar.

3.7.5 Expansion/Diversification Units

Existing/New Micro, Small and Medium Manufacturing Enterprises and Small Industries (including Manufacturing/Information Technology/Biotechnology (IT/BT)) Units, qualifying as Expansion/Diversification Units, will also be eligible to get the incentives for Expansion/Diversification, equivalent to 80% of the incentives admissible for New Units. The eligibility period for availing of the incentives will however be reduced by one year than that admissible to a New Unit in case of Expansion/Diversification Units.

3.7.6 Industrial Promotion Subsidy (IPS)

The eligible New/Expansion Micro, Small and Medium Manufacturing Enterprises, which are set up in different parts of the State, will be eligible for Industrial Promotion Subsidy (IPS), as per their taluk Categorization. Eligible Micro, Small and medium enterprises shall be offered IPS on 100% Gross State Goods & Services Tax (SGST) payable by the unit on the first sale of eligible products billed and delivered within Maharashtra. The modalities for disbursement of incentives shall be as per the guidelines issued vide GR dated 12.06.2018, 20.12.2018 and 08.03.2019 and the guidelines issued from time to time.

3.7.7 Interest Subsidy

All eligible new Micro, Small and Medium Manufacturing Enterprises will be eligible for interest subsidy in respect of interest actually paid to the Banks and Public Financial Institutions (excluding unsecured loans, private loans/borrowings, loans from Non-Banking Financial Company (NBFC) etc.) for claim period, on the amount of term loans taken for acquisition of new Fixed Assets required for the project accepted by the implementing agency. The amount of interest subsidy will be calculated @ effective rate of interest, after deducting the interest subsidy receivable from any agency of the State Government or under any Govt. of India Scheme and the penal/ compound interest or 5% per annum, whichever is less. The quantum of interest subsidy payable to the eligible unit every year will not exceed the bills paid for electricity consumed during the relevant year.

3.7.8 Exemption from Electricity Duty

Eligible New Units in developing areas (such as Group C, D, and D + which are less developed areas, lesser-developed areas of the State, not covered under Group A/Group B/Group C and least developed areas, not covered under Group A/Group B/Group C/Group D, respectively; Group A and B are industrially developed areas and areas where some industrial development has taken place, but are less developed than the areas under Group A, respectively), No Industry District(s), Aspirational Districts and Naxalism affected Area will be exempted from payment of Electricity Duty during applicable eligibility period.

3.7.9 Power Tariff Subsidy

Eligible New Micro, Small and Medium Enterprises (MSME) and Small Industries will be eligible for power tariff subsidy. The subsidy will be to the tune of Rs. 1/ - per unit for the Units located in Vidarbha, Marathwada, North Maharashtra and the Districts of Raigad, Ratnagiri and Sindhudurg in Kokan Region, No Industry Districts, Naxalism Affected Areas, and Aspirational Districts and Rs. 0.50 per unit for the Units in other areas of the State for a period of 3 years from the date of commencement of commercial production, for the energy consumed and paid. The Units in Group "A" areas will however not be eligible for this incentive. Units will be eligible for a basket of incentives mentioned in Table 6. The total quantum of which will be linked to the actual eligible Fixed Capital Investment. The total quantum of incentives and the Eligibility Period will be as given in Table 6. However, aggregate fiscal incentives provided by various departments or agencies of the State Government shall not exceed the basket of incentives (as percentage of FCI). The incentives will be granted to the units on first-cum-first serve basis. Maharashtra also provides additional incentives for strengthening MSME units.

Table 6 Basket of incentives for MSMEs

Taluka/area classification	Maximum permissible fixed capital investment (INR crore)	Maximum ceiling of basket of incentives as % of FCI	Eligibility period (Years)
A	For the purpose of this policy, MSME shall include units as per the MSMED Act, 2006, as well as the units with FCI of upto INR 50 crore	–	–
B		30%	7
C		40%	7
D		50%	10
D^{+}		60%	10
Vidarbha, Marathwada, Ratnagiri, Sindhudurg & Dhule		80%	10
No industry Districts, Naxalism affected areas* and Aspirational Districts**		100%	10

* Naxalism affected areas as per Government Resolution No.: PSI-2013/(CR-54)/IND-8 Dated 1st April 2013 issued by Government of Maharashtra Industries, Energy and Labour Department
** Aspirational Districts are Osmanabad, Gadchiroli, Washim and Nandurbar
Note
Group A: Denotes industrially developed areas
Group B: Denotes areas where some industrial development has taken place, but are less developed than the areas under Group A
Group C: Denotes areas, which are less developed than those covered under Group B
Group D: Denotes the lesser-developed areas of the State, not covered under Group A/Group B/ Group C
Group D+: Denotes the least developed areas, not covered under Group A/Group B/Group C/Group D

Growth potential of the industry

Today, India is seen as a reliable sourcing destination for high quality leather, leather products and footwear. Considering the business opportunities, the sector is looking to increase its exports from the present USD 5.3 billion in 2019 to USD 13.7 billion by 2030 and the domestic market potential from the present USD 12.3 billion in 2019 to USD 33.4 billion by 2030 as shown in Fig. 1. For achieving the desired results, CLE has been requesting the Government to implement the Production Linked Incentive Scheme for the footwear and Leather Sector and also provide plug-and-play factories.

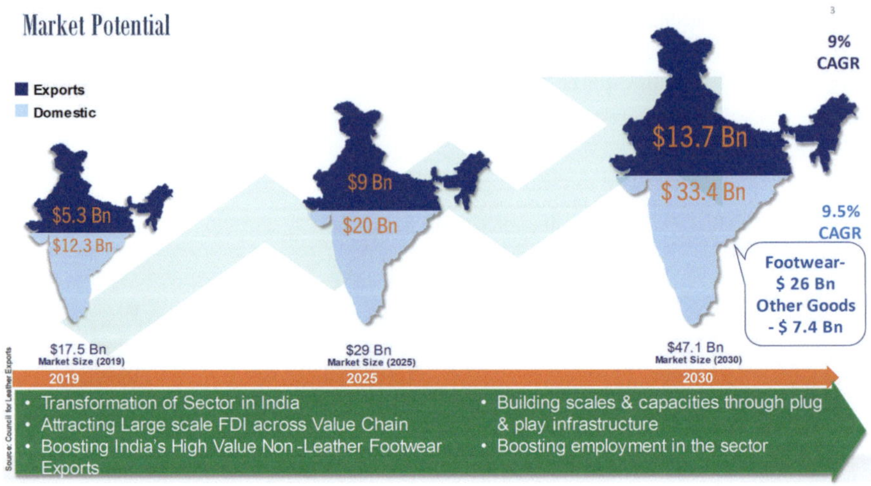

Fig. 1 Growth potentials of the Indian leather and products industry

4 Conclusion

The footwear and leather industry in the country has come a long way but has not reached its optimum potential for growth. Our country has the inherent strengths namely a huge raw material base, availability of skilled manpower, support of technical and training institutions, traditional knowledge of tanning etc. but we need to focus on growth areas namely scaling-up of our production. moving-up the value chain by enhancing our design capabilities, and attracting both domestic and overseas investments in high growth segments like non-leather footwear, ladies footwear etc. to achieve the envisaged turnover of USD 47.1 billion by 2030.

Overview of the Certifying, Regulatory and Restrictive Organizations in the Leather Industry

Luis A. Zugno and Fernando Bellese

1 Introduction

The leather industry has evolved over the last 100 years, and many chemicals that were used in the past such as arsenic, mercury, lead, cadmium, pentachlorophenol, toxic solvents and dyes are no longer allowed. New regulations are continuously introduced that impact chemicals, application methods, exposure limits, discharges, transportation, handling and occupational hygiene. In the 1970s, the industry made great strides in improving pollution control after local, national and international rules and standards were implemented and enforced. Tanneries that do not comply with the basic principles of environmental control are finding themselves excluded from the global trade.

Europe was first to take a great stand on restricting chemicals that can be used in the chemical industry and in the tanneries; the process of restrictions is dynamic and is continuously evolving. Today there is a significant number of chemicals that are not allowed to be used in the tannery or be present in the leather or leather articles. Restriction limits can be based on occupational exposure, consumer exposure in leather articles, or environmental exposure from waste or discharges. The restrictions and limits can affect all the countries that export leather or leather articles. Because the leather industry is global, it is not sufficient just to consider local regulatory requirements. For example, leather may be produced in one country, the leather articles manufactured in another country, and the finished articles exported to other parts of the world resulting in a complex regulatory landscape.

In addition to national and international regulations, some sectors, corporations and non-governmental organizations (NGOs) have created voluntary standards that

L. A. Zugno (✉)
Secretary & Past President IULTCS, Basel, Switzerland
e-mail: drzleather@gmail.com

F. Bellese
Sustainability Specialist, Seabrook, NH, United States

S. K. Janardhanan and L. A. Zugno (eds.), *Emerging Trends in Leather Science and Technology*, https://doi.org/10.1007/978-981-99-9754-1_12

have been adopted and enforced in their supply chain, adding an extra level of complexity. Those standards, and the verification schemes connected to them, have helped to support the adoption of better practices around the world, including in regions where national regulations are not as advanced. NGOs and brands often set high standards that go beyond legal requirements, including aggressive targets for reducing hazardous substances in leather and leather manufacturing process as well as comprehensive programs for supply chain documentation and transparency. While this movement contributes significantly to the long-term sustainability of the leather industry, it can also pose challenges for those companies that are not prepared for the changes.

The information on this publication has been compiled from the organization's websites in September 2022, which will be updated in the future. Therefore, readers are encouraged to use this publication as a reference, but to refer to the specific websites for the most updated information.

2 Standards, Certifications and Other Verification Tools

The leather industry certifications and voluntary standards have progressed enormously, and today they help demonstrate that leather can be made responsibly, safely and compliant with the most demanding requirements. These standards and certifications are important qualification tools that give peace of mind to brands that purchase or specify leather as a material for their consumer products. Many brands are 100% committed to those standards and certifications and will not use leather from tanneries that have not been verified against those initiatives.

Verification tools and certifications are an important way to communicate to consumers all the aspects of sustainability related to the leather industry. They are also an efficient way of assessing performance and managing risk. This brings transparency from the origin of the raw material, through production practices, leather properties and article production.

Current initiatives available in the leather industry focus on different aspects. For example, some evaluate manufacturing environmental performance, while others assess the chemicals used in the industry and their impacts. There are also initiatives targeting product level compliance to ensure that products are not only made responsibly but also safe for use by consumers.

Some of the main certification and standard setting organizations are listed below. Their scopes may differ, and currently there is limited cross acceptance among those initiatives. This is expected to evolve, as standards are periodically reviewed and revised to conform with new information or requirements.

The first group of initiatives listed in this document has developed protocols to allow the evaluation and reporting of social and environmental performance at the production level:

3 Certifications

Leather Working Group (LWG) (www.leatherworkinggroup.com) is headquartered in the UK and composed of leather manufacturers, leather traders, brands, chemical suppliers, machinery suppliers and professionals. It began in 2005 with the LWG Leather Manufacturer audit standard evaluating the environmental impact of leather production. The protocol has expanded to water and energy usage, solid waste and effluent management, air and noise emissions, traceability, health and safety, chemical management, restricted substances, compliance and chromium VI management. As of August 2022, over 1000 leather manufacturers have received the LWG certification. Additionally, the LWG has the Trader Audit Standard, Subcontractor Audit Standard and Commissioning Manufacturer Audit Standard, which includes other important stakeholders involved in leather trading and processing. As a result of the audits, tanneries are rated in three levels: Gold, Silver and Bronze.

The LWG is developing a chain of custody tools to promote best practices in the leather upstream supply chain. It has the goal of 100% deforestation-and conversion-free leather by 2030. It will address animal welfare, environmental and social impacts of production, deforestation and conversion of natural landscapes, biodiversity and climate change.

Sustainable Leather Foundation (SLF) (https://sustainableleatherfoundation.com) is headquartered in the UK and aims to provide the structure and support to enable leather manufacturers and associated facilities to work on the reduction and improvement criteria necessary to be successful. SLF is concerned with the main aspects of sustainability—environmental, social and governance. As a nonprofit organization, the SLF provides a transparent and inclusive approach to demonstrating sustainable good practices through the value chain, giving consumers a clear mechanism to see and understand the sustainable attributes of leather as a material and the work done by the industry to ensure better industrial practices.

The two initiatives above have a wide international reach. However, there are regional initiatives that look at providing similar approaches, in addition to addressing issues that are more specific to their locations. The ICEC in Italy and CSCB in Brazil are two good examples.

The CSCB (www.cicb.org.br), Brazilian Leather Certification for Sustainability covers areas similar to the ones mentioned above but carries an additional social and governance module. It goes deeper into local regulations and challenges such as labor conditions and deforestation.

The ICEC (www.icec.it), Institute of Quality Certification for the Leather Sector in Italy, identifies tanneries in the areas of environmental, ethical-social, economic and product areas.

There are also other initiatives that are not specifically developed for the leather industry, but that have become more common in tanneries due to the engagement from brands in the apparel and footwear sectors with those initiatives.

The Higg Factory Environmental Module (FEM) and Higg Factory Social and Labor Module (FSLM), were developed by the Sustainable Apparel Coalition (https://apparelcoalition.org/). The former, Higg FEM, consists of 7 different modules covering areas such as Environmental Management System (EMS), energy, water, wastewater, air emissions, waste and chemical management. In addition to providing a tool for assessment in those areas, the Higg FEM allows benchmarking and comparison among different factories, making it easier for brands to assess their supply chain.

The Higg FEM is a good option for an initial, high-level assessment, but because it was created to be applied to different sectors, and the verification process is run by more generalist professionals, it is less precise and may not be representative of the real performance in tanneries.

The second module, FSLM, is a social and labor assessment developed to harmonize the efforts of different brands and reduce the number of overlapping initiatives. The questionnaires were developed based on the Converged Assessment Framework (CAF) from the Social & Labor Convergence Program, covering areas such as recruitment and hiring practices, working hours, wages & benefits, worker treatment, worker involvement and health and safety. This alignment has helped to reduce the number of social audits a tannery has to go through, while creating efficient reporting channels that promote transparency and better social and labor practices.

In addition to facility level certifications, the leather, fashion and apparel sectors have seen in the last decade the appearance of initiatives to promote best practices, efficiencies and transparency related to chemical management and use.

Two examples are the ZDHC and Chem-IQ. The first one is a multi-stakeholder initiative, and the second is a private program developed by VF Corporation.

The ZDHC (https://www.roadmaptozero.com) was created in 2011 and is based in the Netherlands. With nearly 200 members, divided between chemical suppliers, brands, solution providers and manufacturers, the initiative has the mission to "enable brands and retailers in the textile, apparel, and footwear industries to implement sustainable chemical management best practices across the value chain. Through collaborative engagement, standard setting, and implementation, [advancing] towards zero discharge of hazardous chemicals".

ZDHC brought an innovative perspective to safer chemical usage by creating standards and verification process that look at the process chemicals used in leather manufacturing, resulting in cleaner discharges. Unlike most initiatives, the tools developed by ZDHC look not only at the practices followed by a specific factory, but also the practices of upstream suppliers in creating a more sustainable supply chain. ZDHC also offers important transparency tools that allow factories to share their compliance performance against MRSL (Manufacturing Restricted Substance List) and Wastewater Guidelines with their customers, driving transparency to a new level.

VF's Chem-IQ (https://www.vfc.com/responsibility/product/chemistry) is another initiative promoting positive changes in the fashion and leather industry. The program "seeks to achieve responsible chemical management" throughout VF's supply chain. Similarly, to ZDHC, the Chem-IQ works with its own MRSL as a way of restricting more hazardous chemicals from their supply chain. Additionally, they look at annual chemical screenings of their suppliers to ensure the products being used are in compliance with their program through the testing of effluent discharge, promoting transparent and responsible management of treatment plants, and limiting possible impacts to the environment.

The last group of standards and certifications covered in this section consists of tools aimed at the testing and labelling of products to attest to the quality and safety of the products to final consumers.

Blue Angel (https://www.blauer-engel.de/en) is a consumer facing ecolabel that was created over 40 years ago with the support of the German government, to support the communication of products that were made responsibly and safely for consumers. While more relevant to the German market, it is recognized as an important certification, linked to responsible processes and safe products. The label covers different products including leather.

Oeko-Tex (https://www.oeko-tex.com/en/our-standards/leather-standard-by-oeko-tex) is another label example that was initially created as a certification for textiles. It currently carries a leather standard intended to test, verify and communicate to consumers the performance of "high human-ecological product safety".

While both Blue Angel and Oeko-Tex are niche labels that impact a small part of the leather industry, they are good examples of increasing requirements that tanneries must consider.

In addition to the above, there are numerous global regulatory bodies that have a direct or indirect impact on the leather industry. Some of these are summarized below.

4 Regulatory Bodies and Regulations

There are also governmental bodies that regulate the use of chemicals in the leather industry or in the final product (leather) and in the waste.

ECHA (https://echa.europa.eu/) is the European Chemicals Agency, an agency of the EU, tasked with protecting health and the environment. Their work also contributes to a well-functioning internal market, innovation and the competitiveness of Europe's chemicals industry. The following three regulations are implemented by ECHA:

(1) **REACH** (https://echa.europa.eu/regulations/reach/understanding-reach) means Registration, Evaluation, Authorization and Restriction of Chemicals in the European Union, adopted to improve the protection of human health and the environment from the risks that can be posed by chemicals, while enhancing the competitiveness of the EU chemicals industry. It also promotes alternative methods for the hazard assessment of substances in order to reduce the number of tests on animals. In principle, REACH applies to all chemical substances; not only those used in industrial processes but also in our day-to-day lives, for example in cleaning products, paints as well as in articles such as clothes,

furniture and electrical appliances. Therefore, the regulation has an impact on most companies in the EU.

REACH places the burden of proof on companies. To comply with the regulation, companies must identify and manage the risks linked to the substances they manufacture and market in the EU. They must demonstrate to ECHA how the substance can be safely used, and they must communicate the risk management measures to the users. Companies need to register their substances and to do this they need to work together with other companies who are registering the same substance.

ECHA receives and evaluates individual registrations for their compliance, and the EU Member States evaluate selected substances to clarify initial concerns for human health or for the environment. Authorities and ECHA's scientific committees assess whether the risks of substances can be managed. Authorities can ban hazardous substances if their risks are unmanageable. They can also decide to restrict a use or make it subject to a prior authorization.

In general, under REACH you may have one of these roles:

Manufacturer: If you make chemicals, either to use yourself or to supply to other people (even if it is for export), then you will probably have some important responsibilities under REACH.

Importer: If you buy anything from outside the EU/EEA, you are likely to have some responsibilities under REACH. It may be individual chemicals, mixtures for onward sale or finished products, like clothes, furniture or plastic goods.

Downstream users: Most companies use chemicals, sometimes even without realizing it, therefore it is important to your obligations if you handle any chemicals in your industrial or professional activity. You might have some responsibilities under REACH.

Companies established outside the EU: If you are a company established outside the EU, you are not bound by the obligations of REACH, even if you export their products into the customs territory of the European Union. The responsibility for fulfilling the requirements of REACH, such as registration lies with the importers established in the European Union, or with the only representative of a non-EU manufacturer established in the European Union.

(2) **SVHC** (https://echa.europa.eu/candidate-list-table) ECHA has published the Substances of Very High Concern (SVHC) Candidate List. Once a substance is added to the Candidate List, REACH imposes immediate obligations on manufacturers and importers to declare the substances if present. Article 33(1) of the REACH Regulation states that manufacturers and importers of articles (products) are required to notify their customers of the presence of any SVHC in their products exceeding 0.1% by weight and provide instructions on the safe use of the product. Listing of a substance as an SVHC is the first step in the procedure for authorization or restriction of use of a chemical.

(3) **BPR** (https://echa.europa.eu/regulations/biocidal-products-regulation/understanding-bpr) The Biocidal Products Regulation (BPR, Regulation (EU) 528/2012) concerns the placing on the market and use of biocidal products, which

are used to protect humans, animals, materials or articles against harmful organisms such as pests or bacteria, by the action of the active substances contained in the biocidal product. This regulation aims to improve the functioning of the biocidal products market in the EU, while ensuring a high level of protection for humans and the environment. Biocidal actives and biocidal products must be approved (or under review) under the BPR to be used in Europe. According to the regulation, articles can only be treated with biocidal products containing active substances approved (or under review) in the EU.

EPA (https://www.epa.gov/) The Environmental Protection Agency is an independent executive agency of the United States federal government tasked with environmental protection matters. EPA's mission is to ensure that Americans have clean air, land, and water. This includes ensuring that chemicals in the marketplace are reviewed for safety. Two EPA regulations relevant for the leather industry are discussed below.

(1) The **Federal Insecticide, Fungicide, and Rodenticide Act (FIFRA)** provides for regulation of pesticide distribution, sale, and use. All pesticides distributed or sold in the United States must be registered (licensed) by EPA. Before EPA registers a pesticide under FIFRA, the applicant must show, among other things, that using the pesticide according to specifications "will not generally cause unreasonable adverse effects on the environment".
(2) The **Toxic Substances Control Act(TSCA)** of 1976 provides EPA with the authority to require reporting, recordkeeping, and testing requirements related to chemical substances. TSCA also grants the authority to EPA to set restrictions on the use of chemicals. Pesticides, discussed above, are excluded from TSCA.

5 State Requirements

Proposition 65 (https://oehha.ca.gov/about/home) is valid in California, USA, and requires businesses to provide warnings to Californians about significant exposures to chemicals that cause cancer, birth defects or other reproductive harm. These chemicals can be in the products that Californians purchase, in their homes or workplaces, or that are released into the environment. Proposition 65 requires California to publish a list of chemicals known to cause cancer, birth defects or other reproductive harm. This list, which must be updated at least once a year, has grown to include approximately 900 chemicals since it was first published in 1987. The official name of Proposition 65 is the Safe Drinking Water and Toxic Enforcement Act of 1986. The Proposition 65 list contains a wide range of naturally occurring and synthetic chemicals that include additives or ingredients in pesticides, common household products, food, drugs, dyes, or solvents. Listed chemicals may also be used in manufacturing and construction, or they may be byproducts of chemical processes, such as motor vehicle exhaust.

Warning label: By law, a warning must be provided for listed chemicals unless the exposure is low enough to pose no significant risk of cancer or is significantly below levels observed to cause birth defects or other reproductive harm. Leather containing chemicals on the Proposition 65 list—above Safe Harbor limits must have a Warning Label.

6 Restricted Lists

RSL (Restricted Substances Lists) are made by brands, manufacturers and retailers to ensure that their products comply with regulations and standards that govern which substances and amounts of substances can be used in products. The RSLs are always changing –most of the time increasing the number of restricted chemicals and reducing the limits of chemicals. Leather buyers only purchase products from a tannery when leather is compliant with their RSL. The RSLs are not standardized but most of them have similarities, some are more stringent than others.

The objectives of the RSLs are:

- Protect the consumer, consumer health and environment
- Comply with regulations and standards (including NGO's)
- Help protect revenue by reducing cost of lawsuits, returns and recalls
- Enhance brand image by promoting health and environmental stewardship
- Enhance sustainability image of the brands.

Example: https://afirm-group.com/wp-content/uploads/2022/02/2022_AFIRM_RSL_2022_02161.pdf.

7 Other Relevant Leather Organizations

IULTCS (International Union of Leather Technologists and Chemists Societies) (https://iultcs.org/), a world-wide organization of professional societies, was established in London in 1897. There are currently 19 Member Societies, 6 Associate Members and 1 Supporting Member—representing some 3000 individual members.

According to IULTCS statutes, the aims of the Union are to foster cooperation between member societies, to hold congresses to further the advancement of leather science and technology, to form commissions for special studies and to establish international methods of samples and testing of leather and materials associated with leather manufacture.

IULTCS has a special agreement with **ISO** (https://www.iso.org/standards.html) (International Organization for Standardization) which means that there are special ISO Standards unique to leather testing that take into consideration all the material's properties. ISO recognizes IULTCS as the international standardizing organization for leather test methods.

The ISO Standards are adopted globally as the basis for local national standards. Almost 100 leather specific procedures have been developed and approved as ISO Standards.

Examples include standards for measuring area, tensile and tear strength, dimensional shrinkage, heat resistance, water repellence, fogging, flame resistance and color fastness, as well as a range of chemical tests for leather.

Other important national standards institutions for leather are ASTM (https://www.astm.org/) (United States) and **DIN** (https://www.din.de/de/ueber-normen-und-standards/din-norm) (Germany).

Leather Naturally (https://www.leathernaturally.org/) is a not-for-profit industry members association that focuses on education and the promotion of leather. It promotes the use of globally manufactured sustainable leather and seeks to inspire and inform designers, creators, and consumers about its beauty, quality and versatility.

Leather Naturally's educational and promotional activities focus on leathers made from hides and skins that are a by-product from the food industry that would otherwise mostly be turned into waste.

Responsibly made leather is a valuable alternative for oil-based products such as plastics—including PVC and Polyester—adding value to an important renewable material, via efficient and modern processing techniques. Leather Naturally believes that the promotion of sustainably made leathers supports a continuous effort and the development of best practices supporting a circular economy. Moreover, it facilitates dialogues between the leather upstream supply chain, manufacturers and consumers.

One4 Leather (https://www.one4leather.com/) is a group of manufacturers and suppliers, active in the automotive leather supply chain, who have set themselves the task to dispel myths, rumors and incorrect information regarding the resourcing, production, and usage of leather in automotive interiors. Their mission is to rekindle the appreciation for leather as a high-value, renewable material and the preferred choice for car interiors, by sharing its full story. By providing verified and complete information, they aim to offer the general public and industry the facts and insights about leather.

UNIDO Leather Panel (https://leatherpanel.org/) is the UNIDO global forum for technical assistance programs dealing with the leather-based industry sector. The Leather Panel provides information on good practices in small-scale manufacturing as well as professional training and pollution control procedures in the leather, footwear and leather products industries. There are important publications on Hides and Skins, Leather Processing, Cleaner Technologies, Solid Wastes and Tannery Effluent treatment. They also have E-learning modules on Leather Manufacture, Tannery Effluent Treatment and Occupational Safety and Health.

8 Conclusions

Certifications are powerful tools to standardize best practices and reward tanneries that comply with several parameters measured through certifications. The certifications help the tanneries, brands and consumers to have a transparent and effective communication tool.

Constantly new chemicals are being restricted and/or limited in use; others require additional measures of occupational hygiene to be handled.

The regulations, restrictions and recommendations that limit the use of chemicals in the leather production are dynamic and always evolving. They bring more safety and confidence to consumers, and ultimately help to promote the use of leather. We always say that leather industry is safe, regulated and responsible; this is the result of certifications, norms, restrictions and best practices. This work outlines the best information we have in September 2022.

Acknowledgements The authors want to express gratitude to Mrs. Laura Pearson for the technical review and suggestions.

Call for a Unique North-South-South-Triangular Co-operation for Education for Leather and Allied Sector

J. Raghava Rao, G. C. Jayakumar, and T. Ramasami

Abbreviations

ACF	Academy of Costume and Fashion
ADS	Arsutoria Design School
BCSIR	Bangladesh Council of Scientific and Industrial Research
BDU	Bahir Dar University
BLS	Bristish Leather School
BTU	Bir Tikendrajit University
CAIS	College of Applied and Industrial Sciences
CLFI	China Leather and Footwear Industrial Research Institute
CLS	Chennai Leather School
DEI	Dayalbagh Educational Institute
DeKUT	Dedan Kimathi University of Technology
DLT	Department of Leather Technology, Anna University
E&T	Education and Training
EFLED	Engineering Faculty Leather Engineering Department
ELRC	Exotic Leather Research Centre
ESSUTM	East—Siberia State University of Technology and Management
FDDI	Footwear Design and Development Institute
FILK	Freiberg Institute GmbH
FLYERS	Frontline Leaders and Young Entrepreneurs' Resources for Start-Ups
GCELT	Government College of Engineering and Leather Technology
GILT	Government Institute of Leather Technology

J. Raghava Rao (✉) · G. C. Jayakumar
CSIR-Central Leather Research Institute, Adyar, Chennai 600020, India
e-mail: jrrao61@gmail.com; jrrao@clri.res.in

T. Ramasami
Department of Science and Technology, Government of India, New Delhi, India

S. K. Janardhanan and L. A. Zugno (eds.), *Emerging Trends in Leather Science and Technology*, https://doi.org/10.1007/978-981-99-9754-1_13

HBTU	Harcourt Butler Technical University
HRD	Human resource Development
IBTeC	Brazilian Institute of Technology for Leather, Footwear and Artifacts
ICLT	Institute for Creative Leather Technologies
ICPI	Leather and Footwear Research Institute
ILET	Institute of Leather Engineering and Technology
ILRI	Italian Leather Research Institute
INCDTP	National Research and Development Institute for Textile and Leather
KLLCE	Key Laboratory of Leather Chemistry and Engineering
LASRA	Leather and Shoe Research Association
LACOURO	Laboratory of Leather and Environmental Studies
LCF	London College of Fashion
LCI-B	LCI Barcelona
LGR	Leather Institute German Tanners' School Reutlingen
LIRI	Leather Industries Research Institute
LIZ	Leather Institute of Zimbabwe
LRI BCSIR	Leather Research Institute, Bangladesh Council of Scientific and Industrial Research
LRL	Leather Research Laboratory
LSRI	Leather and Shoes Research Institute
MIT	Muzaffarpur Institute of Technology
NIFT	National Institute of Fashion Technology
NILST	Nigeria Institute of Leather and Science Technology
PATK	Politeknik ATK Yogyakarta
PFS	Polimoda Fashion School
SCB	School of Chemical and Bio- Engineering
SDG	Sustainable Development Goals
SDT	School of Design Technology
SFC	Saruwaka Footwear College
SME	Small Medium Enterprise
SUST	Shaanxi University of Science and Technology
UdL	Universitat de Lleida. Campus d'Igualada
UoN	University of Nairobi
WWEH	World Wide Education Hub
WWWP	World Wide Web Portal

1 Introduction

Education has been part of a cultural activity. Education for leather sector is to be viewed as a part of an economic activity. It serves the equity and empowerment needs in certain techno-social environments. Economic viability of the leather and allied sector in the new material world is being restricted to some specific social-economic situations. Manufacturing activities have been shifting from continent to continent during the last 100 years. **Education and Training (E&T) infrastructure and talent supply chain are needed for the human resources, however, not so mobile**. There is a need to map the landscape of educational and training institutes' infrastructure globally and assess its supply and demand gaps of a shifting manufacturing base (Schmél, 2010).

A case is made in this work for creating a sharing and caring learning environment for meeting the needs of global leather and allied product sector by establishing a learning platform for **World Wide Education Hub (WWEH)** under a global collaboration. Social environment, pedagogy and educational practices, information and communication technology are expected to impact the scenario of professional education and training applicable for leather and leather products manufacture. The trade on leather being global, time is ripe for globalizing talent supply chain as well as educational and training infrastructure for higher education for leather and allied sector.

2 Evaluation of Leather Education in Global Platform

Early leather education started way back in 1891 as a training course in British Leather School (BLS). Since then, it has transformed into various programs like B.Sc., M.Sc., B.Tech., M.A., B.Des., M.Des., M.Tech. and Ph.D. in different countries depending on the needs of the local industry. The global portfolio of the leather and leather product courses offered in various countries is shown in Fig. 1.

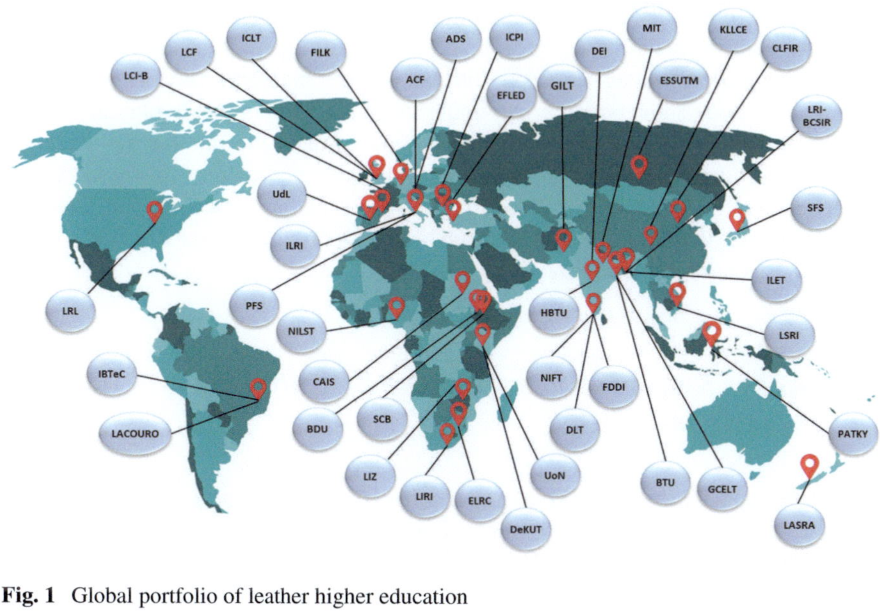

Fig. 1 Global portfolio of leather higher education

3 Context Defines Content in Professional Education

Professional education for any industrial sector is generally integrated into the national educational system. This means personal development of a professional occurs within formal conditions of the local context. Education, by definition, is the preparation of unprepared minds. Training by character is about preparation of underprepared minds to meet specific tasks and activities. Training in leather sector is relevant to critical functions to be performed by employees in leather and/or product manufacturing. Needs for professional education and training are always associated with economic, managerial and technical development. Development and application of professional education and training are associated with four parts of building viz. skill, expertise, professionalism and knowledge base. Skill meets the needs of local context. Knowledge is essential for long term sustainability and global requirements. Professionalism helps in stability. Expertise adds values. In case of leather, a few national landscapes of education and training contribute to all four domains namely skill, expertise, professionalism and knowledge.

4 Global Landscaping of Human Resource Development for Leather and Allied Industry

The industry requires a workforce at different levels of organizational structure. Most countries undertake many Human Resource Development (HRD) initiatives from time to time to cope with technology and market trends. In many countries, colleges and universities are engaged in higher education for leather sector.

Various training schools, research institutions and organizations are engaged in meeting the growing needs of the traditional leather industry. The modernization of the leather sector has become inevitable globally. Modern tools play an essential role in the development of leather education to meet the intellectual needs of the industry and for moving from the traditional artisanal activity to an advanced industrial sector. The status of various higher-level leather education and research institutions in various countries is presented below. Details of the courses and services offered are given in Appendix 1.

4.1 Leather Research Laboratory (LRL), University of Cincinnati, USA

The Tanners Lab, which is now known as Leather Research Laboratory (LRL) housed at the University of Cincinnati is a centre of excellence for testing of leather in United States of America. This was established in the year 1924. The testing facility in LRL is focussed on ISO 17025 certification. It provides customized testing for leather and technology training. With limited manufacturing activity within the USA, the ability of the sector to attract talent for higher education in leather is limited. Progressive weakening of higher educational infrastructure for leather and allied sector is a reality. USA, as a major market for leather and allied products, needs expertise in testing, quality appraisal and intelligent procurement of products. E&T system in USA is currently is so designed to meet the needs of user market rather than intelligent manufacturing.

4.2 Brazillian Institute of Technology for Leather, Footwear and Artifacts (IBTeC), Brazil

Brazilian Institute of Leather, Footwear and Artifacts technology (IBTeC) has been functioning for about 47 years. It is located at the largest Brazilian footwear cluster, Vale do Sinos, Rio Grande do Sul, region. IBTeC has well equipped laboratory testing facility and provides services to fulfil the technical standards and quality of the leather products in Brazil. IBTeC also provides testing facilities for the footwear components, PPE and textile industries. IBTeC also works with technical consultancy to train industries and maintain technological and scientific co-operation agreements with technical schools, universities and development agencies, which makes it possible to align its research with market demands. The mission of IBTeC is to create and apply differentiated technological solutions to increase international competitiveness and the sustainable development of Brazilian industry, especially the leather-footwear system. Brazilian leather and allied industry is strong and enjoys proximity to the major market viz. USA. Hence, Brazil needs an E&T system that meets the end-to-end requirements of the entire sector viz. raw material supply to the production of leather and allied products in environmentally sustainable and globally competitive manner. Brazil is emerging as an important centre for education and training for leather and allied industries in the Americas. It is currently a viable system.

4.3 Laboratory of Leather and Environmental Studies (LACOURO), Brazil

LACOURO is linked to the Department of Chemical Engineering of the Federal University of Rio Grande do Sul and its mission is the study, development, optimization and innovation in the leather production, combined with environmental protection and qualified human resources. LACOURO aims at promoting Leather Science and Technology with the intention of providing benefits and important differentials to ensure quality of raw materials, chemicals, and leather products, to increase the competitiveness of companies emphasizing the development and use of eco-friendly technology. (Thanks Note: Dr. Luis A. Zugno, IULTCS Executive Secretary).

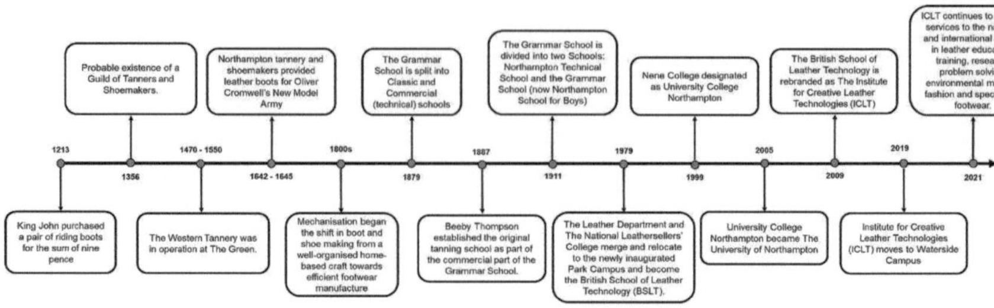

Fig. 2 The evolution of ICLT

4.4 *Institute for Creative Leather Technologies (ICLT), Northampton University, England*

The Institute for Creative Leather Technologies is an educational institute for the development of worldwide leather sector through leather education. The research field of the institute focuses on the automotive, fashion, footwear and allied leather industries. The leather application and research in different fields by understanding the difference in the new and the traditional leather processing is possible. Institute for Creative Leather Technologists, Northampton enjoys a global reputation drawing support from tradition of the past. With the reality of the manufacturing base having shifted out of UK and Europe, most of the users of the educational infrastructure are students and learners emerging from other socio-economics. ICLT is rapidly readjusting to the new norms of the emerging landscape of higher education and training for leather and allied sector. It is able to attract global talent for both learning and teaching. The evolution of ICLT is given in Fig. 2.

4.5 *London College of Fashion (LCF), England*

The MA Footwear course exists within the Craft Program, in the School of Design Technology (SDT) at London College of Fashion. The program also includes MA Artifact, and BA (Hons) Cordwainers Footwear, BA (Hons) Cordwainers Fashion, Bags & Accessories and BA (Hons) Bespoke Tailoring. The SDT also includes MA Fashion Futures, which closely aligns to the MA Footwear course's making and theory elements and provides collaborative opportunities and shared resources. MA Footwear benefits from close links with industry and with other educational institutions and has a strong focus on research. The course is one of the only two MA Footwear courses in the UK and has built a unique core curriculum around conceptual and innovative approaches to footwear, where footwear is defined as any design for the lower extremities. This specialist footwear masters level course gets students to challenge the concept of footwear design and push conventional notions of shoe design. This is an advanced E&T infrastructure that could benefit and form part of the World Wide Web for fashion and footwear education in future (https://www.arts.ac.uk/colleges/london-college-of-fashion).

4.6 LCI Barcelona (LCI-B), Spain

LCI-B has a diverse campus location worldwide, with 23 campuses on 5 continents. The LCI is focused on international collaboration with other institutes and educating over 20,000 learners worldwide each year. Leather fashion and footwear design educational training at LCI-B makes the students shine in the leather market in Spain. With adaptive digital learning options and other technological-based leather, LCI-B education has reached heights in Spain.

4.7 Universitat de Lleida. Campus d'Igualada (UdL), Spain

Since 1957, Igualada has offered technical studies specialized in leather technology. Currently, under Universitat de Lleida, campus Igualada, is still possible to obtain a Master's degree in Leather Engineering and receives students from all over the world. In the same facilities, A3 Leather Innovation Centre offers to tanneries specialized training for companies and doctoral candidates. The Centre focuses on research for leather and new materials, provides technology transfer and it is equipped with a full experimental tanning plant and laboratories for leather chemical and physical tests with UNE ISO / IEC 17,025 Quality Management System (Thanks Note: Dr. Joan Carles Castell, IULTCS).

4.8 Education and Training System in Italy

When Italy enjoyed a dominant position in global leather trade, the E&T system of the country emerged strong on account of SME manufacturing base. With economy of scope benefits and fashion end strong brand strength affording market leadership, the E&T system in Italy has been able to attract talent supply. With changing global trends and with evolving market leadership of Italy, these systems may well be forced into changes by external circumstances. The E&T system enjoys a developed infrastructure for meeting the speciality needs of the fashion and brand products. With other socio-economics in the world, competing for space in the high end and medium price segment, the E&T facilities might well consider embracing a global perspective and embrace global practices in leather education and training in both physical and cyber space.

i. **Italian Leather Research Institute (ILRI), Italy**

Founded in Naples in 1885 by Royal Decree, the Italian Leather Research Institute (SSIP) has gradually broadened its fields of activities, tailored to the needs of the tanning industry. The Institute is currently working on a national and international scale.

ii. **Polimoda Fashion School (PFS), Florence, Italy**

The Master in Shoe Design, operated in partnership with Ferragamo, is a program aimed at developing a professional capable of designing and developing footwear with a strong and innovative creative direction. Starting from the basis of human anatomy, the shoe designer must be a master of both theory and production, with an aptitude for interpreting market needs through the lens of creativity and craft.

iii. **Academy of Costume and Fashion (ACF), Italy**

(Academic Master in Accessories Design) The Master's degree in Accessories Design counts on the participation of leading companies in the field and some of the most important manufacturing companies of Made in Italy leather goods, such as tanneries, heel factory, productions, and laboratories. The Master's will provide students with the conceptual knowledge base and tools necessary to understand new trends in fashion, developing new products, positioning them properly within the market of leather goods, footwear and eyewear, and be able to capture the fashion atmosphere.

iv. **Arsutoria Design School (ADS), Italy**

Since 1947, Arsutoria has been offering its students highly specialized diplomas and training paths, providing them with the skills to express their talent in making and designing shoes and bags all over the world.

4.9 Department of Leather and Leather Processing, Freiberg Institute GmbH Freiberg (FILK), Germany

The institute is an independent, accredited testing laboratory providing a wide range of testing services. These include standardized and normed tests in accordance with German, European and International standardization bodies (DIN, EN, ISO, VDA), technical conditions of delivery of the automotive and furniture industries and other sectors in which flexible polymer materials are used. In addition, there are specific testing and service offers such as analysis accompanying processes or research, proof of usability or functionalities, comparative material tests, pollutant analyses and the examination of the marketability of materials in accordance with legal regulations. Germany is the second largest market for leather and allied products and renowned status as a nation penchant for quality. The E&T system in Germany is designed for testing, evaluation, quality standardization and procurement. The E&T landscape in

Germany suits best for specializing in product quality, clean and lean manufacturing and advanced needs of high end automobile sector.

4.10 Leather Institute German Tanners' School Reutlingen (LGR), Germany

Leather Institute German Tanners' School Reutlingen (LGR) provides the technical management of tomorrow's leather and chemical agent industry with sound knowledge of the theoretical and technological fundamentals of leather manufacture. It enables the graduates to assume responsible, independent functions in the fields of production, development and also application engineering. The practical traineeship in tanning, dyeing and finishing offered in the institute's own teaching tannery is unique anywhere in the world. This application-oriented approach is offered on this scale only by the Lederinstitut Gerberschule Reutlingen.[1]

4.11 Leather and Footwear Research Institute (ICPI), Bucharest, Romania

The Leather and Footwear Institute (ICPI) was established in 1951 in Bucharest under the National Research and Development Institute for Textile and Leather (INCDTP). The institute has a unique facility with sound expertise for the research and development of the leather processing at Romania, where they also focus on footwear manufacturing and related fields. There is a tradition of quality testing and product standardization in this institute.

4.12 Engineering Faculty Leather Engineering Department (EFLED), EGE University, Turkey

Leather and Leather Goods Industry (including; garments, footwear, bags, automotive leathers, upholstery, decorative products, accessories, etc.) is a wide and comprehensive industrial branch. Well educated engineers in this field are in high demand by the industry in Turkey. The Department of Leather Engineering in Ege University, Turkey is one of the few academic institutions in the world giving education in this field. Thus, the graduates have wide job opportunities both in Turkey and all over the world. The department also offers double major/minor programs with the other engineering departments of the faculty that provide an opportunity for multiple diplomas. With the social and personal development activities supported by the department, the students start their career with advantages. Turkey enjoys special

[1] LGR school is no longer in service.

advantages in Europe for supporting the leather and allied sector. It is natural that the E&T system in Turkey enjoys opportunities for a special position in Europe.

4.13 East—Siberia State University of Technology and Management (ESSUTM), Ulan-Ude, Republic of Buryatia, Russia

East Siberia State University of Technology and Management is a leading regional university located in Ulan-Ude (Buryatia, Russian Federation) and it provides leather education through leather and fur technology.

4.14 Education and Training Infrastructure for Leather and Allied Sector in Africa

With Africa emerging as a potential continent of opportunities for leather and allied sector in the latter part of the 21st Century, these institutions could benefit from meaningful collaborations with more advanced E&T infrastructure in Europe and Asia.

i. **College of Applied and Industrial Sciences (CAIS), University of Bahri, Sudan**

The College includes a number of departments in basic and applied sciences. The College offers a five-year academic program leading to the degree of B.Sc. (Honours) and MSc and PhD degrees in various disciplines like textile technology; leather technology; food technology, etc. The college also offers, in collaboration with the Centre for Human Resource Development, a three-year technical diploma of laboratory technology in the fields of biology, biochemistry and chemistry. The policy of the University of Bahri as stipulated in its act has been geared to attain its educational, research and informational goals.

ii. **Nigeria Institute of Leather and Science Technology (NILST), Nigeria**

Provide courses of instructions, training and research in the field of leather and leather product technology and conduct research and development on leather technologies and goods production. It is a centre of excellence in the production of scientists, scientific models and products.

iii. **Bahir Dar University (BDU), Ethiopia**

Cognizant of the fact that the leather sector is not benefiting from the country's strong raw material base and the marketing opportunities given to the developing countries, the Government of the Federal Democratic Republic of Ethiopia (FDRE) established different Institutes with the main objectives of producing trained professionals, conducting research and development activities, rendering consultancy services and providing technical support to the sector.

iv. **School of Chemical and Bio-engineering (SCB), Addis Ababa University, Ethiopia**

Addis Ababa University's mission is to produce competent graduates, provide need-based community service and produce problem-solving research outputs through innovative and creative education, research and consultancy service to foster social and economic development of the country.

v. **Leather Industries Research Institute (LIRI), Rhode University, Grahamstown, Union of South Africa**

Founded in 1904; the University has a well-established reputation for academic excellence. With just over 8200 students, Rhodes is a small University, which enjoys the distinction of having among the best undergraduate pass and graduation rates in South Africa, outstanding postgraduate success rates, and the best research output per academic staff member. This is a testimony to the quality of students that Rhodes attracts and of academic provision and the commitment of Rhodes staff to student development and success. Of the total students, 30% are postgraduates, and over 18% are international students from 54 countries worldwide, making Rhodes a dynamic and cosmopolitan knowledge institute in the offered courses. While Rhode University, South Africa enjoys huge advantages for research leadership in leather sector traditionally from the days of Shuttleworth, it could consider playing a centric role in human resources development of Africa as a whole through some well targeted E&T Programs.

vi. **Exotic Leather Research Centre (ELRC), South Africa**

The Exotic Leather Research Centre (ELRC) approved by the University of Pretoria Senate was established in the year 2013. The ELRC acts as a Technical Collaborator for Exotic Leather in South Africa, which is responsible for developing the best practice industry standards and benchmarks to improve the long term sustainability and global competitiveness of the industry.

vii. **Leather Institute of Zimbabwe (LIZ)**

Leather Institute of Zimbabwe was established on the first of January 1985 by the footwear and leather goods manufacturers of Zimbabwe to address problems relating to shortage of skilled labour, technical and managerial staff and the needs of the industry regarding value addition of hides and skins. LIZ has two laboratories (physical and chemical), a training facility and a standard library with tools that benefit its members. The Institute is in high demand in Zimbabwe and regionally because of its focus on industry needs. Currently, 90% of LIZ is absorbed by the local industry.

viii. **Dedan Kimathi University of Technology (DeKUT), Kenya**

Dedan Kimathi University of Technology is a Chartered Public University of Kenya, which caters the need of human resources requirements at all levels of the country through diploma and master programs in leather technology. The University trains the students in the different aspects of leather processing and product developments thereby enhancing the skilling of students meeting the industry requirements.

ix. **University of Nairobi (UoN), Kenya**

The University of Nairobi at Mombasa, the second largest city in Kenya, is a fast-growing campus with around 2000 students. The University has a strong faculty base with excellent academics in diverse disciplines to train and enrich the students. Kenyan, with 30% GDP of its economy largely dependent on Agriculture, of which 10% is from the livestock sector. Thus the need for better utilization of livestock is significant. The leather education provided at this university helps the country to focus on livestock management and development in technologies suitable for their circumstances to improve the country's GDP.

4.15 Leather and Shoe Research Association (LASRA), New Zealand

Leather and Shoe Research Association is primarily a research association to meet the requirements of leather and allied industries of New Zealand. The research association provides support to the leather industry through modern research and technologies along with the testing services. It provides fundamental education in leather manufacturing and offers a training program.

4.16 Education and Training Landscape in Asian Leather and Allied Sector

Asia is currently the manufacturing base of leather and allied products. The continent enjoys the supply of talent base needed for engaging in the leather and allied sector in countries grouped under the category of Lower Middle income groups or Low income groups. Some of the countries in Asia like India, Pakistan and Bangladesh, benefit also from a developed education and training infrastructure for leather and allied industrial sector, following the legacy of their colonial rulers, UK; who leveraged the supply of raw hides and skins during the late 19th and early twentieth centuries. There are special strengths of the E&T system in Asia arising out of (a) legacy from the 19th—early twentieth centuries, (b) developed educational infrastructure, (c) supply of motivated and talented supply chain of youth engaged in leather sector, (d) developed intellectual infrastructure, (e) strength of manufacturing base, (f) expertise arbitrage for leather and allied sector and (g) good working case studies and models for developing World Wide Web Platform (WWWP) for leather sector.

i. **Government Institute of Leather Technology (GILT), Gujranwala, Pakistan**

This Institute is one of the oldest Institutes serving the export oriented leather sector industries in Pakistan. It was established in 1949 by the Government of Punjab. Till today, the Institute remains the only premier institution for the development of Leather and footwear Technology in Pakistan. However, additional functions such as common facilities, advisory and testing services were introduced later on. Currently, technical services have surpassed education and training activities in the centre.

ii. **Vietnam Leather and Shoes Research Institute (LSRI), Vietnam**

The institute focuses mainly on research, development and application of technologies in the field of leather footwear and environmental protection technology. With the motto of research associated with practice, the application of research results and new products to production has been promoted by the Institute, focusing on the fields of tanning, fashion and environment.

iii. **Institute of Leather Engineering and Technology (ILET), University of Dhaka**

The Institute of Leather Engineering & Technology is an educational institution at Hazaribagh, Dhaka, Bangladesh that provides education related to making leather, leather products and footwear. The Institute offers graduate courses through the University of Dhaka. ILET is an institute of University of Dhaka where admission is given through the admission test.

iv. **Leather Research Institute (LRI), Bangladesh**

Leather Research Institute under the Bangladesh Council of Scientific and Industrial Research (BCSIR) is the only government oriented R&D organization of Bangladesh, playing its role since 2000. The Institute has six research divisions and twelve sections comprising of modern equipment to carry out research and development programs for finding innovation in leather processing technology, developing cost effective chemicals and novel environmental technologies, creative designing of leather products, utilization of slaughter house by-products and tannery wastes. Since its inception, a good number of processes have been developed and are ready for lease out to the entrepreneurs. The Institute also renders technical services to the industrialist to solve their specific industrial problems.

v. **Key Laboratory of Leather Chemistry and Engineering (KLLCE), Sichuan University, China**

The Key Laboratory of Leather Chemistry and Engineering under Ministry of Education, Sichuan University was officially established on August 17, 2000. But the origin of leather education started in China from 1921 through the department of leather engineering of Sichuan University. With surplus investment from industries and government, the institute had achieved the development of basic research in leather research with technological development and achievements in technology transfer. The comprehensive scientific research capabilities have reached the leading level compared with the similar Laboratories at home and abroad.

vi. **China Leather and Footwear Industrial Research Institute (CLFI), China**

China Leather and Footwear Industrial Research Institute (CLFI) is one of the national-level scientific research institutes for the development of the leather and footwear sector. The former China Leather and Shoemaking Research Institute Co., Ltd., now known as China Leather Institute, was founded in 1959. Institute carries out technological research and development and standardization of leather manufacture. The institute also involved in the technical services for import and export of leather and its products under the Ministry of Commerce.

vii. **Shaanxi University of Science and Technology (SUST)**

Shaanxi University of Science and Technology is one of the renowned universities. The university is known for its multidisciplinary field. This University was the first institution of higher learning in light industry.

viii. **Politeknik ATK Yogyakarta (PATK), Indonesia**

Initial leather education started in 1954 to provide the technical resources to the leather industry, which later evolved as a leather high school in the year 1958 which was then structured into Politeknik ATK Yogyakarta, in the year 2015 with three departments; leather processing technology, leather product processing technology and rubber processing technology. Leather processing and leather product processing are diploma programs to meet the human resource requirements for leather processing and product industries.

ix. **Saruwaka Footwear College (SFC), Japan**

Saruwaka Footwear College is based in Asakusa, Tokyo, the centre of the leather shoe industry, and was founded in 1999 by a bespoke shoemaker, Guild, who has been making hand-made shoes using the hand-sewn and welted method for 25 years. The school specializes in hand-made shoes, and has sent out more than 500 graduates since its opening. The name Salwaka comes from the town of Saruwaka, where the school flourished as the centre of theatre during the Edo period.

5 Case Study of India in Education and Training Infrastructure for Leather

It was World War I that was indirectly responsible for setting up schools of tanning for the training of technicians who were essential for the supervision of bulk production of the British army's requirements of leather. The British set up the first tanning school at Madras (present Chennai) in 1914. Later other schools sprang up at Kanpur (1916), Calcutta (present Kolkata) (1919), Jalandhar (1934) and in many places later post-Independence. The beginning of technical education in leather thus corresponds to the period when leather manufacture on an organized industrial scale commenced.

The dawn of Independence saw the launching of the professional degree program in leather technology.

i. **Government College of Engineering and Leather Technology (GCELT), Maulana Abdul Kalam Azad University of Technology, West Bengal**

Government College of Engineering and Leather Technology is one of the pioneering Engineering colleges of our Country established by British Raj in the year 1919. Enabling teaching and research in Leather Technology and Footwear and leather goods making. Providing training and organizing seminars/workshops to disseminate the knowledge to the stake holders. The evolution of various programs in leather technology is shown in Fig. 3.

ii. **Harcourt Butler Technical University (HBTU), Kanpur, UP**

Harcourt Butler Technical University was established in the year 1920. The Department of Leather Technology was established in the year 1978 (then HBTI). Department of Leather Technology at HBTU aims at being a leader of innovation in the area of Leather to produce quality technologists of world standards to deliver the benefits of the developed technologies to the people. The evolution of various programs in leather technology is shown in Fig. 4.

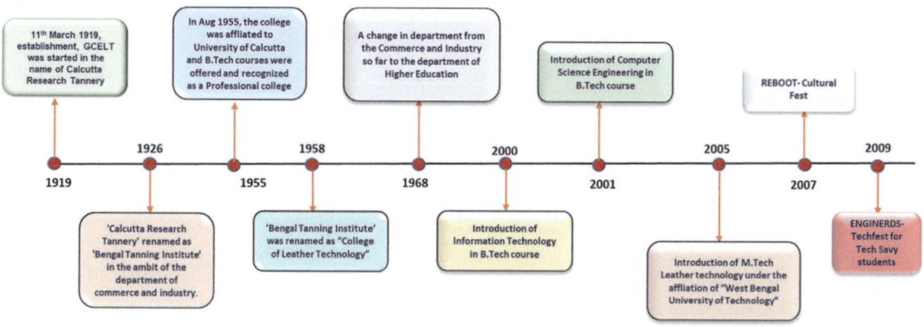

Fig. 3 Milestones of government college of engineering and leather technology

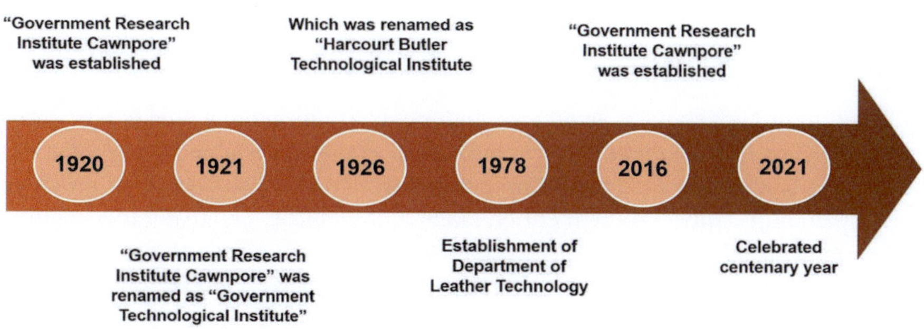

Fig. 4 Milestones of Harcourt butler technical university

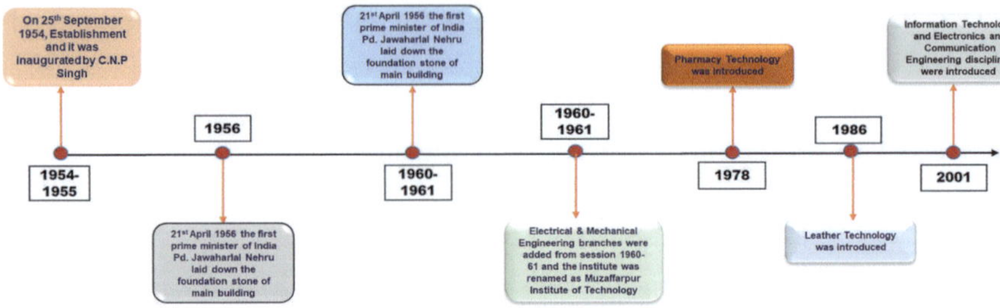

Fig. 5 Milestones of Muzaffarpur institute of technology

iii. **Muzaffarpur Institute of Technology (MIT), Bihar**

Muzaffarpur Institute of Technology came into existence on 25th September 1954. It was inaugurated by Sri C.P.N. Sinha, the then Governor of East Punjab. Leather Technology was introduced in 1986. The evolution of various programs in leather technology is shown in Fig. 5.

iv. **National Institute of Fashion Technology (NIFT)**

NIFT is the pioneering institute of fashion education in the country set up in the year 1986 and has been in the vanguard of providing professional human resource to the textile and apparel industry. It was made a statutory institute in 2006 by an Act of the Indian Parliament with the President of India as 'Visitor' and has full-fledged campuses all across the country. Over the years, NIFT has also been working as a knowledge service provider to the Union and State governments in the area of design development and positioning of handlooms and handicrafts.

v. **Footwear Design and Development Institute (FDDI)**

Footwear Design and Development Institute was established in 1986, under the aegis of Ministry of Commerce and Industry, Govt. of India. FDDI is a premier academic and training institution dedicated to the development and growth of the footwear and products, leather products, retail and fashion merchandise and fashion design sectors. FDDI has Pan India Presence with 12 state-of-the-art campuses spread across the country.

vi. **Bir Tikendrajit University (BTU), Imphal, Manipur**

Bir Tikendrajit University is an educational university in India, which is a new-age University. The University believes in individual empowerment and social progress for training students to face the future competitive world. The education provided at Bir Tikendrajit University will enrich talent and inculcate creative understanding for the students. Bir Tikendrajit University aims at becoming a global institution through innovative programs linking humanities, arts, social sciences, pure sciences and health sciences, which will empower our students and faculty to collectively contribute to the world.

vii. **Dayalbagh Educational Institute (DEI) (Deemed to be University)**

The Department of Footwear Technology supports Under-Graduate, Postgraduate and Doctoral Programs. At the UG and PG levels, a broad-based course structure enables the students to acquire core competence and specialization in the fields of Shoe Designing, Ergonomics of Shoes, Orthopaedic Footwear, Footwear production, testing, fabrication, packaging and marketing. Agra, especially Dayalbagh, has deep roots in the leather and footwear sector. Dayalbagh, as an institution, has been offering a certificate course in Leather and Footwear Technology since 1927. Diploma in leather and footwear technology was started in 2004, and in 2014 a B.Tech. program in footwear technology was introduced.

6 Unique Model for Academy-Research-Industry for Education and Training in Leather: Case Study

Gandhi observed in 1934 that ₹9 crores of raw hides and skins were exported out of India causing not only material drain but also losing an opportunity for creating meaningful employment and contributing to social equity. He called for training and education in leather (Anu, 2021). As early as 1945, the Vice Chancellor of University of Madras, sought the formation of Central Leather Research Institute as a part of National R&D laboratory network. He declared also that such a national laboratory would be part of the University of Madras. He rendered the Department of Leather Technology of University of Madras as a part of CLRI. The institute from its very inception, in 1948 became a part of a vibrant network among the trinity, Academy-Research-Industry. It was established in the then Madras as early as in April 1948. This remains a unique and exemplary model for shaping, education and training infrastructure for the global needs of the future.

The leather technology course was introduced from 1945 in the then University of Madras, Chennai. Anna University, inherited a strong and vibrant educational infrastructure as a part of the Triple Helix in leather right from its inception in 1980. The Council of Scientific and Industrial Research, New Delhi supported the leather technology program and the Department of Leather Technology of A.C. Tech is housed at CLRI since 1948. The evolution of various programs in leather technology is shown in Fig. 6. The Department of Leather Technology offers, B. Tech full time and part time and M. Tech in leather technology and M. Tech footwear science & engineering. The doctoral program offered in leather technology is gaining visibility from the global research domain (Sadulla, 1995).

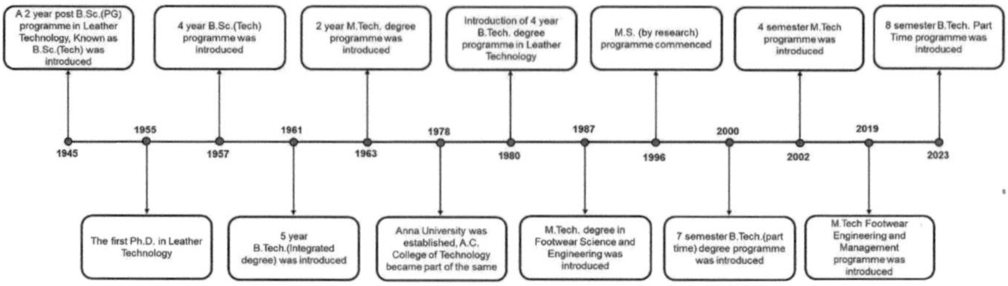

Fig. 6 Milestones of department of leather technology, AC Tech

7 Sustainable Education System

A well-structured education system must have facilities and opportunities for research and innovation. As per the European university association, Sustainable Development Goals (SDG) can be categorized into three broad areas: the well-being of the involved community, environmental protection and National economic development. Academic institutions play a central role in all three areas. A sustainable education system must train its students to bring social cohesion and economic prosperity and ensure environmental protection (Vinuesa et al., 2020).

Technology has broken conventional barriers. The world has become a globally networked community of information assets. The challenges which confront the system of education have evolved rapidly, perhaps too rapidly for our educational system to develop pragmatic solutions to meet them. Infrastructure is one of the key pillars in attaining the sustainability. Industry based education system is a key to reap the success of the training programs. This is primarily achieved through the stakeholder's participation in all the stages of the education programs right from curriculum development to participation in project work and industrial internship (Hassan Mohamed, 2007).

8 Lessons Learnt from Triple Helix of Academy-Research-Industry Model of Chennai School on Leather Domiciled at CLRI

A triple helix model for leather education has been adopted by Chennai Leather School (CLS). The model is depicted in Fig. 7a. The triumvirates created a Trinity partnership model shown in Fig. 7b. A synergizing academy in Anna University and a vibrant, ready-to-absorb leather industry made this trinity research-academy industry a reality.

Chennai Leather School (CLS) seeded in 1948 as a Triple Helix Model has been able to integrate dynamically the changing needs of an industrial sector like leather by adjusting continuously technological and R&D needs. This is achieved by twinning

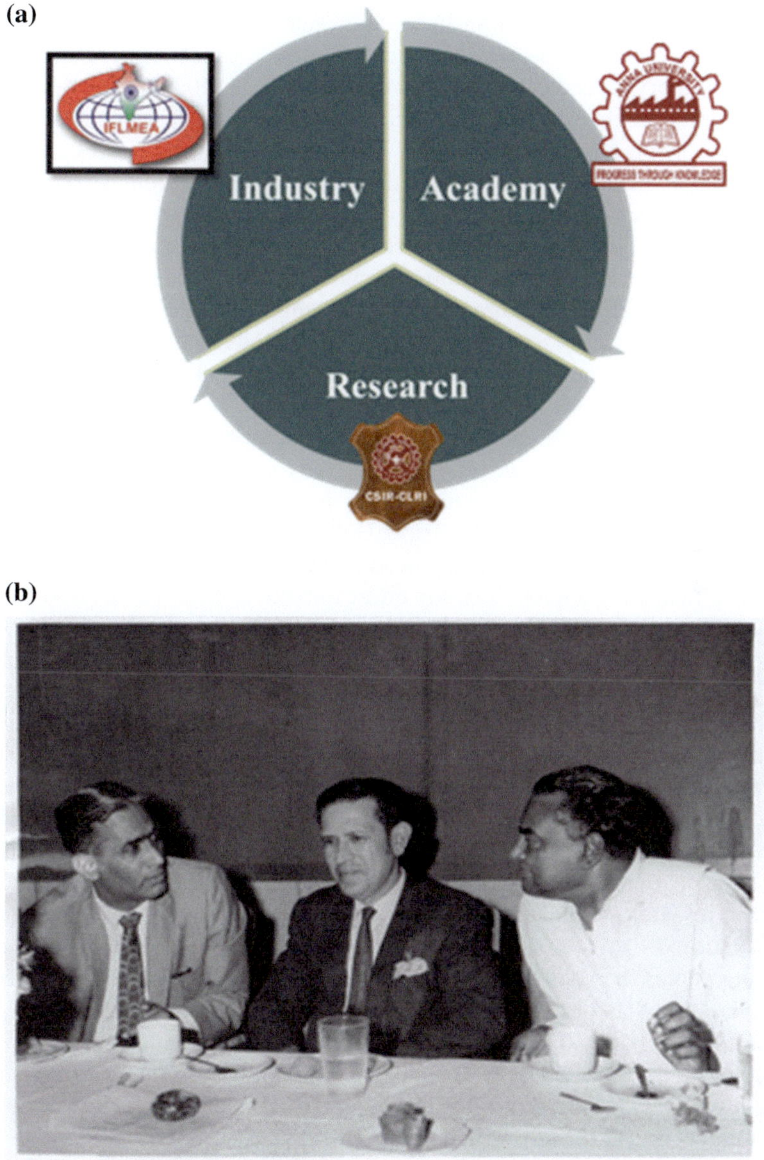

Fig. 7 **a** Trinity model of academy-research-industry. **b** Dr. Y. Nayudamma (Research), Prof. GS Ladha (Academy) and Sri Nagappa (Industry)

under and postgraduate education in the university sector with the largest public funded R&D system of the world namely Central Leather Research Institute, a national laboratory for dedicated R&D for the sector. Today 60% of the Indian Leather Industry in India is manned by the alumni of the University-Research combine. The Triple Helix adapts a relationship rather than a transaction model. It continuously funnels technological developments into the industrial sector through the undergraduate students educated at the Department of Leather Technology of a state university domiciled at the Central Leather Research Institute, a federal R&D laboratory. The continuous modernization of R&D infrastructure benefits also updating of both intellectual and physical infrastructures deployed for undergraduate education. The costs of educating young people are amortized over the various incomes gained by leather research and development of CLRI. Anna University, ranked high in the country and as the leader in engineering education outside of Indian Institute of Technology system is able to attract the best talent for undergraduate education in leather and allied technology. Demand pull for talent for study in leather sector in Tamil Nadu compares well to the information technology in the local context. It is a message in itself. The global footprint of the Triple Helix Model of Chennai Leather School in leather and allied sector is significant and adds merits for wider expansion. It is eminently situated for augmenting any effort to build a World Wide Web platform for higher education for leather for global needs merits study (Ramasami, 2015).

A perceptible change in the quality and quantum of admission in the B.Tech. and M.Tech. programs of Anna University is noticeable. From a mere 10 in the '70s, the B.Tech. admission now stands at 60, with the best of the talents (top 5% of the aspirants for engineering education in Tamil Nadu) opting for Leather Technology. The Triple Helix of Chennai Leather School is currently a vibrant house for meeting R&D needs of the leather and allied sector. The model merits deeper study for adaptation and global adoption (Sadulla, 2022).

The Chennai Leather School of Anna University CLRI combine is one of the most productive research schools in the world with a global share of 14.37% in scholarly journals indexed by Web of Science. Among the top 10 most prolific researchers publishing papers in Web of Science indexed journals in the world, three are from Chennai Leather School (Shanthi & Thanuskodi, 2019).

Anna University CLRI combine has emerged as a dependable partner for developing countries in their intellectual pursuits and training. With its international co-operation and networking strategies, the school is emerging as a global leader in leather engineering education and training. This is illustrated in Fig. 8.

The student's placement record at the Chennai Leather School, is exhilarating. Many have developed into entrepreneurs and technocrats and reached the pinnacle of their careers. Frontline Leaders and Young Entrepreneurs' Resources for Start-Ups (FLYERS) club helps seed start-up culture in the Department of Leather Technology, Anna University, Chennai. The club's objective is to catch and embed the spirit of enterprise. The club emerged as a resource centre for motivated students to join the start-up revolution in leather and leather products.

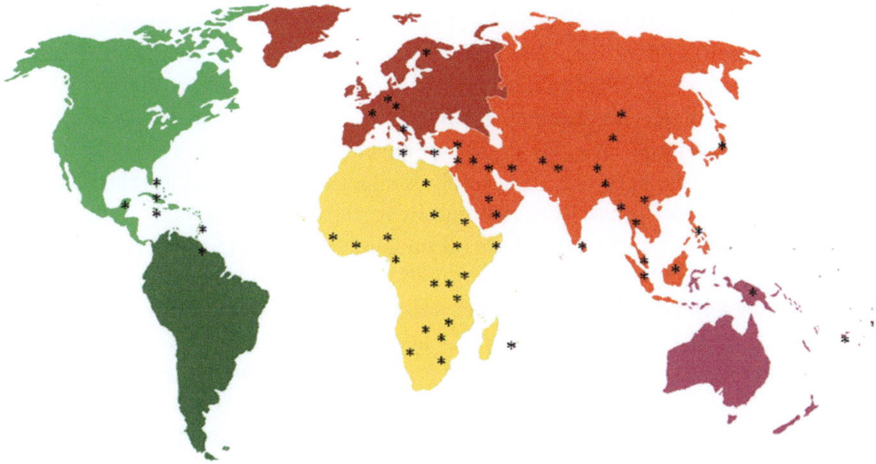

Fig. 8 Countries (*) benefited from training at CSIR-CLRI, Chennai

8.1 Adopting the Imprints from Cambridge Phenomena

Chennai Leather School is promoting the development of an entrepreneurship and innovation ecosystem through the formation of students run Start-up club, Alumni-led entrepreneurship programs and risk financing system, school supported mentoring activities, industry participating in promotional activities by instituting awards and prizes for young entrepreneurs and start-up companies and many more. Currently, the Chennai School is studying the Cambridge Phenomenon more deeply for creating a Chennai Phenomenon in leather and allied product sector in the next five years.

9 An Analysis of Viability of Global Higher Education for Leather and Allied Sector

Global leather and allied sector are challenged by five intrinsic limitations. They are (1) inelastic supply of raw material resources, (2) large mobility of materials across the world contributing to a significant carbon footprint, (3) unbridged demand-supply position in the Education and Training landscape on a global scale, (4) the economy scope model of the Small and medium Enterprise sector not permitting adequate Research and Development efforts and (5) aggressive competitive campaigns of the synthetic product sector in preference to the leather and allied products. Raw material supply position is favourable for high and upper middle income group countries, but the ability to attract local talent for higher education for leather and allied sector is limited. Leather and product manufacturing is most active in Asia. Currently, only some countries in Asia have developed E&T infrastructure for higher education in leather. Chennai Leather School is unique and is able to serve mostly the needs of the local region in Tamil Nadu on account of the mandate of public university under the control of a State Government. Countries in Africa foresee leather and allied sector as an opportunity area for equitable development of their economy and yet they suffer from relatively weak E&T and intellectual and capital infrastructure for serving the needs of Human Resource and Research developments.

Leather and allied sector built on the foundation of SMEs (Small Medium Enterprises) world over is having to compete with aggressive and predatory practices of man-made materials in various product segments on account of their weak R&D backing and ability to invest into R&D and present leather and allied products favourably in terms of its contributions to score Sustainability Development Goals. The man-made material product sector enjoys the benefit of mass scale production, economy of scale, strong R&D back up system, ability to attract talent in high and upper middle income economies and aggressive marketing practices.

Situation is most ripe for the global education and training systems for leather and allied sector to collaborate worldwide and not compete internally to the determinant of the entire sector in global dynamic competition from man-made material sector.

Coronavirus pandemic has taught the world of education and training sector the benefit of online teaching and virtual classrooms. It has become possible for an expert in the North to teach real time learners of the South. Let us face the reality. Manufacturing of leather and allied products would be in the South in future also. The use of high end and imported product segments would in the North. There are countries in the South, which have reached technological developments matching those in the North or even better in leather sector. There is an ideal situation for establishing triangular North-South-South co-operation in Education and Training for leather and allied sector.

10 Call for a Unique North-South-South Co-operative Education for Leather and Allied Sector

The ground reality of global leather sector is such that the main market for leather derived products are in the North and is likely to move further into North in the next 20 years. Education and Training infrastructure for leather and allied sector in the North which has the expertise and professionalism may remain handicapped by the limited ability of local manufacturing to attract adequate supply of talents. Talent supply pipeline challenges are likely to manifest in the North.

There are a large number of countries in the South which plan their equitable developments of reaching sustainable development goals through the leather sector by creation of employment, women empowerment and addition of values to their raw materials. The local education and training infrastructure may be handicapped by the inadequacy of both physical and intellectual as well as expertise infrastructure.

There are some countries in the South which show cultural similarities with the nations in the South coping to develop the leather sector and professional and technical strength with some leaders in the North which languish by talent supply chain problems.

A country like India with a successful Chennai Leather School is in an ideal position to catalyze the formation of noble North-South-South co-operation in Education and Training for leather and allied sector and participate enthusiastically in sharing the lessons learnt from the utility of their unique Triple Helix Model.

Time is ripe for seeking collaborative excellence in education, training and consultation for leather and allied sector. Internal competitions within the sector are likely to favour the growth of synthetic product sector at the cost of leather sector as a whole. Hence, this call for a unique North-South-South co-operation in E&T for leather for collective global good.

To enhance the teaching-learning experience, various pedagogies, such as Project-Based Learning, Activity-Based Learning, Self-Learning, Industry Internships, Project Work etc., should be utilized. Industry Internships and Project Works in different countries, too, as international exposure is about getting quality education in specific fields, adapting to different mindsets, and broadening students' horizons. Paradigm shifts in leather education to prepare well-educated and employable engineers and technologists and instil creativity and passion for innovation and entrepreneurship in them.

11 Summary

An attempt has been made in this work to enumerate E&T related activities in the overall landscape of global leather sector. Several intrinsic features of the global leather and allied sector seem to be rendering progressively the current Education and Training models of the world unviable. Unviability of the E&T system in national economics is likely to elicit huge negative impacts on the health of global leather and allied sector. While manufacturing of leather and allied products in SME sector may help high economy of scope model for fashion oriented market, economy of scale conditions would apply for supporting education, training, consultation and R&D infrastructures with respect to both physical and intellectual capital. An attempt has been made to propose a unique North-South-South co-operative model in higher education and advanced training for leather and allied sector built on the grammar of collaborative excellence and co-operative endeavour among similarly placed subsystem. Relative success of Chennai Leather School in a limited context presents a real-life experiment that merits close study, scrutiny, adaptation and careful adoption on a global scale. This is a call for alignments of the global education community to unite for the cause of a sustainable global leather sector.

Appendix 1: Geographical Distribution of Various Leather Programs

S.no.	Institute name	Establishment	Courses	Services
1	Leather Research Laboratory (LRI), University of Cincinnati, USA	1924	• Leather Orientation Course	Physical and Chemical testing of leathers
2	Brazilian Institute of Technology for Leather Footwear and Artifacts (IBTeC), Brazil	1976	• Research Program	Technical Scientific Services for the leather-footwear, components, PPE and textile industries
3	Laboratory of Leather and Environmental Studies (LACOURO), Porto Alegre, Brazil	2001	• Chemical Engineering	Physical and Chemical testing of leather and degradation studies
4	Institute for Creative Leather Technologies (ICLT), Northampton University, England	1979	• B.Sc. and M.Sc. in Leather Technology • Diploma in the higher education in leather technology • Certificate of higher education in leather technology • M.Phil. and Ph.D. of Leather Technology	Research in Scientific and Technological needs of the automotive, fashion, footwear and allied leather industries
5	London College of Fashion (LCF), England	1906	• MA Footwear • MA Artifact • BA (Hons) Cordwainers Footwear • BA (Hons) Cordwainers Fashion, Bags & Accessories • BA (Hons) Bespoke Tailoring	Courses in all things fashion, from business to design and fashion curation. With over 60 undergraduate and postgraduate degrees, and 165 short courses

(continued)

(continued)

S.no.	Institute name	Establishment	Courses	Services
6	LCI-Barcelona, Spain	1959	• Bachelor's Degree in Fashion Design (English or Spanish) • Postgraduate in Footwear Design (in Spanish only) • Master in Design of Footwear, Bags and leather goods collections (in Spanish only)	• Certificates of Higher Education (CFGS) • Official, Dual Degree and Private Master's and Postgraduate Degrees • Specialization Courses • Pattern Making Courses • Summer Courses
7	Universitat de Lleida. Campus d'Igualada (UdL), Spain	1957	• Master in Leather Engineering	Training: Company training, Doctoral candidate training Research: New technology and materials in the leather sector Technology transfer Experimental tanning plant services Laboratory analysis quality assurance
8	Italian Leather Research Institute (ILRI), Italy	1885	• Higher Technical Diploma with the certification of the skills corresponding to the V level of the European Qualification Framework	Polytechnic of leather, R&D activities, training, product and process certification, analyses, consultancy Services
9	Polimoda Fashion School (PFS), Florence, Italy	1986	• The Master in Shoe Design	Designing and developing footwear with a strong and innovative creative direction
10	Academy of Costume and Fashion (ACF), Italy	1964	• Master in Accessories Design	Provide students with the conceptual knowledge base and tools necessary to understand new trends in fashion, developing new products, positioning them properly within the market of leather goods, footwear and eyewear

(continued)

(continued)

S.no.	Institute name	Establishment	Courses	Services
11	Arsutoria Design School, (ADS) Italy	1947	• Highly specialized diplomas making and designing shoes and bags	Specialized skill programs
12	Department of Leather and Leather Processing, (FILK) Freiberg Institute GmbH Freiberg, Germany	1897	• Internship/Ph.D. Programs • Lecture at Technische Universität Bergakademie Freiberg • Lecture at University of the Arts in Bern	Testing of polymer materials, Research and development of Projects, professional training
13	Leather and Footwear Research Institute (ICPI), Bucharest, Romania	1951	• Mentoring in European Project Semester • Research mentorship for Masters' studies within ERASMUS program for Ege University, Izmir, Turkey; • Postdoc research internships for students from University of Salerno, Italy	Developing new chemical auxiliaries, green technologies or for recovery of collagen and keratin based by-products
14	Engineering Faculty Leather Engineering Department (EFLED), EGE University, Turkey	1960	• B.Sc. and M.Sc. of Leather Engineering • Ph.D. of Leather Engineering	Research and development of leather and leather products

(continued)

(continued)

S.no.	Institute name	Establishment	Courses	Services
15	East—Siberia State University of Technology and Management (ESSUTM), Ulan-Ude, Republic of Buryatia, Russia	1962	• Bachelor Degree in Environmental Engineering and Water Consumption • Bachelor Degree in Technology of Light Industry Products	Leather and Fur Technology. Water Resources and Commodity Research of Non-Food Goods
16	College of Applied and Industrial Sciences (CAIS),University of Bahri, Sudan	2011	• B.Sc. of Leather Technology • M.Sc. of Leather Technology • Ph.D. of Leather Technology	It is an educational University. It's act has been geared to attain its educational, research and informational goals to empower
17	Nigeria Institute of leather and Science Technology (NILST), Nigeria	1964	• Higher diploma and Diploma in Footwear Technology • Higher diploma and Diploma in Footwear Design & Innovation • Higher National Diploma and National Diploma in Leather Technology • Advanced certificate and certificate program in Leather Manufacturing • Advanced certificate and certificate program in hides and skin Improvement	The institute focuses on training and research in the field of leather and leather products technology

(continued)

(continued)

S.no.	Institute name	Establishment	Courses	Services
18	Bahir Dar University (BDU), Ethiopia	2000	• B.Sc. in Leather Product Engineering • M.Sc. in Leather Product Design & Engineering • Ph.D. programs through Leather Processing and Product Research and Innovation Centre	It is an educational university, producing trained professionals, conducting research and development activities, rendering consultancy services and providing technical support to the sector
19	School of Chemical and Bio-Engineering (SCB), Addis Ababa University, Ethiopia	1960	• B.Sc. in leather technology • M.Sc. in leather technology • Ph.D. programs in leather technology	The major activities are research, community services and teaching–learning
20	Leather Industries Research Institute (LIRI), Rhode University. Grahamstown, Union of South Africa	1904	Leather Program • M.Sc. • Ph.D.	The institute providing professional education and training
21	Exotic Leather Research Centre (ELRC), South Africa	2013	Leather Program • Master Degree • Ph.D. • Post-Doctoral Program	Training, research and diagnostic resource for the crocodile and ostrich industry nationally and Africa

(continued)

(continued)

S.no.	Institute name	Establishment	Courses	Services
22	Leather Institute of Zimbabwe (LIZ), Zimbabwe	1985	National Diploma-3 years • Leather Manufacture Technology • Footwear Manufacture Technology • Leather Goods Manufacture Higher National Diploma-1 Year • Leather Manufacture Technology • Footwear Manufacture Technology • Leather Goods Manufacture	Operative Course (240 h) • Basic Leather Manufacture • Basic Footwear manufacture • Basic Leather Goods manufacture • Basic Shoe Design • Stitching Operative Course • Hides Improvement
23	Dedan Kimathi University of Technology (DeKUT), Kenya	2012	• Diploma in Leather technology • B.Sc. in Leather technology • M.Sc. in Leather technology	Training and research
24	University of Nairobi (UoN), Kenya	1970	• B.Sc. in Leather Science and Technology	Training and research
25	Leather and Shoe Research Association (LASRA), New Zealand	1928	• Leather Manufacturing and Training Program	The institute aims to provide research facilities for innovations in leather/collagen sciences with a view of improving the New Zealand leather manufacturing industry. It also provides testing for industries and training programs on leather manufacturing for students

(continued)

(continued)

S.no.	Institute name	Establishment	Courses	Services
26	Govt. Institute of Leather Technology (GILT), Gujranwala, Pakistan	1949	• DAE Leather Engineering Technology • DAE Footwear Engineering Technology • Certificate courses in Auto Cade, Computer Applications, Leather Garments Stitching, Shoe Upper Stitching, Leather Manufacturing	DAE Degree programs in Leather and Footwear Technologies and short time certificate courses on various skills aligning with the leather and footwear industries
27	Vietnam Leather and Shoes Research Institute (LSRI), Vietnam	1973	• Research and Training Programs	Training and research

(continued)

(continued)

S.no.	Institute name	Establishment	Courses	Services
28	Institute of Leather Engineering and Technology (ILET), University of Dhaka, Bangladesh	1949	• B.Sc. in Leather Engineering • B.Sc. in Footwear Engineering • B.Sc. in Leather Product Engineering • B.Sc. Engineering in Leather Engineering • B.Sc. Engineering in Leather Product Engineering • B.Sc. Engineering in Footwear Engineering • M.Sc. in Leather Engineering • M.Sc. Engineering in Leather Engineering • M.Sc. Engineering in Leather Product Engineering • M.Sc. Engineering in Footwear Engineering	The institute aims to initiate, organize and undertake research in the field of Leather Engineering and Technology, provide quality control and laboratory facilities for testing of leather and allied materials and leather products
29	Leather Research Institute (LRI), BCSIR, Bangladesh	2000	• Leather Research Program	Physical and Chemical Testing

(continued)

(continued)

S.no.	Institute name	Establishment	Courses	Services
30	Key Laboratory of Leather Chemistry and Engineering (KLLCE), Sichuan University, China	2000	• UG in Leather Chemistry and Engineering • UG in Design of Leather Goods • UG in Leather commerce and trade • PG in Leather Chemistry and Engineering	The Institute provides laboratory facilities for testing in the areas of Chemistry of tannage; Protein chemistry and histology of the skin; Method, theory of tannage and clean leather-making technology; Utilization of leather waste resources; Leather chemicals; Modern leather analysis and testing techniques etc
31	China Leather and Footwear Industrial Research Institute (CLFI), China	1959	• Leather Research and Training Program	The institute aims to provide research facilities for various aspects of leather and footwear industries
32	Shaanxi University of Science and Technology (SUST), China	1958	• UG in Light Industry and Chemical Engineering (Leather Engineering) • PG in Light Industry Technology and Engineering • Ph.D. in Light Industry Technology and Engineering	Shaanxi University of Science and Technology established as Beijing institute of light industry to promote the development of china's light industry
33	Politeknik ATK Yogyakarta (PATK), Indonesia	1954	• Polytechnic in Leather Treatment Technology • Polytechnic in Leather Product Processing Technology • Polytechnic in Rubber and Plastic Processing Technology	The institute aims to provide leather education and training in accordance with the needs of the leather industries. It also provides research facilities in improvement of machineries and equipments
34	Saruwaka Footwear College (SFC), Japan	1999	• Training in hand-made shoes	Specialized in hand-made shoes using the hand- sewn and welted method

(continued)

(continued)

S.no.	Institute name	Establishment	Courses	Services
35	Government College of Engineering and Leather Technology (GCELT), West Bengal, India	1919	• B.Tech. in Leather Technology • M.Tech. in Leather Technology	The institute provides programs in various fields including leather, computer science and information technology. It also provides advanced certificate course in Boot, Shoes and Leather Goods manufacture
36	Harcourt Butler Technical University (HBTU), Kanpur, Uttar Pradesh, India	1978	• B.Tech. in Chemical Technology (Leather Technology)	It is an educational university which produces a good amount of leather technologies every year with a greater knowledge in leather science and latest technologies
37	Muzaffarpur Institute of Technology (MIT), Bihar, India	1954	• B.Tech. in Leather Technology	This educational institute strives to emerge as a national leader in graduate level studies in all sub areas of leather field and to make significant contribution to the development of the society, industry, nation and the world
38	National Institute of Fashion Technology (NIFT), India	2010	• B.Des. in Leather Design	NIFT, a fashion oriented educational institute offers a learning experience of the highest standards in fashion pertaining to design, technology and management
39	Footwear Design and Development Institute (FDDI), India	1986	• B.Des. in Footwear Design & Production • M.Des. in Footwear Design & Production • Bachelor of Design (B.Des.) in Leather, Lifestyle & Product Design	FDDI provides services such as testing services, design services and consultancy & project management services

(continued)

(continued)

S.no.	Institute name	Establishment	Courses	Services
40	Bir Tikendrajit University (BTU), Imphal, Manipur, India	2020	• B.Tech. in Leather Technology & Cosmetology	Engineering's field of leather technology focuses on the synthesis, manufacture, and refinement of leather for practical application. It also covers the creation of synthetic leather and how effectively commercial products may be made using it
41	Dayalbagh Educational Institute (DEI), Agra, Uttar Pradesh, India	1950	• B.Tech. in Footwear Technology • M.Tech. in Footwear Technology • Ph.D. in Footwear Technology	The institute facilitates collaborative research activities
42	Department of Leather Technology (DLT), Anna University Chennai, India	1945	• B.Tech. in Leather Technology • M.Tech. in Leather Technology • M.Tech. in Footwear Engineering and Management • Ms. By Research in Leather and allied science • Ph.D. in Leather and allied Science	The department of Anna University aspires to ensure high quality education in leather and leather products manufacture through the academic collaboration with CSIR-Central Leather Research Institute

References

Anu, B. (2021). *Bahuroopee Gandhi*. Prabhat Prakashan, India.

Hassan Mohamed, H. A. (2007). *Promoting South-South and North-South cooperation in education and research: A question of responsibility. globalization and education: Proceedings of the joint working group, the pontifical academy of sciences, the pontifical academy of social sciences,* 19–17 November 2005, Casino Pio IV, edited by Marcelo Sánchez Sorondo, Edmond Malinvaud and Pierre Léna, De Gruyter (pp.79–90).

Ramasami, T. (2015). *Education in leather science and technology: way to discoveries and solutions.* ALFA Compendium (pp. 5–18).

Sadulla, S. (1995). *HRD-Education & training, Kothari's Deskbook on leather industry*, SC Kothari Group, Chennai (pp. 247–265).
Sadulla, S. (2022). *Repositioning leather-professional education in the context of sustainable development of global leather industry.*
Schmél, F. (2010). *Professional education and training in the leather-based industries.*, UNIDO.
Shanthi, B., & Thanuskodi, S. (2019). Leather technology research output: A scientometric analysis on web of science database (2009–2018). *Library Philosophy and Practice (e-journal)*, 2940.
Vinuesa, R., Azizpour, H., Leite, I., et al. (2020). The role of artificial intelligence in achieving the sustainable development goals. *Nature Communications, 11*, 233.

Country Status and Case Studies

Emerging Scope for Vegetable Tanning in Eco-Leather: A Case of Kenya

Arthur Onyuka and Alvin Asava Sasia

1 Introduction

1.1 Overview and Structure of the Kenyan Leather Industry

Kenya is endowed with an abundant natural resource of livestock and is the third largest holder of livestock in Africa. The raw material base consists of a livestock population of about 18.8 million cattle, 19 million sheep, 27 million goats, 2 million donkeys and 3 million camels (Kenya National Bureau of Statistics [KNBS], 2019). There is also an emerging new source of supply of raw materials consisting of exotic livestock (ostrich, crocodiles and other reptiles) and fish. Despite this resource, hides and skins value addition has been relatively low. This situation led to the blooming of the hides and skins export market. However, over the last decade, the government through policy intervention sought to reverse the negative trend by the imposition of punitive tariffs (80% tax) on exports of raw hides and skins (Mwinyihija, 2014b; Wayua & Kagunyu, 2012). The rationale for this intervention is to encourage local manufacturing and increase the capacity of domestic manufacturers. Overall, the hide, skins and leather industry is estimated to contribute about 4% to the country's agricultural GDP with a net contribution of about 1.5% to the national GDP (Mwinyihija, 2014a).

Even though value addition on raw hides and skins has been minimal, the leather industry in Kenya is considered one of the oldest manufacturing sectors, dating back to the colonial period. Its transformation has undergone several metamorphosis and upheavals along its development. Currently, the sector is estimated to be worth $500 million annually, and contributes close to 3.5% to the African continent's production output. Over 90% of Kenyan exports worth $140 million are predominantly wet blue leather and accounts for about 0.14% of the world leather exports. The Kenyan leather

A. Onyuka (✉) · A. A. Sasia
Kenya Industrial Research and Development Institute, P.O. Box 30650-00100, Nairobi, Kenya
e-mail: arthuronyuka@hotmail.com

S. K. Janardhanan and L. A. Zugno (eds.), *Emerging Trends in Leather Science and Technology*, https://doi.org/10.1007/978-981-99-9754-1_14

industry consists of about 14 operational tanneries, which are mainly family-owned micro, small and medium-sized enterprises (World Bank Group, 2015). As a key agricultural sub sector, the industry has been identified for commodity development with high potential to impact rural development, employment and wealth creation.

Despite its potential, the Kenyan leather industry has not been fully exploited for productive economic gain and remains uncompetitive in the global market. According to a study by the World Bank Group (2015), key challenges identified as being responsible for the slow growth of the sector include: poor husbandry practices, poor quality of raw materials, obsolete manufacturing technologies, lack of skilled manpower, weak research and development (R&D) initiative, poor systems for pollution control and waste management, and introduction of cheap used shoes and leather articles into the local market. Nonetheless, interventions through policy frameworks have been enlisted to rejuvenate the leather sector in Kenya. The legal and policy frameworks for the transformation of the industry have been enacted. Requisite infrastructure and facility development are in place.

1.2 Structure of the Kenyan Leather Sector

Broadly, the industry has been divided into the formal and informal sectors, coordinated through a weak and unbalanced supply link. At the lower lever, there is micro and small tanneries which engage in subcontract tanning, supplying mainly wet blues to the large-scale tanneries. The annual production of leather in Kenya is mainly made up of semi-processed tanned wet blue and crust leather with about 2% being finished leather. The finished leather finds many applications in the downstream sectors of the consumer product industry such as in the production of footwear, belts, bags, garments, upholstery and gloves. Close to about 80% of the country's leather production is destined for export as semi-processed material consisting largely of wet blues. The major destinations of the country's exports are China, India, Vietnam, Korea and Japan (World Bank Group, 2015); supporting the generation of foreign exchange and attracting foreign direct investments into the country.

The biggest proportion of output of finished leather in Kenya is vegetable re-tanned leather. The largest domain for end-use applications for this leather is the belts, hand bags, travel ware and cases. The market for these products has been on the rise. This has been attributed to the growth of the Kenyan middle class, whose appetite and demand for high quality leather products has been elicited, presenting an opportunity for scaling production for the domestic market. On the other hand, the leather hand bags sub-sector is the country's most competitive *vis-à-vis* global markets (World Bank Group, 2015). Whereas the sub-sector has attracted the interests of both high end tanneries and mid-size companies, there is an ever-growing number of local producers that operate informally who are making a mark in this segment. Their production of handmade, African distinctive designs is growing in the safari tourism niche. The country seeks to build its reputation by improving the quality of the leather articles and establishing the 'Made in Kenya' brand distinction (World Bank Group, 2015).

1.3 Socio-economic Impact of the Leather Industry in Kenya

The growth and expansion of Kenya's leather manufacturing sector have come with positive impacts on the country's socio-economic development. The industry being a labour-intensive enterprise, accounts for significant numbers of employment in the agricultural sector by offering employment to about 14,000 people. Of this, 5,000 people are employed directly in the formal sector, while the informal sector accounts for the balance of the employment numbers. This number is expected to rise higher considering the emphasis the Kenyan government has put in the manufacturing sector through the 'Big Four' development agenda. Further, the increase in employment is expected to continue, especially with the trend in world leather production indicating a shift from western to Asia, and recently into Africa because of less stricter environmental regulations and lower labour costs (Mwinyihija, 2015).

The leather industry as an industrial sector is highly regarded as a driver of industrial development and economic growth. Its potential for commodity development is the critical ingredient for impacting rural development, employment and wealth creation. Consequently, the sector has been designated as a priority sector under the manufacturing segment of the country's economic blue print (Onyango et al., 2019). Through manufacturing, the country seeks to increase the proportion of locally made products. A focus on the local production of footwear, belts, bags, garments, upholstery and gloves plays a vital role in generating wealth and improving the well-being of the people. Currently, the earnings of local dealers in commodities reached an estimated $20 million (Ogolla & Wanjau, 2013). It is expected that both vertical and horizontal growth of the sector will thrive with the proliferation of the leather value-chain actors.

1.4 Leather Processing

Leather processing involves the conversion of hides and skins, which are putrescible outer covering of animals, to stable non-putrescible forms which can be utilized for making footwear, leather goods, garments and apparels. The processes involved in the transformation of hides and skins into leather are generally categorized into three: beam house, tanning and the finishing operations. The tanning operation is the most vital and fundamental phase in the transformation. It transforms permanently and stabilizes the collagen protein against degradation by heat and thermo-mechanical stress. Thus, it improves leather's innate properties by imparting definite physical, chemical and biological properties (Sivaram & Barik, 2018). Generally, two types of tannages are identifiable, broadly categorized as inorganic and organic tannages.

Inorganic tanning refers to the use of mineral salts as primary tanning agents. The important mineral salts are basic chromium (III), aluminum (III), zirconium (IV), titanium (IV), silica and both cations of iron (II & III). Among these mineral tannages, chrome tanning accounts for more than 80% of global tannage. This is due to the unique and excellent physical and chemical properties that it imparts to the resultant leather. Moreover, chrome tanning confers superior handling quality, hydrothermal stability and excellent use performance characteristics. Commercially available as basic chromium sulphate (BCS), chrome is a strong tanning agent. It easily binds onto the collagen, forming coordinate interactions between carboxyl sites of the collagen. The tanning by this method is fast, thus reducing the turnaround time for the production of leathers. However, chromium needs to be properly managed in the tannery regarding liquid and solid wastes, chrome-tanned leather shavings, trimmings, buffing dust and chromium sludge. Today, there are complementary technologies available for the tanneries to work with chromium salts and comply with national, international regulations and Restricted Substances Lists (RSL).

With the rising environmental regulations, the Kenyan leather industry has been put into sharp focus. The enforcement of the domestic environmental regulations by the Kenya National Environment and Management Authority (NEMA), has led to the closure of some tanneries for non-compliance. It has been established that the cost of non-compliance for small firms in developing countries is the major barrier to trade and remaining competitive in the face of environmental sanctions. In this respect, the shift to organic tanning materials represents one of the commendable strategies for the leather industry to remain economically viable. On the other hand, organic tanning implies the use of tanning materials such as vegetable tannins, aldehydes, oil and synthetic tanning materials as the primary tannage. The application of these tannages continues to gain popularity due to their considered ecological benefits.

1.5 Principles of Vegetable Tanning

Vegetable tanning involves the use of vegetable tanning materials (more commonly referred to as tannins) as the primary tannage in an aqueous infusion with the pelt for a considerable amount of time. Tannins are so called because of their property to bind and precipitate proteins. Vegetable tannins are extracts from the leaves, nuts, bark and wood of various plant sources that contain commercially viable concentrations. Extraction is by means of leaching with water (hot water preferably) yielding mixtures of polyphenolic compounds with molecular weight in the range of 500–20,000 Daltons (Bule et al., 2020; Falcão & Araújo, 2018). The extracts have three fractions separated with regard to the molar mass. The different molar weight fractions play significantly different roles in the reactions with collagen.

The low molecular weight fractions below 500 Daltons are termed as non-tans. They lack astringency and have very low tanning power. Their presence contributes to the solubility of the tans and hence the penetration into the pelt. The medium molecular weight fraction between 500 and 3000 Daltons is designated as tans and is the fraction responsible for the tanning function. The phenolic hydroxyl content determines the astringency as well as the tanning function. Tannins bearing high molecular weight fraction, above 3000 Daltons are termed as gums. They consist of polyphenols complex with some carbohydrate species and thus their high molar mass. Gums are undesirable since their presence not only impedes penetration into the pelt, causing surface reaction but also affect the quality of leather.

Vegetable tannins are broadly categorized as hydrolysable and condensed tannins based on their structural characteristics. Hydrolysable tannins are compounds with a saccharide-based core (of glucose or its derivatives) whose hydroxyl groups are esterified by the carboxylate species on the pyrogallol group. Based on their further classification as gallotannins or ellagitannins, their properties are dependent in part on the ease of hydrolysis of the ester bonds and the reactivity of the phenolic hydroxyl groups. Gallotannins feature a glucose core which is (i) esterified only with gallic acid and (ii) has presence of other depside linked galloyl units. Depside esterification refers to the intra esterification of the gallate groups by the phenolic hydroxyls of the same molecule (Jourdes et al., 2013). Ellagitannins are characterized by the presence of biarylic units of hexahydroxydiphenoyl (HHDP) which confers its unique structural identity. They also display pronounced structural variability and have strong tendency to aggregate forming dimeric and oligomeric derivatives (Jourdes et al., 2013).

Condensed tannins also known as proanthocyanidins are flavonoid oligomers whose monomeric units are connected through C–C and C–O–C bonds (Bule et al., 2020). The flavonoid ring structure is made up of both aromatic and alicyclic ring structures with varying patterns and positions of hydroxyls. The monomeric units in the ring structure are stereoisomers that link through polymerization at the alicyclic ring, consequently, different conformations are formed which are either linear or branched polyphenols. Small variations in fundamental substitution patterns give rise to various subgroups. Generally, the main subclasses are flavonols, flavones, flava-3-ols, isoflavones, chalcones, flavanones, catechins, leucoanthocyanidins, and anthocyanidins. The condensed tannins are ubiquitous in nature and exhibit strong interactions with proteins and metal ions owing to the presence of several hydroxyl groups in their structure (Bule et al., 2020).

The primary objective of the protein–tannin interaction is cross-linking reaction of the protein with the purpose of conferring stability. These interactions are mainly through hydrophobic and hydrogen bonding (Bule et al., 2020). Though other mechanisms such as covalent bonding have been studied (Covington, 2009, pp. 310). At the onset, the dominant interactions are that of hydrophobic bonding, since the tannin molecules are relatively insoluble in water. The reactions proceed to hydrogen bonding. Both condensed and hydrolyzable tannins form hydrogen bonds. The presence of reactive hydroxyl group on the polyphenols induces reaction with the peptide oxygen radical on the collagen. The basic amino and carboxyl groups of the polypeptide act as active binding sites for the vegetable tannin fixation. In addition, the flavonoid macromolecule of the condensed tannins undergoes oxidization, giving quinones that bind covalently with the amino groups of the collagen.

Several factors influence the interactions of tannins with collagens. Tannin molecule penetration should precede any interactions, to avoid surface fixation. The size, stereochemistry, type of the protein, and appropriate medium for interaction (such as pH and ionic strength) have all been explored in relation to the interaction between tannins and proteins. Other than the mechanism of interactions, it has also been shown that the quantity of bound tannins and the quantity of repeating amino acid sequences both affect the stability of the tannin-protein complex. The tannin-protein interactions are both dynamic and time dependent. Complimentary factors such as conformational flexibility of both the tannins and protein increase the amount and strength of interaction (Niaz & Khan, 2020). From the chemistry of the tannins, research has shown that the aromatic rings and phenolic groups which act as multi-dentate ligands on the protein surface and the increase in the number of polyphenols galloyl groups have positive effects on the efficacy of the protein binding (Kuria et al., 2016; Ncube & Van Staden, 2015).

Vegetable tannins in leather processing are utilized as either primary tannage or as a retanning agent. Their introduction in leather processing is meant to achieve product-specific characteristics. Ideally, vegetable tannins are incorporated to improve the feel and handle of the leather. Because of their high molar mass, tannins confer a filling effect on the looser and softer parts of the leather. Filling produces leathers with more enhanced physical properties and a uniform cutting area. Moreover, vegetable tannins are the preferred material in the production of corrected grain leathers. They also impact the resistance of leather to both alkali and perspiration. As a retanning agent, vegetable tannins improve the wetting back property which is essential in the subsequent processes such as dyeing. Based on the plant sources, vegetable tannins are known to confer unique attributes to leather. They produce compact, full and easily embossable leather which possess better water permeability, stability, strength and molding properties.

1.6 Historical Perspective of Vegetable Tanning in Kenya

There is no notable literature on vegetable tanning in Kenya. However, the history of vegetable tannins in Kenya dates back to colonial times with geopolitical connections and varying commercial interests from abroad. At the heart of the Kenyan tannin industry is the acacia wattle bark, which has dominated both production and use in the local tanning industry. The acacia wattle tree was introduced in the country in 1903 by the colonial white settlers. The objective was the production of fuel for the Kenyan and Ugandan Railway. Its use was found unsatisfactory for this purpose opening up the opportunity for the sale of Acacia barks (Leckie, 1938; Maxon, 2001). The first sale of acacia dry barks for tannin was recorded in the year 1910 as exports from the colony. At the time, acacia was the main source of tannins on the world market. Its demand rose after world war I, and the sale of acacia tannin was profitable spurring cultivation of acacia trees in Kenya on a large scale. The growth of the Acacia trees in Kenya invited the commercial interest of the two leading companies at the time involved in the processing and sale of vegetable tanning agents worldwide (Maxon, 2001). The commercial exploration culminated in the setting up of the East African

Tanning Extract Company (EATEC), a subsidiary of the South African-based Natal Tanning Extract Company (NTEC).

Kenya grows the wattle trees, mimosa (*Acacia dealbata*), which is originally from Australia and is known for its high tannin content (Maxon, 2001). Because of its origin during the colonial period, it was majorly cultivated by European settler farmers. The country's stake in the world tannin market grew gradually over the years, leading Kenya to be the second largest supplier after South Africa. However, in the early 2000, EATEC ceased operations. The exit of EATEC led to the depletion of the wattle farm estates and loss of interest in small holder farmers who no longer saw the commercial value of wattle trees (Cheboiwo et al., 2018). Consequently, the interest in wattle tree farming waned, depleting the share of the country in the world tannin industry. Today, only a meagre fraction of the estate is still in operation with only three small enterprises in operation and being sustained on the local bark supplies and little imports from Tanzania.

The country's export for the black wattle extract fell to just about 50 tonnes, in the first decade following the exit of EATEC. Whereas this suppression of the export potential has been attributed to the falling production of the acacia bark, one significant factor has been the increasing local demand for vegetable tannins attributed to the expanding leather tanning sector. It is estimated that close to 10,000 tonnes of tannins will be consumed by the tanning industry annually. In addition, the demand for tannins in the country is expected to increase exponentially with the growth of the domestic leather sector. According to a study by Cheboiwo et al. (2018) Kenya's demand for tannins could outstrip the local supply if the current growth rates persist.

At the turn of the millennium, the country embarked on an ambitious policy framework, that was geared towards revitalizing the manufacturing sector. The Kenyan government under the "Big Four" initiative and the Third Medium-Term Plan (MTP) prioritized the development of the leather sub-sector as part of efforts to generate more employment opportunities and grow incomes through industrial development. The key policy reforms to support the development of the leather sub-sector included the establishment of a leather council, to execute the sector reforms and build the public–private partnerships; imposition of high export taxes (80%) on exports of raw hides and skins, that was aimed at encouraging local value addition and product differentiation and improved institutional framework and increased budgetary allocations to the sector which would spur growth of the local leather industry. These interventions have been noteworthy, leading to the increase in the number of operating tanneries, the balance of trade between raw and crust leathers and increase in the number of small and medium enterprises dealing with leather products and articles.

1.7 Traditional Rural Tanning in Kenya

The practice of rural tanning has been the communities' mainstay for decades now. Hides and skins from rural slaughter slabs and homestead slaughters have been utilized for local processing and value-addition in the make-shift community tanneries. Whereas the scale and volume of production of veg-tanned leather through rural tanning is not well documented, its widespread practice among the local communities in Kenya has been captured as an economic activity (Kenya National Bureau of Statistics, 2019). The choice of vegetable tanning material is always dictated by what is locally available, the community's cultural beliefs and what has been used before by the older generations. Kenya is home to many indigenous tree species with potential for tannin extraction. Baseline surveys by the Kenya Forestry Research Institute has recorded the exploration of a variety of acacia species such as *A. hockii, A. kirkii, A. mellifera, A. mearnsii, A. nilotica, A. polyacantha, A. sayel, A. senegal, A. sieberiana* and *A. xanthophloea* for tannin production (Kuria et al., 2016; Mugedo & Waterman, 1992). Although, these vegetable tanning materials have shown potential for tannin extraction, their exploitation for commercial vegetable tannins has not been low.

While the commercial focus has largely been laid on *Acacia dealbata*, the use of other indigenous Kenyan species such as the aforementioned acacia family is only limited to rural tanning, mostly by communities in the pursuit of handicraft leather articles for the preservation of heritage and culture. The acacia family is varied and is said to probably contain in excess of 1000 different species. Of this, close to 100 species have been recorded in Africa, out of which 37 are native to Kenya (Mugedo & Waterman, 1992). Whereas only *Acacia dealbata* is the major source of commercial tannin extract, several others are used, although usually only in those areas where they occur naturally (Mwinyihija, 2011). The focal point that has been raised through research, is the pursuit of these species as alternative sources of tannins. The merit in this endeavor is the adaptation to propagation of these species to grow in arid and semi-arid lands in Kenya, which have little agricultural output.

Traditionally, hides and skins from most arid and semi-arid lands are sun-dried. Since they cover a vast area, proper collection channels for the hides and skins to reach tanneries hardly exist. In addition, the prices offered per piece for either hide or skin coupled with long distances to collection centres upset the motivation by locals to engage in hides and skins as an economic activity. Therefore, the aspect of rural tanning in these areas has gained traction. The practice is simple tanning devoid of any technical skills and equipment. Value addition is learnt through apprenticeship and passed on from one generation to the other. Consequently, the resultant leather is sometimes of low quality. However, the persons engaged in these activities have mastered the art and to some degree can produce good leather with some degree of reproducibility. The leather is used in making traditional containers, covers for beds, clothing, sandals and other leather articles. Some other pastoral Kenyan communities, notably the Maasai, have embraced rural tanning enterprises creating employment through the establishment of successful handicraft businesses that produce and supply leather artifacts to the tourism sector (World Bank Group, 2015).

1.8 Current Trends in Vegetable Tanning in Kenya

Today, the use of vegetable tanning materials in leather processing continues to draw interest because of the commercial value of non-mineral tannages. This is attributed to adoption of organic tanning which is highly regarded for its non-toxic form and the generation of less environmental pollution, thus, its prospectus for perceived ecological credentials to the leather industry. The rising consumer preference for eco-friendly materials has specifically elevated the importance of vegetable tanning materials. Its use is considered a green alternative to mineral tannages such as chromic oxide given its renewable nature, natural abundance and to an extent its environmental benefits (Das et al., 2022). This has however drawn mixed reactions since vegetable tanning effluents have a high load of chemical oxygen demand (COD) arising from incomplete exhaustion of chemicals and presence of non-tannins (Balakrishnan et al., 2018). The effluents also exhibit low biodegradability, carry a high phenol content and pose a recalcitrant color stream (European Commission [EU], 2003), which in principle affects the surface watercourse. Moreover, comparative water consumption per kilogram of leather processed is high for vegetable tanning than is for chromic oxide.

Presently, mimosa (*Acacia dealbata*) is the most used commercial vegetable tannins in Kenya. Although the use of quebracho has been registered, its use is very limited to specific contracts and periods over the last five years. According to the Kenya National Bureau of Statistics, the country imported wattle extract worth about $0.5 million over the last five years. The origin for these imports are South Africa (82%), Brazil (12%) and Tanzania (5%). Though the later refers to the wattle bark imports for supplementing local production of mimosa. The balance of supply for the leather industry is complemented by the local supply from the wattle extract processing from EATEC. Locally, the sulfited mimosa is preferred by most tanneries and accounts for the great majority of leather crust output. Sulfitation reduces the astringency and improves solubility, hence improved penetration to the vegetable tannins. The colour output of mimosa is reddish tinge to red-brown, while the sulphited one gives an extremely light colouring to the crust leather. The latter can be dyed to any other shade of colour with ease.

The conventional techniques in organic tanning are pit tannage, drum tannage and a hybrid of these two techniques. Whereas pit tannage bears the hallmark of classical method of leather processing, modifications such as concentration gradients, mechanical action and counter-current flow have elevated its importance in modern tanning systems. The technique to be deployed for vegetable tanning in a tannery is solely the choice of the individual tannery. However, several factors may influence the choice, key among them is: level of float required, water consumption optimization, the time of tannage required, desired end product type and environmental impact relating to the pollution load of the resultant effluents. The vegetable tanning systems that pose the least environmental impact are considered to be those that discharge the lowest volume of spent liquors, at a minimum load of the tanning material.

The drum tannage is the most preferred technique by the Kenyan tanneries and is used in the production of medium to light leathers. It affords the tanner great control over a number of variables, enabling the production of satisfactory leather in a much shorter time. The drum technique is characterized by concentrated liquors, low floats and mechanical action which accelerates the vegetable tanning process. Although mechanical action increases the speed of tannin penetration, it has been shown to result in pebbling of the grain. As a caution, some tanneries subject the pelts to a pre-tannage in pits to ensure partial penetration of the tannins and a degree of tannage before drumming in stronger concentrations of the tanning material. This trend is also noticeable for thicker hides, in which case, hides are tanned with up to 10% of the total offer of tannins first, followed by splitting operation before full penetration of the tannins is achieved. Interestingly, there is less demand for full veg i.e., (E.I) leathers by the formal local commodity sub-sector players. A growing demand for semi-chrome leathers has seen the industry pre-tan with chromium (up to 8%) followed with mimosa retannage (about 30% offer). In this case, both chrome tannage and the mimosa retannage are carried out in the drums, relegating the pits from the industry.

The pit tannage is based on the principle of counter-current flow, in which the pelts are applied to a series of cascading pits that differ from each other in concentration. Fresh pelts are applied to the pit with the lowest concentration of tan, and will stay there until the concentration of the tan is denuded. They will then be moved across to other pits with higher concentration of the tan than the preceding pit until full penetration is achieved, while the exhausted liquors are pumped down the series of pits to be used on the fresh pelts. The great advantage of pit tannage is that it results in effluents with less environmental impact. However, its major drawback is the volume of liquors used and the time of tannage required for full penetration. It takes considerable time, ranging from a few weeks to over a year for the completion of tanning in pit tannage system. The variation in time depends on the tanning agent employed. Locally, this technique is rarely used. Although some tanneries carry out the hybrid technique involving a combination of both the pit and drum systems. As earlier mentioned, in vegetable tannage, pits could be used for a shorter time first, just enough to give enough resistance to the grain against mechanical action before pursuing complete penetration of the tannins in the drum.

1.9 Contract Tanning

The aims espoused under the country's economic blueprint relate to creating a viable and sustainable industry to propel the country toward inclusive prosperity. To this end, various interventions have been put in place. At the manufacturing level, the country seeks to increase the proportion of finished leather from the current minimum of 2%. According to the World Bank Group (2015) report, the first strategy under this goal, is the establishment of industrial parks which will not only combine a broad spectrum of manufacturing resources under one roof but also integrate the various mid-sized plants in the same region. Secondly, increase product differentiation by mainstreaming Small- and Medium-Size Enterprises (SMEs) in the formal value chain channels while increasing market access for local products. To achieve the former, the country seeks to leverage productivity and quality by the informal sector by encouraging contract manufacturing through establishment of common manufacturing facilities. Bearing in mind that the focal area of interest by the informal sector is in leather articles and handicraft items whose domain is vegetable-tanned leather, it is expected that this intervention would lead to an increase in the proportion of finished leather, exploration of other locally available materials for commercial bark extracts and the proliferation of subcontracting networks in leather tanning and product manufacturing.

2 Conclusions—Sustainability and Future Prospects

An important merit in achieving value addition and product quality is the adoption of clean and sustainable technologies. The ever growing global demand for eco-friendly materials coupled with the rising consumer preference for greener materials is the biggest driver for the emerging trends in leather processing. In this regard, Kenyan tanneries have little choices other than be compliant to access global brands and markets for their products. The concept of sustainable production is not well developed in the local leather industry. The industry's segmentation into the formal and informal sectors presents a unique challenge. However, interventions through collaborative efforts by local institutions such as Kenya Industrial Research and Development Institute, Kenya Leather Development Council, Kenya National Cleaner Production Centre and Kenya Climate Innovation Centre have continuously extended technical support to the tanning and leather products manufacturing industry in driving sustainable production processes.

The basis of this support is geared towards increasing investments by the SMEs through formal channels of value addition, while improving the social and environmental impact of their business models. The goal has been in the upgrading of skills, technology, waste management and upscaling of organically tanned leather. This is expected to substantially contribute to the leather industry's long term viability and global competitiveness.

References

Balakrishnan, A., Sri Bala Kameswari, K., & Kalyanaraman, C. (2018). Assessment of aerobic biodegradability for vegetable tanning process wastewater generated from leather industry. In *Water quality management* (Vol. 79, pp. 151–160). Springer.

Bule, M., Khan, F., Nisar, M. F., & Niaz, K. (2020). Tannins (hydrolysable tannins, condensed tannins, phlorotannins, flavono-ellagitannins). In *Recent advances in natural product analysis* (pp. 132–146).

Cheboiwo, J. K., Mutta, D., Kiprop, J., & Gattama, S. (2018). Public private partnerships opportunities for forestry sector development in Kenya: Synthesis of primary and secondary production actors, and trade. *Journal of Environment and Earth Science, 8*(1), 47–69.

Commision, E. (2003). Integrated pollution prevention and control (IPPC) reference document on best available techniques for the tanning of hides and skins. *European Commision,* 1–1023.

Covington, A. D. (2009). *Tanning chemistry: The science of leather*. RSC Publishing.

Covington, A. D., & Wise, W. R. (2020). Current trends in leather science. *Journal of Leather Science and Engineering, 2*(1), 1–9.

Das, R. K., Mizan, A., Zohra, F. T., Ahmed, S., Ahmed, K. S., & Hossain, H. (2022). Extraction of a novel tanning agent from indigenous plant bark and its application in leather processing. *Journal of Leather Science and Engineering, 4*(1).

Falcão, L., & Araújo, M. E. M. (2018). Vegetable tannins used in the manufacture of historic leathers. *Molecules, 23*(5), 8–10.

Jourdes, M., Pouysegu, L., Deffieux, D., Teissedre, P., & Quideau, S. (2013). Hydrolyzable tannins: Gallotannins and ellagitannins. In *Natural products: Phytochemistry, botany and metabolism of alkaloids, phenolics and terpenes* (pp. 1975–2010).

Kenya National Bureau of Statistics. (2019). Population and housing census. https://www.knbs.or.ke/

Kuria, A., Ombui, J., & Onyuka, A. (2016). Tannin analysis of selected plants from Laikipia County, Kenya. *Journal of the Society of Leather Technologists and Chemists, 100*(2), 73–76.

Leckie, W. G. (1938). The growing of wattle and production of wattle bark in Kenya. *The East African Agricultural Journal, 4*(1), 51–62.

Maxon, R. M. (2001). Where did the trees go? The wattle Bark Industry in Western Kenya. *The International Journal of African Historical Studies, 34*(3), 565–584.

Mugedo, J., & Waterman, P. (1992). Sources of Tannin: Alternatives to Wattle (Acacia Mearnsii) amog indigenous Kenyan species. *Economic Botany, 46*(1), 55–63.

Mwinyihija, M. (2011). Essentials of ecotoxicology in the tanning industry. *Journal of Environmental Chemistry and Ecotoxicology, 3*(13), 323–331.

Mwinyihija, M. (2014a). A prognosis of the leather sector in Kenya; the upheavals and antidotes associated with value creation. *Management, 2014*(1), 21–29.

Mwinyihija, M. (2014b). Analysis of the trading regime of the leather value chain Strata in Kenya. *Business Management and Strategy, 5*(2).

Mwinyihija, M. (2015). Africa's renaissance and potential in the leather sector with refection to the global performance. *Journal of Africa Leather and Leather Products Advances, 2*(1), 1–17.

Ncube, B., & Van Staden, J. (2015). Tilting plant metabolism for improved metabolite biosynthesis and enhanced human benefit. *Molecules, 20*(7), 12698–12731.

Niaz, K., & Khan, F. (2020). Analysis of polyphenolics. In A. S. Silva, S. F. Nabavi, M. Saeedi, & S. M. Nabavi (Eds.), *Recent advances in natural products analysis* (pp. 39–197).

Ogolla, P., & Wanjau, K. (2013). Factors affecting value addition in the leather industry in Kenya. *European Journal of Business and Innovation Research, 1*(3), 45–55.

Onyango, C., Musyoka, P., Shibia, A., & Laibuni, N. (2019). Towards revitalizing Kenya's skins, hides and leather products industry (221/2019). The Kenya Institute for Public Policy Research and Analysis (KIPPRA).

Sivaram, N. M., & Barik, D. (2018). Toxic waste from leather industries. In *Energy from toxic organic waste for heat and power generation* (pp. 55–67).

Wayua, F., & Kagunyu, A. (2012). Constraints and opportunities in the hides and skins value chain in pastoral areas of northern Kenya. *Livestock Research for Rural Development, 24*(8).
World Bank Group. (2015). Kenya—Leather industry: Diagnosis, strategy, and action plan.

Benchmarking Best Global Practices in Ethiopian Leather Sector

Wondu Legesse and Mohammed Hussien

1 Introduction

The tanning industry in Ethiopia was started in 1918. Since then, the leather industry in Ethiopia has reached a relatively high stage of development and now enjoys a reputation as a producer of good quality leather, recognized in international markets. The country has comparative advantages in the large resource base in hides and skins and an internationally recognized high quality sheep and goatskins.

Despite comparative advantages, tanning industry in Ethiopia shows low level of competitiveness in the international market. This is due to low level of capacity utilization and production, low off take ratio, low recovery rates, deterioration of quality of raw hides and skins, low level of technology of leather manufacturing, low level of productivity of manpower, material and machinery, poor marketing strategies, non-availability, or difficulty in getting specialty chemicals, etc.

In order to strengthen the competitiveness of Ethiopian tanning industries, the government of Ethiopia under Ministry of Industry (MoI) represented with its Leather Industry Development Institute (LIDI) signed an agreement on 21st January 2010 with Council of Scientific and Industrial Research (CSIR) under the Ministry of Science and Technology of the government of India, represented by Central Leather Research Institute (CLRI). The title of the project was "Implementation of Benchmarking (Technology upgradation) for selected seven Ethiopian tanning industries". The duration of the first phase was six months and was launched on 14th March 2010. The seven tanneries selected for this program were Batu, Addis Ababa, Dire, Hafde, Colba, Wallia, and Bahirdar.

W. Legesse (✉)
United Nations Industry Development Organization (UNIDO), Addis Ababa, Ethiopia
e-mail: wonduleg@gmail.com

M. Hussien
Leather and Leather Products Industry Research and Development Center (LLPI-RDC), Addis Ababa, Ethiopia

S. K. Janardhanan and L. A. Zugno (eds.), *Emerging Trends in Leather Science and Technology*, https://doi.org/10.1007/978-981-99-9754-1_15

The project cost was more than USD 560,000 (five hundred sixty thousand US dollars). One project coordinator, one project director, seven scientists specialized in leather technology, two environmental engineers, two machine engineers, and seven consultants with industrial experience participated in the project from CLRI side. From LIDI side, 14 experts and 2 project coordinators were involved in the benchmarking implementation.

On completion of total duration of 6 months of direct interventions in the selected tanneries, the project was reviewed by the Steering Committee in a Meeting held on 7th September 2010. In the steering committee meeting, it was decided to scale up the project in other tanneries in Ethiopia and four more tanneries have been added to the program in the extension program.

2 Objective of Benchmarking Program

2.1 Broad Objective

The primary objective is to turn around the overall performance of the Ethiopian tanning sector, in particular, to improve the efficiency and effectiveness of seven selected export-oriented tanneries enabling them to produce finished leather that can be tradedin the international market in a competitive and sustainable manner.

2.2 Specific Objectives

- To identify the causes/effects of the performance in the selected tanneries regarding production, technology, human skills and quality of the product.
- To design a comprehensive solution strategy to transform the performance of the tanneries to the level of international standards
- To plan and implement technology upgradation and benchmarking measures in terms of Product development, Quality management, Standardization, Cost control, Productivity, Environmental safeguards, Occupational safety, etc., in order to bring about international level competitiveness at firm level
- To develop human skills in the selected tanneries for performance improvement and build capacity through training and documentation to enable sustained transformation
- To ensure sustainable performance by developing documentary tools such as manuals, skill level monitoring and upgrading plans, product and process inspection manuals, planning guidelines.
- To establish and implement quality management system within the framework of benchmarking and targeting transformational turnaround for the selected companies.

3 Scope of the Project

- Preparing an action plan to transform the performance of the selected firms by identifying bottlenecks that hinder the production process.
- Implement performance improvement measures in the production process to transform the productivity level of each firm and produce competitive finished leather in close collaboration with the management of the firm and with the active involvement of the key personnel.
- Monitor progress on the performance improvement activities, develop useful documentations to be used by the tannery and provide biweekly progress and monthly report/documentations to MoI-LIDI.
- Cover the requirements of the international standards of product technology as applicable for the tanneries including the systems of treatment of effluents.

4 Methodolgy of the Program

4.1 Phase 1: Gap Analysis

The first step was to assess the performance of the existing production system and to identify the gap in the existing production system in terms of production technology, human skills and quality of the products and then to propose viable and acceptable technology solutions for effective management.

4.2 Phase 2: Design & Implementation of Action Plan

Design and implement all activities like technology upgradation, quality enhancement, standardization of process, productivity gain, environmental safeguards, occupational safety, and ensure participation/direct involvement of the company staff and thereby build capacity of the human resource working in the production area.

Other activities in this phase include organizing training sessions to inculcate capacity building, skill development and system development to sustain the technology upgradation initiatives. Developing manuals and documentation were also carried out in this phase.

4.3 Phase 3: Monitoring and Evaluation

To follow and monitor benchmarking activities, a follow-up and monitoring committee composed of project coordinator from CLRI, project coordinator from

LIDI and representative from Engineering Capacity Building Program (ECBP) was formed. Activities were reported by the Performance Improvement Team (PIT) every two weeks to LIDI, and then they are communicated to the national export steering committee by LIDI coordinator every month. The structure and contents of the action plan have also been approved by the steering committee. In addition, the assessment was done by the steering committee after the completion of 4 and 6 months of benchmarking program.

5 Structure and Function of Benchmarking

Floor level operators, Technical Working Group (TWG), Performance Improvement Team (PIT), Monitoring and follow-up committee, steering committee were organized to plan and design, implement, monitor evaluate benchmarking activities. The floor level operators facilitate the execution of action plan. The Technical Working Group (TWG) understands gaps and action plans and communicates to the work floor about the corrective actions. The Performance Improvement Team (PIT) is composed of the Tannery Manager, key personnel from the respective tanneries, CLRI and industrial experts, and two experts from LIDI. The function of PIT is to identify and analyze gaps and to prepare action plans. The role of the follow-up and monitoring committee is to guide, follow up, and give feedback to PIT. The steering committee, chaired by state minister of Ministry of Industry, assesses, evaluates, and gives direction about the execution of benchmarking.

6 Actions Carried Out in Benchmarking

Actions carried out in benchmarking can be broadly categorized into system development, product development, maintenance management, environmental management, training and capacity building, manual preparation and documentation, and Occupational Safety and Health (OSH) system improvement. Except OSH, all other activities have been carried out in the seven selected tanneries.

System development efforts include process re-standardization in the raw yard, beam house and tanning operations, quality control systems in all yards, development of article-specific machinery operations, the system for productivity improvement and better capacity utilization, article-specific production planning, infrastructure/layout management and good housekeeping.

Product development activities include developing new products, upgrading techniques to lower grades and rejects, demonstrating on a larger scale and bulk trial and preparing process technology documents.

Maintenance management includes assessment of existing machinery and layout, developing maintenance system, developing preventive maintenance schedule for critical machines, troubleshooting and maintaining machinery, training of TWG maintenance team.

The environmental management system includes an assessment of existing effluent treatment methods, an assessment of solid waste disposal system, training of TWG of the existing treatment methods and finally preparing a blueprint for Effluent Treatment Plant (ETP).

Training and capacity building activities include training of supervisors/workers in different yards/sections in technology/culture change, training of machinery operators for specific products/increasing productivity/quality improvement, training of supervisors in quality control/process control, training in color matching in dyeing and finishing, training in product development and training of personnel for overall capacity building.

Manual preparation and documentation include process technology documents for the products developed, quality/process control manuals, training manuals for workers and supervisors in different yards/sections/machine operations, manuals for trimming and assortment/selection, manual for color matching and ModEuropshade cards, manual for maintenance of machinery, blueprint for ETP training, manual for ETP supervisors/workers and OSH manual.

7 Changes Brought About After Intervention

Tannery A

See Table 1 and Fig. 1.

Tannery B

See Table 2 and Fig. 2.

Tannery C

See Table 3 and Fig. 3.

Tannery D

See Table 4 and Fig. 4.

Tannery E

See Table 5 and Fig. 5.

Tannery F

See Table 6 and Fig. 6.

Tannery G

See Table 7 and Fig. 7.

Table 1 Actions carried out in Tannery A and achievements

Area/operation	Unit of measurement	Before intervention	After intervention
Buffing operation	Qualitative	Non-uniform buffing	Uniform buffing system for goat suede
Setting out operation	Qualitative	Improper setting	Effective mechanism of setting out for sheep skin
Spray machine	Qualitative		Developed effective use of Auto spray machine
Finishing	Qualitative	Mostly exported as crust leather	More than 80% of our export is finished leather
Product diversification	Qualitative	Available	Increased by 30% in sheep and cow
Level of specialization	Qualitative	Crust leather for local market and export	Finished sheep upper (aniline &semi-aniline) leather Cow upper & lining leather from lower grades Goat suede and printed suede leather

Fig. 1 Some of the product developments carried out in Tannery A

8 Constraints Faced

When implementing such a big project, it is natural to expect challenges. Problems faced in benchmarking implementation can be categorized as input shortages, mindset problems, and coordination failures.

It is known that in product development, a range of variety of chemicals are required. However, a variety of chemicals for dyeing, post tanning, finishing and

Table 2 Actions carried out in Tannery B and achievements

Area/operation	Unit of measurement	Before intervention (Dec–Feb)	After intervention (Mar–May)	Remarks
Cow waxy	Sq.ft./month	527,010	573,105	Finished leathers
Cow Asko softy	Sq.ft./month	1,769	36,665	
Cow full grain	Sq.ft./month	2,197	12,355	
Sheep upper	Sq.ft./month	15,000	105,602	
Products developed from cow	No of products	1	16	9 products exported; 7 products market linkage is required (56%)
Products developed from goat	No of products	1	4	3 products accepted by the customer (75%)
Products developed from Sheep	No. of products	1	3	3 products exported (100%)
Trimming in liming	%	Trimming loss of more than 15%	Reduced to 3–4%	Management appreciated and strictly followed the suggestions provided by Indian expert
Trimming in Wet Blue	%	Trimming loss of more than 7–8%	Reduced to 3–4%	
Trimming in dyed crust	%	Trimming loss of more than 10–12%	Reduced to 2–3%	
Packing		Wrong method of packing	Trained packing of finished leathers by keeping the grain outside, which prevents the formation of looseness and folding	

specialty chemicals were not available to a sufficient level in almost all tanneries in which benchmarking was implemented. Another constraint faced during implementation is non-availability of required machines for finished leathers in some tanneries. This includes buffing machine, ironing or plating machine, roller coater, hand spray machine with hood, and Finiflex machine. The above-mentioned problems challenged the possibility of producing diversified and marketable products. It also constrained the possibility of upgrading rejects and lower grades of hides and skins.

Fig. 2 Some of the product developments carried out in Tannery B

Table 3 Actions carried out in Tannery C and achievements

Area/operation	Unit of measurement	Before intervention	After intervention	Remarks
Cow hides	Quantity & quality	Only up to Wet Blue	started making finished articles (25,000 Sq.ft.)	Management liked the samples and received orders
Beam house operation of hides	Quality	Wrinkles and growth marks	No more wrinkles and growth marks and wet blue hides are more open	Resulted in better area yield and final product quality
Preventive maintenance		Lack of annual preventive maintenance plan	The annual maintenance plan put in place	CLRI machine engineer has given training to the mechanical department staff

Another challenge faced was spare parts for machines were not available at the desired level. The maintenance of machinery was constrained by the non-availability of spare parts. These problems affected the quality of produced leathers. It also affected the productivity of labor due to downtime.

Another factor that challenged the project was the human factor. It is known that when producing good quality suede and glove leather, the skill of operators for buffing and polishing machines is very critical. However, skilled operators for buffing and polishing machines were not available. In some tanneries, qualified personnel for research and development, quality control and product development were not available. It is not only the non-availability of personnel but also the mentioned departments were not existent.

One of the mindset problems faced during the execution of benchmarking is the initial resistance to change from some tanners. Nevertheless, this problem diminished over time. The source of this resistance was tanner's expectation from the consultant was too much. Another mindset problem was tanners are comfortable with the local

Fig. 3 Some of the product developments carried out in Tannery C

market and export of crust leathers. They did not want to take risks in the international market as well as in the export of finished leather.

Another problem was some of the tannery owners lack sufficient involvement in the benchmarking process. Some local technicians were not interested in product development. The last but not the least challenge in connection with the mindset problem was tanners' poor attention to environmental aspects. In some tanneries, there were no permanently and exclusively assigned personnel to the effluent treatment plants.

Coordination failures include additional time taken for the preparation of action plan. This was expected to be finalized in two weeks, but it was extended up to two months due to several disparities among various stakeholders involved in the process. Another challenge was the extension of project execution time owing to the ill health of experts and poor performance, though at the end of the project period, it was compensated. Another challenge was most tanneries considered benchmarking as product development and gave minimal attention to system development. Further, PIT members did not have a clear and uniform understanding of system development.

Other problems include insufficient floor space for system improvement, layout rationalization in some tanneries, and difficulty in getting working capital in some tanneries resulting in slowing down/incompletion of some activities.

Table 4 Actions carried out in Tannery D and achievements

Area/operation	Unit of measurement	Before intervention	After intervention	Remarks
Goat crusting process	Area improvement	Wrinkles, fold marks and rough grain	Fine, flat and smooth grain—no wrinkles	About 10% increase in the area achieved More profit
Sheep nappa—black	Process time	1.5 shift	1.0 shift	Productivity per month increased Power saving Cost saving on man, machine and materials
New setting method	Physical increase in area	Drawn and wrinkle grain	Flat, fine and smooth grain	An increase in area is proven and achieved 10% gain in area and value Reduction in rejection quantity
Finishing—Sheep	Upgradation	Domestic sale	Export	Value addition Additional earnings to the country
Finishing—Cow	Full grain	Domestic	Export	Increase in profit margin
Finishing—Cow	Corrected grain	Domestic	Export	Full utilization of crusts Improvement in grade Reduction in rejections

9 Conclusion and Recommendation

The system development effort mainly focuses on the wet end part of the tannery. Particularly it concentrates in raw hide/skin yard, beam house yard and tanning yard. Post-tanning and finishing were given less attention and less time. There were great achievements in trimming, the proper way of piling pickled skins, wet blue and crust leather. In some tanneries, productivity is increased by direct processing from pelt to wet blue without piling and selecting at the pickle stage. And also, in one tannery productivity is increased by solving the problem of water pipelines for cold and hot water. However, quality control efforts and re-standardization of the beam house process and tanning yards should be strengthened. Tanners should have dedicated quality control personnel and department. Moreover, capabilities in post-tanning and finishing of export quality should be strengthened.

Fig. 4 Some of the product developments carried out in Tannery D

Table 5 Actions carried out in Tannery E and achievements

Area/operation	Before intervention	After intervention	Remarks
Development of gloving leathers from sheep, goat and cow	Goat and cow were not processed No washed look, surface dyed and pigment dyed sheep gloves	A wide range of articles developed	Process standardized Provided a document on “Art of Glove Leather Making”
Process control	pH and Baume checks were not carried out for all the batches	pH and Baume checks were carried out for each lot and documented	Provided a document on “Manual on Leather Making”

Many products were developed in all tanneries except one tannery. Some of them scaled up the sample trial to bulk production. But as compared to the number of sample trials developed, bulk trials were few. Some sample trails were exhibited in Shanghai leather fair, and they were able to create market linkage. However, most of the sample trials were constrained by non-availability of specialty chemicals. In addition, evaluation of sample trials was not properly done. Sending samples to the customers and getting feedback was not exercised. Utilization of splits, upgradation of lower grades and rejections were not linked to the market. Thus, the product developed should be market linked, specialty chemicals should be availed, capability in

Fig. 5 Some of the product developments carried out in Tannery E

Table 6 Actions carried out in Tannery F and achievements

Area/ operation	Unit of measurement	Before intervention	After intervention	Remarks
Direct wet blue process	Drum running time-electric power	24 h	12 h	Approximately 50% saving in electric power *(Yet to be implemented)*
Crust selection	Degree of upgradation	–	20% upgradation	Selectors trained to select based on cutting value 20% upgrade of grade V to grade IV observed E.g.: Approx. 0.50 $USD per sq.ft.; for one million sq.ft. / month the company will get 500,000$USD per month
Goat suede crust	Upgradation	Grade V	Grade III	Grade V suede crust upgraded to Grade III after buffing operation Suggested doing buffing operation
Sheep crust	Value addition	Grade VI	50% grade IV 40% grade V	The upgradation obtained by employing different chemical and mechanical operations Suggested doing dyed crust

Fig. 6 Some of the product developments carried out in Tannery F

Table 7 Actions carried out in Tannery G and achievements

Area/operation	Unit of measurement	Before intervention	After intervention	Remarks
Goat & cow raw to pickle	Time	3 days	1.5 days	Standardized
Raw direct wet blue	Time	3 days	2 days	To be standardized
General production	Sq.ft./month	375,000	800,000	(50% intervention)

dyeing and finishing should be strengthened, and evaluation of the sample developed should be carried out.

With regard to environment, in most of the tanneries although provision for treatment of lime and chrome liquor was provided, they were not segregating lime and chrome liquor and treating liquors properly. In one tannery, the ETP was not properly designed with physio-chemical treatment process. And in another tannery, there was no ETP present. In some tanneries, although there was a filter press to dewater the sludge generated in primary settling tank, it was observed that filter press was not used. In some tanneries, although they had the facility to recover chrome, they were not recovering it. In some, there was no exclusive operator available for ETP. Also, when there were failures in ETP are encountered, serious regulatory actions not enforced. In some tanneries, there were no basic chemicals such as alum, lime and magnesium oxide for chemical dosing in ETP and chrome recovery system.

Thus, it is evident that there are huge gaps in the environmental management systems. Tanners should provide more attention to it. Chrome and sulfide liquors

Fig. 7 Some of the product developments carried out in Tannery G

should be separated and treated properly. Those tanneries that have the facility to recover chrome should do so. They should also operate the filter press. Dedicated and exclusive personnel for ETP should be appointed. They should strive to upgrade their ETP's with secondary and tertiary treatment plants.

Overall, tanners should strive to sustain attained achievements in system and product development. They should strive to improve their system continuously. In order to be competitive in the international market, the quality of their product should be better than their competitors; their cost of production should be lower than other competitors, and finally, in technology, they should be ahead of their competitors.

Challenges and Opportunities for Sustainable Development of Leather Sector in Vietnam

Tran Vu Anh Le, Van Huan Bui, Chi Thanh Nguyen, and Trinh Bui

1 Introduction

1.1 An Overview of Vietnam's Leather Sector

The footwear industry plays an important role in the Vietnamese economy. In recent years, Vietnam is the second-largest exporter of shoes and leather products in the world. Currently, Vietnam's leather and footwear industry has over 2,600 enterprises, employing over 1.4 million people (Statistical yearbook of Vietnam, 2021). Vietnam's footwear products have been present in more than 100 countries on all continents. The main export markets are the countries of the European Union (EU), the US, Japan, Korea and China. In recent years, Vietnam's leather and footwear industry has continuously developed and achieved many great achievements. In 2021, the export turnover of the leather and footwear industry will reach nearly 21 billion USD (United States dollar) (Statistical yearbook of Vietnam, 2021). However, in the coming period (to 2030), besides the advantages, Vietnam's leather and footwear

T. V. A. Le (✉) · C. T. Nguyen
Leather and Shoe Research Institute, No. 160 Hoang Hoa Tham Street, Tay Ho District, Hanoi, Vietnam
e-mail: anhltv@lsi.com.vn

C. T. Nguyen
e-mail: thanhnc@lsi.com.vn

V. H. Bui
Hanoi University of Science and Technology, No. 1 Dai Co Viet, Hai Ba Trung District, Hanoi, Vietnam
e-mail: huan.buivan@hust.edu.vn

T. Bui
University of Art and Design, No. 360 De La Thanh, Dong Da District, Hanoi, Vietnam
e-mail: maitrinh.bui@polimi.it

S. K. Janardhanan and L. A. Zugno (eds.), *Emerging Trends in Leather Science and Technology*, https://doi.org/10.1007/978-981-99-9754-1_16

industry is forecasted to face many challenges related to the United Nations' sustainable development goals, as well as finding a sustainable development direction for the industry.

1.2 Sustainable Development in the Leather Sector

Sustainable development is "development that meets the needs of the present generation without compromising the ability of future generations to meet their own needs" (Keeble, 1988). Recently, sustainability is no longer considered a choice but an important criterion in the production and business of enterprises operating in all fields (Khan et al., 2018). Author Elkington (1997) proposed three key points of sustainability, 3Ps, including People (society), Planet (environment) and Profit (economy). Likewise, the "3E" (economy, ecology and equity) criteria need to be addressed simultaneously in the context of sustainable development. For the industry/business, these three aspects are Employee, Environment and Profit. These aspects have both relative independence and interdependence, promoting each other as a unified whole. When the environment and people are taken care of, there will be positive effects on profits, increasing competitiveness in the market. Therefore, businesses have better profits and can pay more attention to human development and reduce environmental impact (Svensson, 2015a, 2015b; Dwivedi et al., 2021).

The concept of sustainable development is expanded and applied from the macro to the micro level in each specific area of the leather sector, from the supply chain, production and business models to specific solutions related to the life cycle of the product/service. Wenjie (2019) refers to the sustainable development of the supply chain of the Chinese shoe industry on the basis of the analysis of the sustainable development of Nike products. Ville Matinaro et al. (2019) also studied and proposed a sustainable business model for small and medium enterprises in Taiwan. In the book "Leather and Footwear Sustainability" (2020), the authors looked at the manufacturing process, supply chain and supply chain sustainability of footwear products. In addition, sustainability is reflected in the production, distribution and use of leather products (Deselnicu et al., 2014; Rashed-Ul-Islam, 2022), in the company's supply chain management leather industry (Islam et al., 2020), in the use of sustainable materials and design for leather and footwear products/services (Hervaa et al., 2011), in shoe manufacturing (Herva, 2011; Subic et al., 2012), and product reviews (Pearson King, 2010). Many footwear brands such as Nike and Adidas have been transparent in their reports, they have sustainable development programs and strategies with specific goals, plans and actions (Aleksandar Subic, 2012; Sustainable Development Goals, 2019; Adidas Annual sustainability reports). The main contents of the sustainable development program are specified as follows.

- Standardization and certification systems;
- Sustainable materials and accessories for footwear;
- Product ecological design;

- Sustainable production technologies and processes;
- Environmental regulations and standards;
- Health and safety in the workplace of workers;
- Safe products for consumption, sustainable packaging;
- Supply chain management and logistics;
- Green marketing; and
- Practice corporate social responsibility.

1.3 The Need to Develop Vietnam's Leather Sector Towards Sustainability

Sustainable development is the main trend for production, business and consumption activities in the world because this not only has a positive impact on the environment and society, but also brings economic benefits to related parties. Recently, in Vietnam, this trend has attracted attention from the government to the business community. There are three main reasons to answer the question of why it is necessary to develop the leather and footwear industry in a sustainable way.

Firstly, we are facing urgent problems such as resource depletion, waste pollution, climate change and socio-economic instability. In particular, the footwear industry operates on the use of a large number of raw materials and chemicals, and most of these chemicals remain on the products or are discarded causing environmental pollution (Staikos et al., 2006).

Secondly, trade agreements and the concerns of brands, especially international brands, as well as the awareness of consumers and stakeholders about the environmental impacts and social responsibility of this industry are increasing.

In this context, in order to sustainably develop the economy—society, the Government of Vietnam has planned sustainable development strategies and promulgated legal documents on environmental protection, production and consumption such as Decision 889/QD-TTg dated June 24, 2020: National action program on sustainable production and consumption for the period 2021–2030; Decree 08/2022/ND-CP dated November 17, 2020: Detailed regulations on a few documents of the Law on Environmental Protection.

1.4 Approaches to Develop Vietnam's Leather Sector Toward Sustainability

This research is based on identifying and solving "hot" issues with the participation of stakeholders throughout the entire product life cycle and the footwear supply chain, from research to product development, raw material supply, design, manufacture, distribution, use and disposal. The identification is done based on: (1) analysing the current situation of the leather and footwear industry in Vietnam, clarifying the drivers and barriers to the industry's sustainability; (2) point out the factors affecting the sustainable development of Vietnam's leather and footwear industry; and (3) analysis and forecast of development trends of the world leather and footwear industry in relation to sustainable development goals.

There are many criteria related to the "3Es", which should be based on the assessment of the current situation of the leather and footwear industry in Vietnam and the socio-economic conditions of Vietnam to determine the major criteria to find effective solutions. In the supply chain of the leather and footwear industry, Vietnam is focusing mainly on the manufacturing sector, other sectors are limited. Therefore, the main issues that need attention are: to improve production and business efficiency such as minimizing waste; to improve labour productivity, improving product quality; to improve working conditions and raising incomes for workers; to minimize the environmental impact of the product manufacturing process.

In addition, attention should be paid to the role and cooperation of stakeholders such as governments, non-governmental organizations, designers, manufacturers, service providers, users and media.

Indeed, to build a sustainable development model for the leather and footwear industry, a green ecosystem and a prosperous community, satisfying the "3E" criteria of economy, people and environment, specifically as follows.

(1) Effective business model: Enhancing production and business efficiency for enterprises and enterprises must be profitable;
(2) Using resources efficiently, renewable, safe and non-toxic;
(3) Producing and supplying products of high quality, affordable prices, minimizing environmental pollution;
(4) Creating better working conditions, raising incomes, protecting and ensuring the rights of workers;
(5) Extending the life cycle of leather and footwear products, create truly sustainable and sustainable foods, along with recycling and reuse actions, satisfying the "3R" criteria (Reduce, Reuse, Recycle);
(6) Increasing transparency in the supply chain about origin, material and production conditions; and
(7) Increasing the awareness for the new generation's consumers to become green consumers/smart consumers.

It should be emphasized that, in order to achieve the goals of developing a sustainable footwear industry in a sustainable economy, Vietnam needs to act promptly with the participation of stakeholders. Therefore, this chapter focuses on the following content: (1) Assessing the current situation of Vietnam's leather and footwear industry, related to enterprise structure, production and market, equipment and technology, methods, organization of production and labor management, product quality management, logistics, application of automation and enterprise digital transformation, as well as solid waste and solid waste treatment issues; (2) Analysis of domestic and foreign factors affecting the development of the industry; (3) Assess the strengths and limitations of the industry; (4) Assessment of opportunities and challenges for Vietnam's leather and footwear industry based on the perspective of sustainable development; and (5) Proposing strategies to develop Vietnam's leather and footwear industry towards sustainability.

2 Assessment of the Current Situation in Vietnam's Leather Sector

2.1 The Enterprise Structure in Vietnam's Leather Sector

Currently, there are many forms of ownership of leather and footwear enterprises in Vietnam, such as state-owned enterprises, non-state enterprises (limited companies, private companies and joint stock companies) and foreign-invested enterprises (joint venture companies, foreign direct investment—FDI).

In 2021, according to statistics from the Vietnam Leather, Footwear and Handbag Association (Lefaso), Vietnam's footwear industry has above 2,600 manufacturing enterprises, accounting for about 2.4% of the enterprises in the processing and manufacturing industry of Vietnam. Compared to 2010, the number of additional enterprises is over 1,500 enterprises, equivalent to an increase of 168 enterprises per year. The number of enterprises with above 200 employees is 514, accounting for 19.7% of the total number of enterprises in the industry. About 10% of the total enterprises have 3,000 or more employees. By capital scale, there are 230 enterprises, accounting for 8.8% of the total number of enterprises in the leather and footwear industry with a capital of over 8 million USD. This is a rather high index in the group of processing and manufacturing industries of Vietnam. Domestic enterprises are usually small and medium-sized, accounting for above 70% of the total number of leather and footwear enterprises, but accounting for just over 20% of the industry's total export turnover (Statistical yearbook of Vietnam, 2021).

About 80% of leather and footwear enterprises are in the South of Vietnam, the rest are distributed in the Central and Northern regions of Vietnam (Fig. 1).

FDI enterprises from Taiwan and Korea account for the majority of Vietnam's leather and footwear export market share. The production capacity of Taiwanese shoe manufacturers accounts for 50% of Vietnam's total production capacity (Lefaso Annual Report, 2021). In general, the structure of Vietnamese leather and footwear enterprises has the following advantages and limitations.

Advantages: The number of enterprises in the leather and footwear industry is quite large and increases every year. Although FDI enterprises have a small number (over 10%) they are large enterprises, attracting the labour force (about 75%) and making a significant contribution to the export turnover of the industry (about 77%). FDI enterprises also play an important role in improving the level of technology and management of the leather and footwear industry in Vietnam.

Limitations: Businesses are unevenly distributed and concentrated in some key economic regions of Vietnam, so recruiting local workers is and will face many

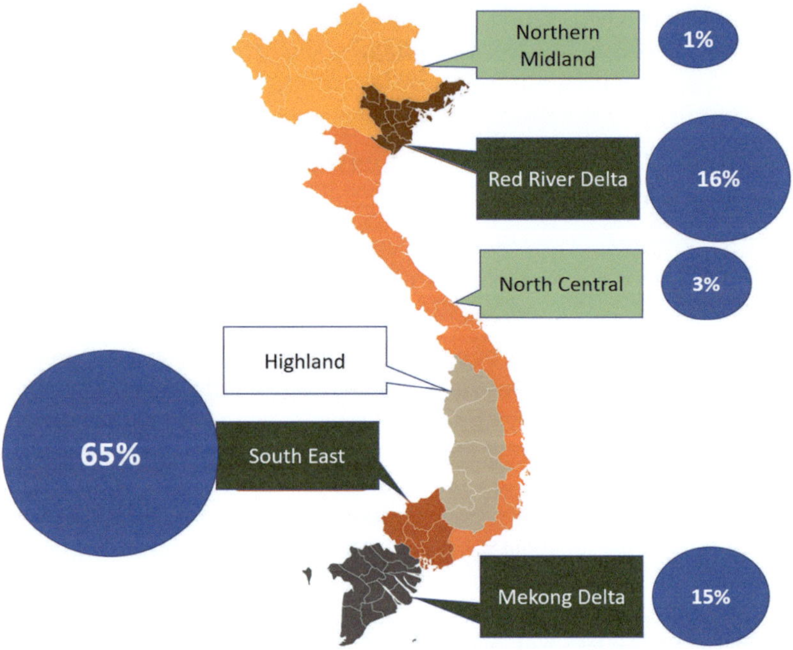

Fig. 1 Current capacity of footwear manufacturing in each economic zone

difficulties. Besides, the number of small and medium enterprises accounts for quite a large number. These are usually Vietnamese enterprises and are much less competitive than FDI enterprises, as well as large Vietnamese enterprises.

2.2 Production and Export Market in Vietnam's Leather Sector

Vietnam is the third country in the world after China and India in terms of shoe production each year. In 2021, Vietnam produced 1.32 billion pairs of shoes, accounting for 6.4% of the total world shoe production (Table 1).

Currently, about 95% of Vietnam's footwear production is for export. Vietnam keeps the second-largest footwear export position and accounts for over 10% of the world's exports (Table 2).

Besides, sports shoes are the major export products of Vietnam's footwear industry, accounting for nearly 70% of output. Followed by leather shoes at 23.5% and canvas shoes at 6.5%. The major products of the leather and footwear industry in recent years are shown in Table 3.

For many years, this industry has experienced stable growth and accounts for a large proportion of Vietnam's exports. In the years 2020 and 2021, due to the impact

Table 1 Top 10 footwear production countries (World Footwear, 2021)

Rank	Country	Pairs (Millions)	World share (%)
1	China	11 118	54.3
2	India	2 080	10.2
3	Vietnam	1 320	6.4
4	Indonesia	1 036	5.1
5	Brazil	764	3.7
6	Turkey	487	2.4
7	Pakistan	483	2.4
8	Bangladesh	423	2.1
9	Mexico	165	0.8
10	Iran	165	0.8

Table 2 Vietnam is the world's second largest exporter of footwear (World Footwear, 2021)

Rank	Country	Pairs (millions)	World share (%)
1	China	7 402	61.1
2	Vietnam	1 233	10.2
3	Indonesia	366	3.0
4	Germany	301	2.5
5	Turkey	280	2.3
6	Belgium	243	2.0
7	India	228	1.9
8	Italy	165	1.4
9	Netherlands	162	1.3
10	Spain	131	1.1

Table 3 Some major products of footwear industry in Vietnam 2016–2020 (Annual report, Lefaso, 2021)

Products	2015	2017	2018	2019	2020	Growth 2016–2020, %/per year
Leather shoes	253.0	263.4	282.5	301.8	289.7	2.7
Cloth shoes	61.5	67.8	72.7	79.7	81.6	5.8
Sports shoes	683.3	771.3	821.2	880.0	859.8	4.8
Bags and other leather goods	350	380	402	450	430	8.7

of the COVID-19 epidemic, the export turnover of the leather and footwear industry has decreased. However, in 2022, the production and export situation showed good signs of recovery (Fig. 2).

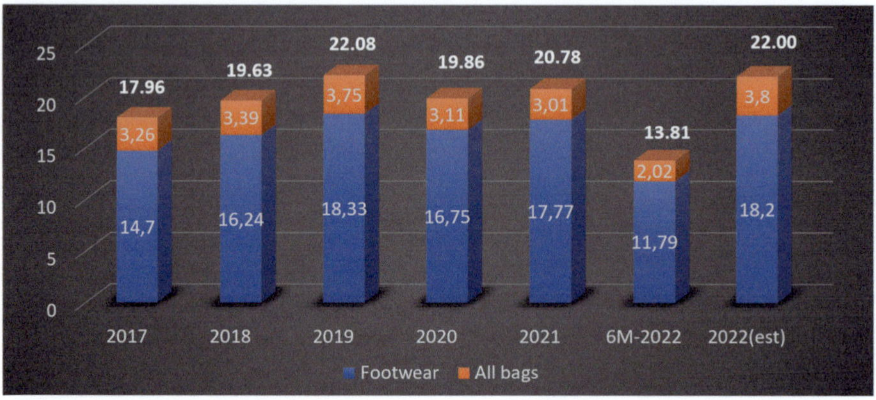

Fig. 2 Export of footwear and bags from Vietnam by years (Annual report, Lefaso, 2022)

The main export markets of Vietnam's footwear industry are North America, the European Union and Asian countries (China, Japan and Korea). Export turnover to the US, EU, China, Japan and South Korea markets accounts for over 80% of the industry's export value (Figs. 3, 4 and Table 4).

The US market is still the largest footwear export market in Vietnam. Next is the EU market. A remarkable point is that in recent years, Vietnam's leather and footwear exports to China accounted for about 5% of the total export turnover of the whole industry. In the first 6 months of 2022, the export turnover of leather and footwear products into China increased sharply to 12.5% (Table 4).

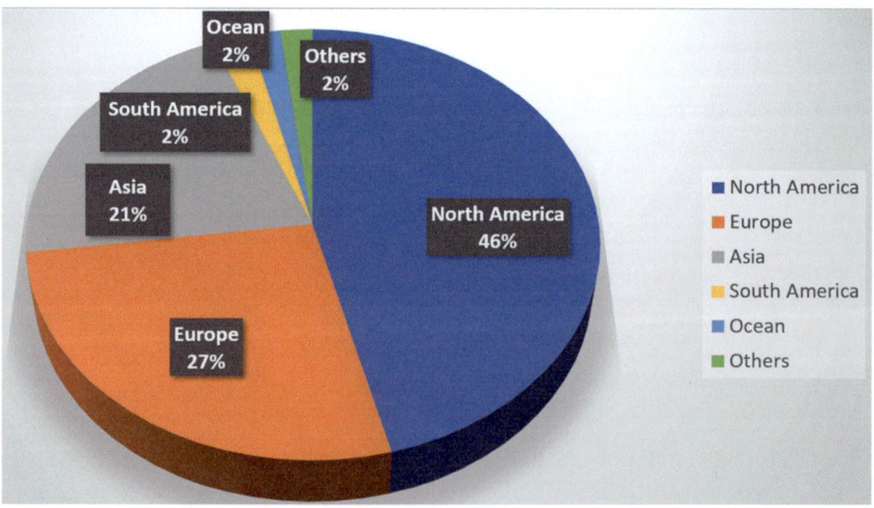

Fig. 3 Export by market of whole Industry, first 6 months/2022 (Annual report, Lefaso, 2022)

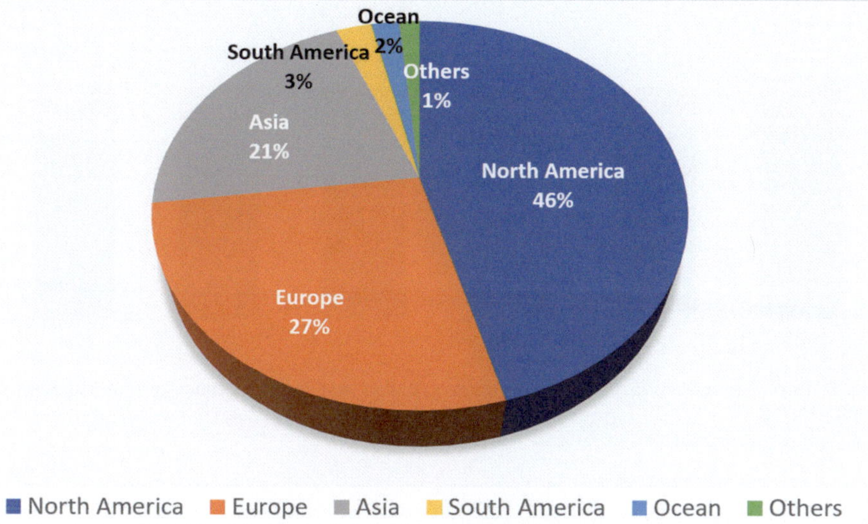

Fig. 4 Export by market of footwear, first 6 months in 2022 (Annual report, Lefaso, 2022)

Table 4 Five biggest countries import from Vietnam, first 6 months in 2022 (Annual report, Lefaso, 2022)

Country	Total import (Mil USD)	Footwear		All kinds of bags	
		Mil USD	Percentage	Mil USD	Percentage
USA	8764.6	7436.8	41.8	1327.8	44.1
EU (27)	4702.7	4042.8	22.7	659.9	21.9
China	1718.3	1595.7	8.9	122.5	4.1
Japan	1066.7	807.3	4.5	259.4	8.6
Korea	614.3	503.5	2.8	110.8	3.7

Exports of FDI enterprises still account for a large proportion of the total export turnover of Vietnam's leather and footwear industry. FDI enterprises are often large and very large enterprises, so although the number of enterprises accounts for just over 10%, export turnover accounts for over 77% (Fig. 5).

Besides, the export price of Vietnam's shoes ranks 6th in the group of high-priced exported shoes in the world. The average price of shoes exported from Vietnam is 17.9 USD, much higher than the world average export price of 10.47 USD/pair (Fig. 6). Most of the major international shoe brands such as Nike, Adidas, Puma, Reebok, New Balance, Salomon, Clarks and Sketcher have ordered production in Vietnam (Annual Report, Lefaso, 2021).

Thus, through an overview of the production and export status of the footwear industry in Vietnam, the advantages and limitations of the industry can be assessed as follows.

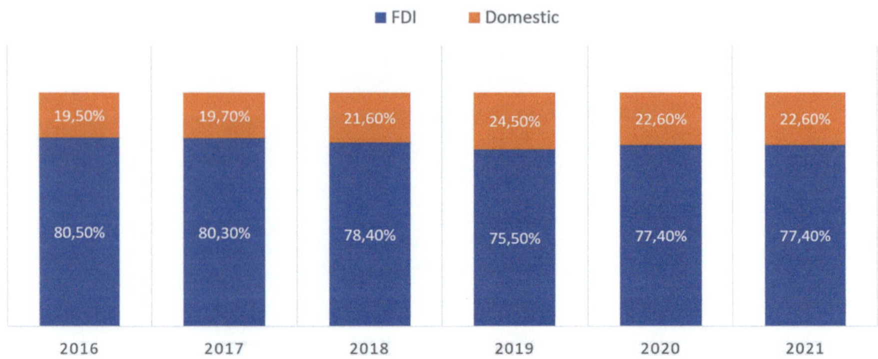

Fig. 5 Export by enterprise sector of Vietnam's leather and footwear industry (Annual Report, Lefaso, 2021)

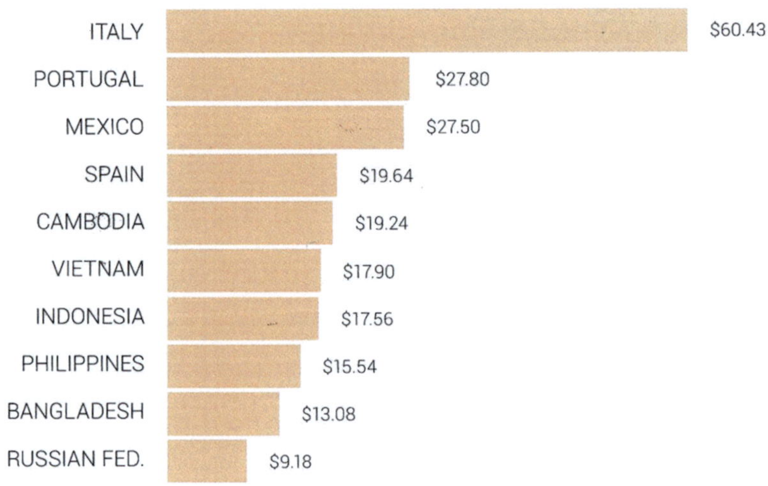

Fig. 6 Vietnam is in the top 6 countries with the highest shoe export prices

Advantages: Vietnam's footwear industry has made a breakthrough in export in the past 10 years. From 2015 to now, Vietnam has become the second-largest footwear exporter in the world. In the last 5 years, this industry has contributed steadily from 7.5 to 9.0% to the total export turnover of Vietnam.

The export market of Vietnam's leather and footwear industry has expanded to above 100 countries, including about 50 countries with a value of over one million USD. The industry's export activities have flourished over the years, leading to maintaining competitive advantages in traditional markets.

The gap in the proportion of the export value of domestic enterprises with FDI enterprises tends to decrease gradually over the years. By 2021, the export turnover of domestic enterprises will account for about 24–25% of the total export turnover of Vietnam's footwear industry.

Limitations: Footwear products made in Vietnam are mainly low-priced shoes. For example, in the case of sneakers and canvas shoes, sports shoes mainly use cold glue sole technology, while the production technology of canvas shoes is mainly boiler vulcanized rubber soles. These are kinds of shoes that require a lot of labour, but the labour productivity is lower than leather shoes, not to mention the fact that workers have to be exposed to many toxic chemicals in the making process. Although the annual export turnover of Vietnam's footwear industry is over 20 billion USD, enterprises benefit only a small part of the export turnover. Since the production method is mainly outsourcing products to other countries, in addition, the localization rate is low (about 50%).

2.3 *Labour Status and Working Environment in Vietnam's Leather Sector*

2.3.1 The Number of Employees

According to statistics, by 2021, the number of employees in the leather and footwear industry will reach over 1.4 million, an increase of above 1.95 times compared to 2010 and accounting for 9.3% of the total number of employees in Vietnamese enterprises. From 2010 to now, on average, the number of employees in Vietnam's leather and footwear industry has increased by about 73,800 people annually (Statistical yearbook of Vietnam, 2021). Besides, according to Lefaso's statistics, the average number of employees/enterprise in the leather and footwear industry is quite high, about 527 employees/enterprise. Of the total number of employees in the footwear industry, footwear production accounts for over 87%; leather production accounted for only 0.65% and handbag production accounted for 12.34%. The proportion of female workers in the leather and footwear industry is about 80%. The foreign-invested sector attracts about 75% of employees.

2.3.2 Labour Quality

Vietnam has a relatively large labour force and low average age, who is able to learn quickly and meet the training requirements to use new technology and machinery. However, for the footwear industry, due to low wages, only unskilled workers from rural areas can be recruited to work in industrial zones. Therefore, the level of culture and social awareness of many workers in this industry is still low. The number of workers graduating from high school accounts for 70%, the rest just finished lower secondary school.

Currently, in Vietnam, there are no vocational schools to train workers in the footwear industry. In enterprises, the training and development of human resources have not been adequately invested. Most workers only learn theory in a quick time, then practice directly on production lines. The average technical level of workers is 2.5/6 (according to Vietnam's occupational ladder), and the majority of workers are aged 20–35. Therefore, the labour productivity of the industry is low (Lefaso Statistics, 2021).

It should be emphasized that, although people are an important factor, in determining the success of an enterprise, the quantity and quality of human resources in the leather and footwear industry are lower than in other industries. Human resources at college, university and graduate levels who are trained in design, technology and management in this industry are being evaluated weakly in terms of skills and lacking in number. For example, compare with the human resources of the textile and garment industry in Vietnam. Highly qualified human resources for the textile industry in Vietnam have trained 65 years ago. Every year, dozens of universities and colleges train about 6000 bachelors, engineers and masters in the textile and garment industry. Meanwhile, up to now, in Vietnam, there are 2 colleges training bachelor's degrees in leather and footwear with a number of about 150 students/per year. These schools have been training for 15 years. Since 2015, there is a university training leather and footwear industry with about 20–30 students/per year.

According to our research results in the topic "Research and evaluate the impact of the Fourth Industrial Revolution on the leather and footwear industry in Vietnam, strategic orientations, policies and solutions for the development of the footwear industry in Vietnam by 2030" (Lefaso, ĐTL.XH.15/18, 2020) in 2020, less than 20% of the total number of direct workers in the leather and footwear industry can use automation equipment, this rate is in FDI enterprises are 20–30%. About 65% of employees indirectly use administrative software.

2.3.3 Labour Productivity

A recent report by the International Labor Organization (ILO) shows that Vietnam's national labour productivity is among the lowest in the Asia-Pacific, only slightly higher than Cambodia, Laos and Myanmar, and lower than Singapore nearly 15 times, lower than Japan 11 times and Korea 10 times. Compared with middle-income Association of Southeast Asian Nations (ASEAN) countries, Vietnam's labour productivity is only 1/5 of Malaysia's and 2/5 of Thailand's.

According to a survey by Lefaso (DTDL.XH.15/18, 2020), the average labour productivity of shoe manufacturing enterprises in ASEAN countries is 8 pairs of shoes/person/day, while, in Vietnam is 6 pairs of shoes/person/day, 75% of the region's average productivity.

2.3.4 Income Level

In the past 5 years (2017–2021), the labour income level of the leather and footwear industry has grown at a rate of about 8.8%/year, higher than that of other processing and manufacturing industries in Vietnam (reaching 7.3%/year). However, the average monthly income of workers in the footwear industry is still much lower than the average income of workers in other industries in Vietnam (Statistical yearbook of Vietnam, 2021).

2.3.5 Working Environment

Research results of the authors Hien (2020) at leather and footwear enterprises in Ho Chi Minh City, one of the largest manufacturing centres for leather and footwear products in Vietnam, show the quality of the working environment. The activity of shoe factories is lower than that of garment factories. There are also indicators of occupational hygiene that do not meet the occupational hygiene standards of Vietnam. Specifically, the temperature in the factory is often high above 30 °C. The wind speed is much different at different locations, where the wind speed is much higher (3.5 m/s). The areas in the shoe factory mostly have substandard light intensity. Low dust content in all regions (0.1–0.3 mg/m^3). The noise parameter has many locations of high intensity (>75 dBA). The solvent vapour content is high. There are locations where the standard is exceeded, in the locations where glue is used. In the summer in Vietnam, most areas are hot and humid, so the occupational hygiene criteria, such as temperature and humidity in the factory are even higher than in Ho Chi Minh City. Meanwhile, shoes are manufactured using cold-soled glue technologies, boiler vulcanized rubber soles and solvent glue, so besides affecting workers' health, they also emit emissions large amounts of solvents and chemicals into the environment. In the coming time, this is also a common difficulty for Vietnamese leather and footwear enterprises, especially small and medium-sized enterprises, when the Free Trade Agreements (FTAs) take effect and environmental and labour issues to be adhered to. In addition, the use of equipment with low automation, many manual jobs and not applying Lean production are also the causes for the slow improvement of the working environment in the leather and footwear industry. Thus, based on the above data, human resources and the working environment of Vietnam's leather and footwear industry have the following advantages and limitations.

Advantages: The leather and footwear industry employs a large labour force and creates many jobs for society, especially since the industry employs many untrained and untrained workers in rural areas. In addition, Vietnamese workers are assessed as hardworking, eager to learn and quickly absorb work and easily adapting to changes in production technology and production management organization. Currently, the salary paid to Vietnamese workers is still low compared to many countries in Southeast Asia.

Limitations: In the current context, Vietnam's human resources have not yet met the development needs of the leather and footwear industry, especially the human resources in design, and middle and high-level production management. For direct production workers, female workers account for 80%, mostly just graduated from high school or lower secondary school. Although young workers are an advantage for training and improving labour productivity, the majority of workers in this group work in small and medium-sized enterprises without sufficient resources for training, fostering and developing higher-quality human resources. While the income level of workers in the leather and footwear industry is lower than in other industries, it is increasingly difficult to attract workers to work for leather and footwear enterprises. The inferior quality of human resources is also a reason hindering technological innovation and digital transformation in Vietnamese leather and footwear enterprises.

Due to the characteristics of products, technology and production equipment, and production organization, the working environment in leather and footwear enterprises, especially small and medium enterprises, is not good. The amount of waste, especially gaseous waste, is quite high due to the high use of solvents and chemicals. In the coming time, leather and footwear businesses are facing difficulties in sustainable development when FTAs force businesses to comply with environmental and labour requirements and regulations (i.e., when implementing FTA commitments, tariff barriers will be removed, but non-tariff barriers will be erected). For example, Germany will introduce new laws on due diligence of supply chains; or the EU will impose a tax on CO_2 emissions per unit of product. This is also the biggest difficulty, especially for small and medium-sized leather and footwear enterprises (Annual Report, Lefaso, 2021).

2.4 Current Status of Equipment and Technology

2.4.1 Equipment and Machinery for the Production of Shoes

According to a survey by Lefaso (DTDL.XH.15/18, 2020), footwear manufacturing equipment in Vietnam is mainly equipment from Taiwan (50%), Korea (20%) and China (10%). In general, the equipment has a high level of mechanics. However, the proportion of semi-automatic or automatic equipment is not high, including programmable cutting machines, milling systems, printing and embroidery systems, and programmable sewing.

Equipment at the material cutting stage is rated as medium to advance in the region. Some large-scale domestic and FDI enterprises have invested in new generation programmable cutting equipment. Currently, about 60% of equipment in the cutting stage has been used for more than 10 years.

The equipment used in the shoe upper sewing process is mainly imported from Taiwan, Korea and some from China. There are a few businesses that use semi-automatic specialized sewing machines, automatic thread trimming and sewing according to the programs in Japan and Germany. About 65% of the equipment in the sewing stage has been used for more than 10 years.

The shoe assembly and finishing equipment systems are mainly from Taiwan and Korea. Some enterprises import retail equipment from Italy, mainly mound machines, shoe sole presses. Currently, about 80% of equipment at this stage has been used for more than 10 years.

Vietnam's leather and shoe industry has an average level of production technology. The production process is being mechanized but not automated. The percentage of manual work is still high. This is one of the direct causes of low labour productivity in the industry, so workers have to regularly work overtime. However, in recent years, a few businesses have invested in complete automation systems of design, application of CAD and CAM in prototyping and production management, quickly responding to customer requirements (Bui et al., 2020a).

2.4.2 Technological Equipment for the Production of Bags

Most of the bag production equipment is sewing machines imported from Taiwan, Korea and China. For the cutting stage, many enterprises still use simple cutters and cutting tools such as ring cutters and cutters (Lefaso, DTDL.XH.15/18, 2020).

2.4.3 Technological Equipment for the Production of Leather

In Vietnam, most of the machinery and equipment in leather production are of a new generation, originating from Italy, Taiwan, Korea, India and China. Up to now, the level of mechanization in the domestic leather production sector has been raised. Many manual steps, such as lifting and transferring materials and chemicals, have been mechanized. In general, more than 50% of equipment used in the tanning industry is of good quality, 30% of medium quality and nearly 20% of poor quality, outdated and needs to be replaced (Lefaso, DTDL.XH.15/18, 2020).

In terms of technology, most FDI enterprises and some large Vietnamese enterprises are applying advanced foreign standards and technological processes. The level of tanning technology of these enterprises is rated as medium and advanced in the region, capable of producing good quality leather goods. Although many countries have applied clean and environmentally friendly technologies to reduce environmental pollution and limit the amount of toxic chemicals remaining in leather products, in Vietnam, due to investment costs, only foreign-invested enterprises or domestic enterprises with export products can use this ecological technology (Lefaso, DTDL.XH.15/18, 2020).

With the fact mentioned above, the equipment and technology for manufacturing leather and footwear products in Vietnam have the following advantages and limitations.

Advantages: Most of the production equipment in the leather and footwear industry is imported from Taiwan, Korea and China. This is a device with a cheaper price than equipment of European origin, the conditions for maintenance and replacement of accessories are also more favourable. In the past, the low level of automation of equipment and the use of many manual processes is also suitable for the human resource level of Vietnam's leather and footwear industry, as well as promoting the advantage of low labour costs in Vietnam.

Limitations: Technological equipment in Vietnam's leather and footwear industry is mainly at the level of mechanization, not yet automated. The percentage of manual work is still high. This is one of the direct reasons why the labour productivity of the industry is always low, resulting in employees having to regularly work overtime. Combined with the major production method is outsourcing, so the production and business efficiency of enterprises is still not high. The lack of modern equipment and a high degree of automation is also the reason why it is difficult for enterprises to apply the form of lean production organization, difficult in the digital transformation of enterprises, as well as difficult to improve the working conditions and environment of employees.

2.5 Current Situation of Producing Raw Materials

Raw materials account for up to 70% of the cost structure of footwear production. Meanwhile, the supporting industry for the production of raw materials in Vietnam is not developed, so the localization rate of leather and footwear products has not been improved. The product localization rate of raw materials and accessories produced in Vietnam is currently about 50% (Annual report, Lefaso, 2021). Important materials such as leather, synthetic leather, fabric for uppers must be imported in large quantities and at high prices (Table 5). Each year, Vietnam has to import about 1.5 billion USD of leather for export production (Statistical yearbook of Vietnam, 2021).

Currently, about 70% of Vietnam's leather, footwear and handbag manufacturing enterprises follow the processing method. Therefore, enterprises are usually not actively looking for and ordering raw materials for production but must follow the instructions of customers, often foreign partners. This is one of the barriers to the development of the production of raw materials and accessories in the leather and footwear industry in Vietnam.

In addition, most of the quality of raw materials and products have not met export requirements related to physical and aesthetic criteria such as color uniformity and color fastness, as well as requirements on biological safety. Some manufacturers meet the requirements for export products but do not meet the delivery schedule, due to the rapid change of samples and short delivery times.

On the other hand, the source of raw materials imported from China with cheap price, rich variety and quite good quality has made domestic enterprises not dare to

Table 5 Satisfaction level of domestically produced materials in Vietnam's leather sector, in 2020

Material	Percentage of meeting the needs of the leather sector (%)	Using purpose
Leather	26.0	Making shoe caps, bags, backpacks
Artificial leather	52.0	Making shoe caps, bags, backpacks
Woven fabric	55.5	Making shoe caps, bags, backpacks
Nonwoven	22.5	Making lining, cushioning shoe uppers, bags
Carton	26.7	Making insoles
Textile material	52.0	Producing shoes, bags, backpacks
Metal and plastic	51.0	Producing shoes, bags, backpacks
Adhesives, solvents, polishing chemicals	53.0	Producing shoes
Shoe's sole	70.0	Producing shoes

invest in the production of raw materials and accessories, due to the fear of not being able to compete in the market price and quality with Chinese goods.

According to Lefaso's survey data (DTKHCN.05/17, Lefaso, 2018), the total number of enterprises producing raw materials, equipment and spare parts (excluding leather tanning) updated by the end of 2018 is about 340 enterprises. In which, foreign enterprises (mainly from Taiwan) accounted for 30%, and Vietnamese enterprises accounted for 70%.

The lack of initiative in the source of domestic materials is also a barrier for leather and footwear enterprises to transform production methods as well as deeply participate in the global supply chain.

2.6 Current Situation of Production Methods, Management, Quality Management

2.6.1 Production Methods

The main model of Vietnam's leather and footwear industry is outsourcing, imported raw materials and accessories up to 50%, produced under international brands, so it hardly participates in the marketing and distribution phase (Fig. 7). Thus, the industry's profit is mainly from manufacturing, accounting for a low proportion of the value-added chain of leather and footwear products (about 3 USD/pair of shoes priced at 100 USD) (Wade and Motawi, 2018).

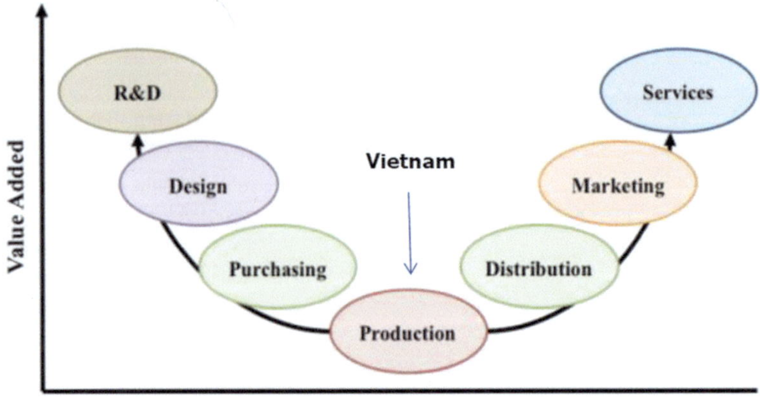

Fig. 7 Model of the value-added chain in the leather and footwear industry (FPT global digital transformation conference, 2019)

Outsourcing production method helps businesses quickly create many jobs for employees, acquire production management skills and advanced technology, save investment capital and avoid risks in the output market. However, outsourcing production also has negative effects such as dependence on foreign partners, enterprises being passive in receiving orders and deploying production; low profits, and no accumulated capital for reinvestment and development; foreign customers completely controlling the management and quality control of products.

The weakness of Vietnam's leather and footwear industry is concentrated in the early stages of "research and development" (R&D), "design" and "supply of raw materials", which has inhibited the growth of the industry. Development of the industry due to dependence on foreign customers for design and supply of raw materials. At the last stages of the value chain, Vietnam has not built a marketing network, distributing and retailing products in the importing country, thus, these are also weaknesses of the industry.

2.6.2 Production Management

Currently, many jobs in the production management process of the leather and footwear industry still have to be done manually, so productivity is low, despite using a lot of managers and executives. This leads to untimely production management, especially in large-scale production enterprises. In small and medium enterprises, the middle management apparatus accounts for 15–20% of the total number of employees. While large enterprises and FDI enterprises have effective management systems, this division only accounts for about 10% (Bui et al., 2022).

Vietnamese leather and footwear enterprises, especially small and medium enterprises, organize production according to the method of traditional long chains:

detailed cutting and preparation lines; upper line sewing, production line and preparation of shoe sole details, sole assembly line and shoe finishing line. These enterprises have not yet applied the lean production method. Lean production has just been applied to some large enterprises and FDI enterprises.

Likewise, the application of new ERP (Enterprise Resource Planning) software and digital management tools are applied in FDI enterprises and a few Vietnamese enterprises with relatively large scales (above 10,000 employees). The reason is that the investment costs are quite expensive and businesses need to change the entire production line system to operate effectively. Besides, in order to apply ERP, requires the determination of leaders and good human resources, 70% of small and medium enterprises in the industry cannot afford to apply (Bui et al., 2022).

2.6.3 Product Quality Management

The surveys of enterprises in the leather and footwear industry show that most of these enterprises have applied the quality management system, according to ISO 9001-2015; environmental management system ISO 14000; SA8000 social responsibility system, ISO/IEC 17,025 laboratory management. Most leather and footwear exporters use the product quality standard system and international standard testing methods according to customers' requirements.

FDI enterprises have paid attention to chemical safety standards and have invested in building testing laboratories with full equipment. Meanwhile, most small and medium-sized leather and footwear enterprises do not have a laboratory fully equipped with equipment to analyse and test the criteria in raw materials and leather footwear products. Some large domestic enterprises have invested in building laboratories, but mainly for physical and mechanical criteria. Many chemical indicators still have to hire foreign testing companies in Vietnam or send them abroad for testing (Lefaso, DTDL.XH.15/18, 2020).

2.7 Current Situation of Application of Automation, Digital Transformation

FDI enterprises and some large domestic private enterprises have basic technology and equipment reaching advanced levels in the region. In the past 5 years, these enterprises have continued to invest in developing and expanding production, through adding more production lines in the existing area and investing in building new and modern factories to welcome new opportunities to export associations to markets that have signed FTAs with Vietnam.

Small and medium enterprises have been replacing and renewing machines and equipment for the production of shoes and bags. The new machinery and equipment lines are arranged according to the short conveyor system, the structure is more complicated than the previous generation of machines and equipment. However, most of these enterprises do not have the financial and technical capacity to invest in expanding and rapidly renewing technology and equipment.

According to research results, the actual application of automation and digital transformation in the leather and footwear industry in Vietnam in 2020 is as follows.

(1) The average degree of automation of shoe and bag businesses reaches about 25%, of which large enterprises reach nearly 50%. Shoe sole manufacturers have about 15% automation because the investment cost for automation is higher than equipment for shoe and bag production;
(2) About 35% of small and medium enterprises apply management software in purchasing and supply. Average enterprises that have applied purchasing and warehouse management software reach about 40–50%, while this rate in large enterprises reaches 80–90%. Enterprises apply software in production management at less than 50%; and
(3) Enterprise management software is applied in about 40% of total enterprises. Integrated systems are applied in about 19% of enterprises and are also concentrated mainly on large enterprises and FDI.

The application of automation in the leather and footwear industry has a great impact on changing today's workforce. By 2020, automation applications have reduced nearly 12% of the industry's workforce, or about 158,000 workers (Bui et al., 2020b).

It can be seen that the application of automation and digital transformation in the leather and footwear industry is currently very limited, due to the following reasons:

(1) Digital technology application is mainly concentrated in large units and foreign-invested units. Small and medium-sized enterprises lack capital, are limited in investment in advanced technology, and are slow to invest in the automation of production stages;
(2) The level of human resources of enterprises is still limited. Enterprises lack high-level technical staff and management capacity is not high;
(3) The precision machinery and mechanical industries in Vietnam are not capable of manufacturing machinery, equipment and moulds for the production of leather and footwear products, so businesses have to spend a lot of money to import machinery and spare parts from abroad;
(4) Domestic enterprises import machinery, equipment and spare parts for the production of shoes and leather products from many sources, so the technological chain is not uniform, lacks synchronization and product quality is therefore not uniform. Thus, the maintenance of equipment faces many difficulties; and
(5) Labour costs are still low, so it is more effective to use more labour, instead of investing in modern technological equipment (Bui et al., 2022).

2.8 Current Situation of Logistic

Logistics is a circle that includes activities such as storage of goods, packaging, packaging, warehousing, freight movement and customs clearance, in order to achieve the ultimate goal of distributing products and goods optimized from supplier to consumer. Most Vietnamese logistics enterprises are still small and lack contacts, only providing transportation services, warehousing and carrying out paperwork related to import and export activities, freight transport. Large logistics enterprises providing package services are often foreign-invested enterprises, leading to increased logistics costs and limited choice of production and business enterprises.

Meanwhile, the characteristic of the leather and footwear industry is to export an enormous volume of products. In 2021, the export value is 20.78 billion USD and the import of raw materials for production is 10–11 billion USD. Therefore, logistics greatly affects production and export costs and costs. According to Lefaso's estimates, in 2021 alone, the cost of logistics activities in Vietnam is much higher than in neighboring countries and in the region. Specifically, Vietnam's logistics costs are 6% higher than Thailand, 7% higher than China, 12% higher than Malaysia and 3 times higher than Singapore's (Annual Report, Lefaso, 2021).

2.9 Current Situation of Solid Waste and Solid Waste Treatment

According to the research results of the Vietnam Leather and Shoes Research Institute (LSI) in the project "Surveying and assessing the consumption of raw materials and solid waste in the leather and footwear industry in Vietnam" (LSI, 4123/QD–BTC, 2015), Vietnam's leather and footwear industry produced 980 million pairs of shoes of all kinds and about 150 million bags. The volume of solid waste in the production of shoes and leather goods in Vietnam in 2015 was estimated at 145,242 tons. According to the estimate of this study, the output of shoes and leather products produced in 2021 is 1.32 million pairs, bags of all kinds are 450 million products and solid waste will be about 245,000 tons.

Actual survey results at enterprises producing shoes and leather products (LSI, 4123/QD-BTC, 2015) show that most of the waste is treated by environmental companies along with waste from other industries. Some types of waste, such as leather, plastic and packaging waste, are sold to the private sector for recycling or reuse. However, this rate is quite low and the use of this waste is not controlled.

During the production of leather, only about 45–55% of raw leather becomes finished leather, and the rest about 55–60% becomes waste, including liquid waste, gaseous waste and most solid waste. Organic and inorganic have a negative impact on the environment. To treat one tonne of raw leather, about 300 kg of chemicals are used in the production process and about 45 m^3 of water is consumed (Lefaso, DTKHCN.05/17, 2018).

In 2015, survey data at Vietnamese leather product manufacturers (LSI, 4123/QD–BTC, 2015) showed that to produce 200 kg of finished leather from about one ton of salted leather, the amount of solid waste generated includes: untapped solid waste about 150–250 kg of leather, 100–150 kg of subcutics; Waste from trimming frills and shavings is about 100–150 kg. In 2015, with an output of 230 million Sqft of leather produced, equivalent to using about 180 thousand tons of salted leather, Vietnam's leather industry emitted 95,600 thousand tons of solid waste to the environment. With an output of 350 million Sqft in 2021, the amount of solid waste from Vietnam's leather industry will increase to 145,500 tons. The management and treatment of solid waste of Vietnamese tannery enterprises have the points to pay attention as follows.

(1) Enterprises are interested in the management and economical use of raw materials in order to minimize solid waste emissions. However, the application of ecological technology and cleaner production is still very limited, so solid waste has not been reduced right from the "beginning of the pipeline";
(2) All non-recyclable and reusable wastes are handled and transported by the enterprise to an environmental company with legal status;
(3) Some organic solid waste before the tanning process is compostable and has high protein content, businesses have implemented reuse solutions to produce some by-products such as fat recovery, manufacturing gelatin, making animal and aquaculture feed, using fat and pre-tanned leather to create biogas as fuel for burning, generating heat and producing bio-organic fertilizers. However, the products from this type of waste are of inferior quality; and
(4) Some other types of solid waste, such as plastic, paper, wood and metal containers of businesses, can also be reused, businesses can be sold to individuals and organizations in need (LSI, 4123/QD–BTC, 2015).

3 Affecting Factors of the Development in Vietnam's Leather Sector

3.1 Market Trends and the Development of the Leather and Footwear Industry in the World

According to the research and synthesis results of Lefaso, Vietnamese organizations and researchers, in the coming time, the following factors will have an impact on the market and development of the world leather and footwear industry as follows.

3.1.1 Fourth Industrial Revolution

The fourth industrial revolution is based on the digitization of the Industrial Revolution 3.0 but differently. It is that machines have begun to replace and support human thinking, or intellectual capacity. This will create unprecedented changes and is expected to create a new turning point in production and business activities.

The industrial revolution 4.0, or more precisely, digitization has fundamentally changed the global production system, completely changing the way people live, work and connect with others; digitization also creates new businesses and industries, changes traditional business models and reshapes the global economy; Applications of big data, real-time, IoTs, or e-commerce are already present on every stage of the value chain.

However, it should be emphasized that core activities are still actual activities such as research, design, production of goods, logistics, and sales. These are fundamental activities for the application of information technology across the entire value chain. With the help of big data, identifying market trends becomes easy, fast and accurate. At the production stage, with the support of real-time management technologies, the production processes are optimized, controlled and reduced wasteful costs in the production process.

3.1.2 US–China Trade War

The trade tension between the US and China that started in early 2018 has been yet ended. The dispute between the two countries has had obvious impacts on the trend of global trade development, including the leather and footwear industry. Currently, China is the world's largest footwear producer and exporter, accounting for 34.9% of the total world turnover. Over the past five years, US footwear imports from China have fallen by about $3 billion a year, roughly equal to the increase in imports from Vietnam. However, this issue is not entirely detrimental to Chinese businesses. Chinese enterprises have adapted to the new strategy of increasing investment and shifting production lines to other countries such as Vietnam, to get Vietnamese origin to avoid punitive taxes from the US. Statistics show that investment from China and Hong Kong in Vietnam's leather and footwear industry tends to increase. The reason is partly due to the impact of the US–China trade war, but on the other hand, it is also to take advantage of market opportunities and anticipate the new generation FTAs that Vietnam has signed and joined.

In addition to Vietnam, Chinese enterprises also shift their origin to countries in South America, the Caribbean and Africa, especially Brazil. In addition, Chinese footwear enterprises have increased their export efforts to Europe and other high-income countries, such as Japan and South Korea.

3.1.3 The COVID-19 Pandemic has Changed the Supply Chain and the Trend of Investment

The COVID-19 pandemic that occurred at the end of 2019 has accelerated the trend of investment shifting of multinational companies from China to other countries due to concerns about too much dependence on a single supply. Due to the impact of the trade war and the pandemic, many international brands have efforts to shorten the supply chain, sourcing locally. The COVID-19 crisis has been a driving force behind the automation of plant operation and management systems in many industries. For example, Nike has plans to digitize its supply chain to reduce delivery times from 60 to 10 days. If successful, this plan could reduce the number of steps in the process by 30%, and based on this simplification, the ability to recover from a crisis will be faster and more efficient (Statistical Report, Lefaso, 2021).

3.1.4 The Conflict Between Ukraine and Russia

The military conflict between Russia and Ukraine is severely disrupting global supply chains, cutting off many important shipping routes, forcing transport companies to suspend service and rising air costs.

3.1.5 Fast Fashion

Fast fashion has dominated the fashion market in recent decades and has caused unsustainable impacts on the environment and society. Along with clothes, shoes and bags are also in this trend. The time from when shoe and handbag designs are introduced during international fashion weeks to their mass sale has been reduced, from one year to just three to four weeks. As such, if production is not on time, delays in the delivery period mean that a retailer's profits can drop from 50% to just 5%. Besides, the footwear retail system is also undergoing many changes. Independent stores specializing in shoes and bags will gradually decrease and increase sales in general supermarkets. In particular, the trend of selling on online platforms is growing due to the superiority of this business model.

3.1.6 The Trends in the World Leather and Footwear Industry

In Lefaso's summary report in 2021, the experts forecasting the world leather and footwear industry have the following notable points.

In the next 10 years, the world's population is forecast to increase by about 10%, from 7.8 billion at present to 8.6 billion in 2030. In which, the increase is mainly in Asia, Africa and South America. Meanwhile, the world's average footwear consumption is about 3.1 pairs/person/year. It is expected that by 2030, this index will increase to 3.5 pairs/per person/per year. It is forecasted that footwear consumption demand will increase by about 2.5–3.0 billion pairs, i.e., about 27–27.5 billion pairs of shoes per year. Specifically, the average shoe consumption of people in North America is quite high, about 5.6 pairs/person/year, 7.0 pairs/person/year in the US alone, and 4.6 pairs/person/year in Europe. However, the population has barely increased in high-income countries such as North America, Western Europe and some Asian countries such as Japan and Korea. Therefore, footwear consumption is forecast to increase in volume only in Asia, Africa and South America. These additional markets often use mid-end to low-end footwear.

Despite the impacts of the COVID-19 epidemic, the footwear industry continues to grow strongly in the Asia region, which produces 90% of the world's shoes and continues to increase by 0.2% in the global production structure. China is still the world's largest footwear producer, accounting for 54.3%. However, in 2020, China has reduced production by more than 2 billion pairs and is on a downward trend in world market share (down 1.0% points), which is reflecting a gradual shift of production from China to other Asia countries.

Over the past 10 years, footwear products with fabric uppers have become increasingly popular, growing 14.6% and now account for a third of the shoes on the market. Footwear with leather, rubber or plastic materials is on the trend of decreasing market share. To date, rubber and plastic footwear accounts for nearly half of the total number of footwear on the market, about 47.5%, while leather footwear accounts for about 38.8%.

From 2011 to now, the export price for each pair of shoes has increased by an average of 3.3% per year. Particularly in 2020, despite being negatively affected by the COVID-19 pandemic, the price increase rate was higher, reaching 6% and reaching an average price of over 10 USD/pair for the first time.

3.2 *Vietnam's Socio-Economic Development*

Despite being heavily affected by the COVID-19 pandemic, Vietnam's economy has still achieved positive results. In 2021, Vietnam's economy will grow by 2.58% compared to 2020. In 2021, export turnover will still maintain a growth of 19% compared to 2020. The signing of free trade agreements has brought about positive signals for the Vietnamese economy, especially the Free Trade Agreement (EU-Vietnam Free Trade Agreement, (EVFTA)) between Vietnam and the EU (Statistical yearbook of Vietnam, 2021).

In terms of population structure, Vietnam is undergoing changes, the population reached over 97.5 million in 2020 and is forecast to increase to 120 million by 2050. In 2019, the population under 35 years old accounted for 55.5%, while the average life expectancy of Vietnamese people is nearly 76 years old, higher than other countries with similar incomes in the region. A middle class is also forming, accounting for 13% of the population and is expected to reach about 26% by 2026. This shows that with a young population structure and a large working-age population, the leather and footwear industry plays a very important role in creating jobs, and promoting the shift of economic structure and labour structure towards industrialization and modernization. Because this is an important industry for workers, especially workers in rural and remote areas.

Along with the high and stable growth of the economy, Gross domestic product (GDP) per capita also increased rapidly. In 2021, Vietnam's GDP per capita will reach 2786 USD, GDP growth in 2021 will reach 2.6%, and the first 9 months of

2022 will reach 8.83% (according to the World Bank). Although from the end of 2019 until now, the COVID-19 pandemic has taken place on a global scale, causing great impacts on the leather and footwear industry, the Vietnamese government has made many timely decisions to support businesses such as reducing taxes, fees, electric prices, and supporting workers. Therefore, the economy is recovering and continues to grow well, so that the spending level on footwear has also increased. The vigorous growth of the domestic market is also an advantage for the development of Vietnam's leather and footwear industry.

3.3 The Impact of Free Trade Agreements in Vietnam's Leather Sector

Since joining ASEAN in 1995, Vietnam has signed 15 multilateral and bilateral free trade agreements and is negotiating two new ones (Table 6).

The member countries of the two new-generation agreements Comprehensive and Progressive Agreement for Trans-Pacific Partnership (CPTPP) and EVFTA account for about 50% of the world market share in leather and footwear. When these agreements have not yet come into force, Vietnam's leather and footwear products are subject to quite high tax rates in these markets, making it difficult to compete with other countries enjoying GSP+ incentives such as Bangladesh, Pakistan and Cambodia. When completing the tax reduction roadmap as committed in CPTPP and EVFTA, Vietnam's leather and footwear products will also enjoy the same 0% tax rate as Bangladesh, Pakistan and Cambodia (Vietnam Ministry of Industry and Trade, 2020).

CPTPP and EVFTA are also factors attracting many domestic and foreign projects to invest in the leather and footwear industry in Vietnam, especially raw materials for leather and footwear to meet the requirements of origin. In particular, in the past five years alone, when the process of negotiating and signing the CPTPP and EVFTA is completed, the number of leather and footwear projects has increased sharply again, accounting for 40% of the number of projects and over 50% of the capital. Registered in the total number of projects and capital from 1988 to the present, investors are mainly from Korea, Taiwan, Hong Kong, China and Japan (Vietnam Ministry of Industry and Trade, 2020).

The growth trend of Vietnam's leather and footwear exports from 1995 to now shows that after each new FTA comes into effect, Vietnam's export turnover has achieved a higher growth rate, which shows that the positive impact of signed FTAs with the trade turnover of Vietnamese leather and footwear. The export of leather and footwear has really increased sharply since Vietnam joined the WTO in 2007, and since then, the trade surplus has increased every year, showing the positive impact of FTAs on the trade balance of the leather and footwear industry in Vietnam.

Table 6 FTA from Vietnam to the world

FTAs has come into effect			
1	AFTA (ASEAN Free Trade Area)	From 1993	ASEAN
2	ACFTA (ASEAN-China Free Trade Area)	From 2003	ASEAN—China
3	AKFTA (ASEAN—Korea Free Trade Area)	From 2007	ASEAN—Korea
4	AJCEP ((ASEAN—Japan Comprehensive Economic Partnership Agreement))	From 2008	ASEAN—Japan
5	VJEPA (Vietnam-Japan Economic Partnership Agreement)	From 2009	Japan
6	AIFTA (ASEAN-India Free Trade Area)	From 2010	ASEAN—India
7	AANZFTA (ASEAN-Australia-New Zealand Free Trade Area)	From 2010	ASEAN—Australia, New Zealand
8	VCFTA (Vietnam-Chile Free Trade Agreement)	From 2014	Chile
9	VKFTA (Vietnam-Korea Free Trade Agreement)	From 2015	Korea
10	VN-EAEU FTA (Vietnam—Eurasian Economic Union)	From 2016	Russia, Belarus, Amenia, Kazakhstan, Kyrgyzstan
11	CPTPP (Comprehensive and Progressive Agreement for Trans-Pacific Partnership)	From 30/12/2018 With Vietnam 14/1/2019	Canada, Mexico, Peru, Chile, New Zealand, Australia, Japan, Singapore, Brunei, Malaysia
12	AHKFTA (ASEAN-Hong Kong, China Free Trade Agreement)	From 12/02/2021	ASEAN—Hong Kong
13	EVFTA (EU–Vietnam Free Trade Agreement)	From 01/08/2020	EU (27 countries)
14	UKVFTA (Vietnam—UK Free Trade Agreement)	From 01/05/2021	United Kingdom

(continued)

Table 6 (continued)

FTAs has come into effect			
15	RCEP (Regional Comprehensive Economic Partnership)	From 01/01/2022	ASEAN-China-Korea-Japan-Australia-New Zealand
FTA under negotiating			
16	Việt Nam—EFTA FTA	Started from May/2012	EFTA (Swizerland—Norway Iceland, Liechtenstein)
17	Việt Nam—Israel FTA	Started from December/2015	Israel

4 Strengths and Weaknesses of Vietnam's Leather Sector

4.1 Strengths

It can be assessed that Vietnam's leather and footwear industry has advantages to develop and attract investment as follows.

(1) The investment environment has been continuously improved positively, along with the process of international economic integration. Vietnam is becoming an attractive destination for investors, including investment in the leather and footwear sector;
(2) The leather and footwear industry is one of the priority sectors of Vietnam's industry. The industry is entitled to participate in the State's encouragement and support programs to modernize and improve competitiveness;
(3) The industry has also built and developed numerous enterprises, their production capacity is quite high; and
(4) Vietnam currently has the advantage of human resources with a young age structure, the ability to quickly adopt new technologies, and the average wage cost in Southeast Asia.

4.2 Weaknesses

Besides the strengths, the leather and footwear industry in Vietnam also has some barriers to the development of the industry, which are:

(1) This industry consists mainly of small and medium enterprises, about 70%. Enterprises of this scale have many limitations in terms of capital, technique and technology, and do not have the initiative to the source of raw materials and auxiliary materials, so most businesses cannot immediately benefit from the advantages of FTAs.

(2) Moreover, enterprises in the industry mainly process products, so they depend a lot on customers and foreign partners for designs, production materials and export markets.
(3) Despite having a large workforce, the industry lacks a workforce with high technical and managerial qualifications. The rapid increase in minimum wage increases labour costs, while the labour productivity of Vietnamese leather and footwear enterprises is still much lower than that of other countries in Southeast Asia;
(4) The supporting industry does not develop in parallel with the development of the leather and footwear industry. There is a lack of synchronous policies to promote the development of supporting industries for the leather and footwear industry, along with limited technological equipment and capabilities, and complete dependence on imports;
(5) Due to the impact of the Industrial Revolution 4.0 trend, fluctuations in world trade and protectionism, the export of Vietnam's leather and footwear industry will face many difficulties. While scientific-technological research and technology application support services have not really been developed; and
(6) Technical barriers to importing countries, labour and environmental requirements and compliance with FTA procedures will also cause difficulties for Vietnamese leather and footwear enterprises.

5 The Sustainable Development of Vietnam's Leather Sector: Opportunities and Challenges

5.1 Criteria for the Development of Vietnam's Leather Sector Towards Sustainability

On the basis of assessing the current situation of Vietnam's leather and footwear industry, clarifying domestic and foreign factors affecting the development of the industry in the coming time, along with the strengths and weaknesses of the leather industry. We analyse opportunities and challenges in the sustainable development of the leather and footwear industry in three aspects: (1) Improving economic efficiency in production and business; (2) Improving working conditions, and raising incomes for employees; and (3) Reducing the environmental impact of the leather and footwear industry. As mentioned above, these three aspects are relatively independent and dependent, closely related to each other. In the current conditions of Vietnam's leather and footwear industry and development trends to 2030, we believe that Vietnam's leather and footwear industry will develop sustainably according to the following important points.

Expanding markets and reducing export taxes: Vietnam's footwear production is mainly for export purposes. Therefore, taking advantage of free trade agreements to reduce export taxes and increase the competitiveness of products is very important.

This is also the basis for improving production and business efficiency and income for employees.

Applying automation and digital transformation to improve labour productivity and product quality: In the next 10 years, shoe production and processing will still be the main activity of Vietnam's leather and footwear industry. In the context of increasing labour costs and strong competition for labour with other industries, the application of automation, Lean application and digital transformation to improve labour productivity, quality products are very important for footwear enterprises, especially small and medium enterprises (Bui et al., 2020a). This might help to improve the working environment, increase income for employees and minimize the environmental impact by reducing waste.

Strongly participation in the global leather and footwear supply chain: Production is the lowest value-added phase in the leather and footwear supply chain. With low processing prices and low labour productivity, enterprises have difficulty paying high wages to workers. Deeply participating in the stages of the chain with high added value is the orientation of leather and footwear enterprises to improve production and business efficiency as well as improve income for employees. At the back end of the chain, such as distribution and retail, businesses have very little, or only in the domestic market. Vietnamese footwear enterprises can participate in the product design and development stages, as well as the supply of raw materials. However, the implementation of these stages depends on many factors, including high-quality human resources and the development of supporting industries. For FDI enterprises and large Vietnamese enterprises, the ability to participate more deeply in these stages of the supply chain has a higher potential with outstanding resources. Vietnam's small and medium-sized leather and footwear enterprises that are manufacturing under CMT (Cut-Make-Trim model) will face many difficulties in participating more deeply in the global leather and footwear supply chain.

Improving working conditions, improving the quality of human resources, raising incomes for workers: Working conditions of workers in the leather and footwear industry are still limited and are underestimated compared to the textile and garment industry, especially in small and medium enterprises. The income level of workers in the leather and footwear industry is lower than that of workers in other industries in Vietnam. With poor working conditions and low income, leather and footwear businesses will find it difficult to attract good workers. Human resources in science and technology with bachelor's degrees and engineers in the leather and footwear industry have not been fully met. Without promoting the increase in the quantity and quality of human resources, it is difficult for Vietnam's leather and footwear industry to successfully apply automation and digital transformation, as well as participate more deeply in the global leather and footwear supply chain. Some research results show that, by 2030, direct and indirect workers need to improve their skills, skills and professional and managerial knowledge. Direct workers need to perform many jobs. Over 40% of employees are trained to operate and use automatic equipment. Indirect employees need to make good use of management software, specialized software, especially enterprises using ERP (Bui et al., 2020b).

Reducing environmental impacts: Vietnam's leather and footwear industry has a significant impact on the environment due to the following reasons: low level of automation in production, using many manual processes; traditional production organization and management; using a lot of chemicals in the technology of glueing cold soles, boiler vulcanized rubber soles, organic solvent adhesives, materials that are difficult to decompose, and cannot be recycled. Without drastic changes, it is difficult for Vietnam to meet the environmental provisions in the new generation of free trade agreements, and the environmental criteria in the sustainable development strategy of footwear brands in the world.

5.2 Opportunities for the Development of Vietnam's Leather Sector Towards Sustainability

5.2.1 Opportunity to Expand the Market and Reduce the Export Tax

Vietnam has participated, negotiated and signed 17 bilateral and multilateral trade agreements, especially the CPTPP, EVFTA, etc., creating more opportunities to expand and access to export markets and promote investment and production development. From the conditions of trade agreements (such as EVFTA), domestic and foreign investors will tend to invest in developing the production of raw materials and accessories to enjoy preferences according to the origin. Thus, Vietnam's leather and footwear industry can improve the domestic supply of raw materials, increase the localization rate of products.

Increased FDI investment brings high export turnover, creates more jobs and contributes to the development of a strong leather and footwear industry.

Production costs in China have increased. This leads to China's exports decreasing, thus, foreign orders and investment shifted to more affordable countries, including Vietnam.

The US–China trade war increases the import tax on Chinese goods into the US, which is also an opportunity for Vietnam to boost exports to the US market.

The rapid change with the variety of products of the domestic and international fashion market creates favorable conditions for businesses to promote product development and brand building activities.

The domestic market with an annual consumption of about 200 million pairs of footwear is a potential market for Vietnamese footwear enterprises.

5.2.2 Opportunities for Automation and Digital Transformation

Change in perception of leaders and business owners: Vietnam's leather and footwear businesses has realized the need to innovate production technology through the application of digitalization and automation technology. In the context of increasing domestic labor costs and import prices of raw materials, their goal is to increase labor productivity, reduce production costs, avoid being left behind and improve product quality and value. Products. Industry 4.0 helps businesses improve the design, production and trade of leather and footwear products, shifting from the CMT model to OEM (Original Equipment Manufacturer) and/or ODM (Original Design Manufacturer). In 2020, according to our research results (Bui et al., 2020a), the three main pillars of Vietnam's footwear business model by 2030 are automation, Lean manufacturing and digital transformation. By 2030, the level of automation application in large enterprises will increase sharply, especially in shoe sole manufacturing enterprises. Enterprises, especially large enterprises, will invest heavily in automatic equipment (cutting machines, sewing machines, rolling mills, pressing soles…). Automatic cutting machines can account for over 90%, sewing machines up to 80% and sole presses up to 85%. This is reasonable because, by 2030, about 80% of enterprises will apply Lean manufacturing, and need to use smart equipment and automation. By 2030, the existing equipment along with the traditional long lines will be depreciated, creating conditions for businesses to invest in automatic equipment.

Also, according to the research results of our project in 2020 (Lefaso, DTDL.XH.15/18, 2020), by 2030, if only automation equipment is applied, the total number of employees in the leather and footwear industry will be will decrease by about 17% (an average annual decrease of about 1.7%), while the labour productivity of the whole industry will increase by 20%. This reduction in the number of employees is not large and is not worrisome compared to the scale of expansion of production investment by about 10% per year of Vietnamese leather and footwear enterprises (Vietnam Ministry of Industry and Trade, 2020).

Technology and software enterprises in Vietnam are developing well: Many technology products and solutions "Made in Vietnam" apply new technologies (AI, Big data, IoT) suitable for Vietnamese enterprises born at a reasonable cost, especially for small and medium enterprises. Vietnam's large technology corporations are capable of consulting and implementing digital transformation roadmaps that are close to the needs and characteristics of Vietnamese businesses, opening up opportunities for digital transformation suitable for Vietnamese footwear enterprises (Ministry of Industry and Trade-UNDP, 2019).

Government's attention: The Vietnamese government is particularly interested in the digital transformation towards building a digital government, a digital economy, a digital society and a digital citizen. Many digital transformation and automation promotion action programs have been promulgated. The Government continues to make efforts to improve institutions, mechanisms and policies to support digital transformation for businesses in general and the footwear industry in particular (Decision No. 749/QD-TTg, 2020).

5.2.3 Opportunity to Participate More Deeply in the Global Footwear Supply Chain

The Vietnamese government is interested in developing supporting industries (e.g., manufacturing raw materials, lasts and moulds) for the leather and footwear industry. By 2030, the localization rate of Vietnam's leather and footwear products will be 65–70% (Resolution No. 115/NQ-CP, 2020).

After more than 30 years of establishment and development, Vietnam's footwear sector has really matured. Compared with other countries in the region, the level of technology and equipment is average and good. Science and technology human resources for the industry are interested in training and supplemented every year, although the number is still modest.

Vietnamese footwear enterprises are determined to change their production model from CMT to OEM and ODM to improve production and business efficiency in the context of increasing labour costs.

Small and medium enterprises have the opportunity to produce according to the OBM model to develop the domestic market.

FDI enterprises and large enterprises in Vietnam will be the driving force for the leather and footwear industry to participate more deeply in the global leather and footwear supply chain.

5.2.4 Opportunities to Improve Working Conditions, the Quality of Human Resources and Incomes for Employees

CPTTP and EVFTA are called new-generation FTAs, characterized by a strong emphasis on labour rights, and at the same time helping workers and businesses enjoy economic benefits equally. The world has reached a consensus on fundamental rights that should be respected by all nations, regardless of their level of economic development. This consensus is articulated in the 1998 ILO Declaration on Fundamental Principles and Rights at Work. The labor-related chapters of the CPTPP and EVFTA aim to enforce these rights.

Vietnam has ratified and committed to many conventions on human rights, principles and fundamental rights at work such as Convention No. 98 in 2019, and Convention No. 105 in 2020. Convention No. 87 is expected to ratify in 2023.

Small and medium-sized business owners and leaders are becoming more and more aware of the implementation of the Law on Occupational Safety and Health as well as the regulations on occupational safety and health issued by the Vietnamese government.

Major footwear brands such as Nike, Adidas, etc. have sustainable development strategies including objectives and implementation plans related to working conditions and income improvement for employees.

Every year, the government adjusts to increase the minimum wage by region—this is also the basis for businesses to increase wages for workers.

Competition in the leather and footwear labor market is becoming increasingly fierce due to the movement of the industry's human resources to other occupations with better income, especially electronics manufacturing or service industries. Vietnam is the country with the highest wage growth rate in the past 10 years, just behind China in terms of labor-intensive, export-oriented countries. This is also a condition for leather and footwear businesses to pay more attention to workers' incomes.

The application of automation, Lean manufacturing and digital transformation will not only increase labour productivity but also improve the working environment, which will help leather and footwear businesses better attract human resources to meet expanding needs of manufacturing. However, the application of automation, Lean and digital management will lead to higher requirements for labour qualifications, especially high-quality science and technology human resources (Lefaso, DTDL.XH. 15/18, 2020). This requires businesses to have a strategy to prepare and improve the qualifications and skills of employees. Enterprises in the industry need to pay attention to and strongly combine with human resource training institutions to increase the quantity and quality of human resources to meet the requirements of applying the achievements of the Industrial Revolution 4.0 in the leather and footwear industry in Vietnam.

5.2.5 Opportunities to Reduce the Environmental Impact of the Leather and Footwear Industry

For free trade agreements such as CPTTP and EVFTA, there are provisions that place great emphasis on protecting environmental sustainability, to ensure that free trade contributes to sustainable development. Strict regulations on carbon tax rates for exports (European CBAM, American Clean Competition Act) have imposed mandatory requirements on businesses in applying technology to reduce emissions and energy savings, such as electricity and water in leather and footwear production (Annual report, Lefaso, 2021).

One of the aspects that are interesting in the sustainable development strategy of major leather and footwear brands (e.g., Nike and Adidas) is to reduce the environmental impact and improve the ecological safety of the product. Environmental goals of footwear brands will be imposed on suppliers of raw materials and processing facilities of leather and footwear products.

The Vietnamese government insists on sustainable green development. The policies focus on attracting investment and giving incentives for the application of new environmentally friendly technologies.

5.3 *Challenges of the Development in Vietnam's Leather Sector Towards Sustainability*

5.3.1 Challenges to Market Expansion and Export Tax Reduction

Increased protectionism and disruption of the global supply chain create a tendency to move production closer to the consumption market, so Vietnam's leather and footwear industry will face difficulties in export activities in the near future.

In markets that use high-cost shoes such as the EU, the US and Japan, the population does not increase, so the output of shoes exported to these markets cannot increase sharply.

Technical barriers imposed from import markets, together with requirements on social responsibility, environmental protection and compliance with procedures to enjoy FTA tax benefits, especially on origin, are challenging Vietnamese leather and footwear enterprises.

Exported products are at risk of being faked by foreign enterprises of Vietnamese origin in order to enjoy preferential tariffs for Vietnam under FTA agreements and avoid being imposed sales tax on the US due to the US–China trade war.

The domestic market is subject to strong competition from cheap products from China and Thailand. Currently, Vietnam has no regulations on the ecological safety of imported leather and footwear products.

5.3.2 Challenges of Strongly Participation in the Global Footwear Supply Chain

When considering the footwear industry in the value chain and participating in the global market, it is necessary to consider that Vietnamese enterprises are mainly small and medium-sized. The market share of the created product lines does not occupy a monopolistic competitive position. Therefore, it is not possible to decide on the price of products, from input materials to output consumer products. Firms are therefore vulnerable to and affected by supply shocks or price shocks coming from external markets.

Analysis of the links in the leather and footwear value chain in Vietnam shows that, after many years of joining the global leather and footwear value chain, although the export turnover is high, the Vietnamese leather and footwear industry still focuses only on CMT. This is the low value-added stage. The biggest limitation of the industry is the uneven development at all stages, especially at the first stages in the leather and footwear value chain, including: product design and development, and raw material supply. Statistics also show the unbalanced development of the leather and footwear industry. While the whole industry has more than 2,600 enterprises, the number of enterprises producing raw materials and auxiliary materials accounts for just over 15%, and the number of enterprises with design activities is less than 20%

of enterprises (Lefaso, DTKHCN.05/17, 2018). Enterprises have not yet paid attention to building Research and Development (R&D) Centers to create added value for products.

Currently, Vietnam's footwear enterprises focus on the export market (accounting for 95% of the production capacity of the entire leather and footwear industry). This market is inherently volatile and difficult to control, and one of its consequences is the neglect of the domestic market (with 95 million people and increasing purchasing power) for neighbouring countries like China. The high dependence on imported raw materials (accounting for about 50% of total demand) makes Vietnam's already weak leather production situation less likely to be improved.

An important factor affecting the transformation of production methods to participate more deeply in the global leather and footwear supply chain is high-quality human resources. As mentioned above, human resources in the leather and footwear industry in Vietnam are both inadequate and weak, and in the coming period, there are no signs of improvement. Producing raw materials, shoe lasts, sole mould, etc., for the leather and footwear industry is a field that not only requires large investment capital but also requires human resources with high scientific and technological qualifications. The current situation of weak science and technology human resources is and will greatly affect the development of the supporting industry for the leather and footwear industry.

5.3.3 Challenges in Automation and Digital Transformation Applications

The Industrial Revolution 4.0 brings drastic changes in productivity, scale and management model, affecting all industrial sectors, including Vietnam's leather and footwear industry. Over 70% of small and medium-sized enterprises in the leather and footwear industry face difficulties in investing and applying new technologies. About 30% of enterprises, including FDI enterprises and large domestic enterprises, have applied automation in each production stage. The application of digital technology in leather and footwear businesses is facing five major challenges as below.

(1) *Low determination from business leaders*: Most businesses do not have a specific digital transformation strategy and roadmap. Many business leaders and footwear business owners focus on operating in the traditional way, and afraid of changing and accepting new things. The lack of leadership with the necessary commitment and will from leadership leads to digital transformation failure.

(2) *Resources for implementing digital transformation in the industry are still weak*: Human resources in the leather and footwear industry are mainly unskilled workers with limited professional qualifications and digital skills, low labour productivity, which directly affects the health and safety of customers' ability to absorb and apply new technologies in production and business activities.

(3) *The scale of leather and footwear enterprises is still small and fragmented*: The machinery, equipment and facilities for the production of leather and footwear are outdated, the production scale is fragmented, the technology and automation content is low. Factors that greatly affect the application of 4.0 technology in production.
(4) *Budget for digital transformation is high*: While the high cost of investment, technology application and automation hinders investment in digital transformation, the limited budget is an invaluable enormous challenge.
(5) *Lack of information technology infrastructure and architecture*: The lack of information technology architecture and corresponding infrastructure makes the deployment of software and systems fragmented, lacking data linkage and connectivity. In addition, the implementation of the latest technologies also requires high requirements in terms of architecture and deployment infrastructure (Bui et al., 2022).

5.3.4 Challenges to Improve Working Conditions, the Quality of Human Resources and Incomes for Employees

Workers in the leather and footwear sector in Vietnam have low qualifications. The main reason is that businesses mainly produce by CMT or OEM processing, so the level of human resources is just enough to meet this production method. High-quality human resources for design, marketing and supply chain management... are very lacking and weak because ODM and OBM production methods have not yet been deployed by businesses.

As analysed above, the leather and footwear industry has a large concentration of female workers, with low qualifications plus less working time than men, so the training process changes jobs or improving qualifications to approach the Revolution Industry 4.0 will be more difficult.

In addition to the environmental risks of water use and pollution, chemical management and greenhouse gas emissions, the footwear industry is a labour-intensive industry and therefore also faces potential challenges in compliance with Vietnam's legal obligations on decent work. The Better Work Program's 2019 annual report shows that the producers participating in the Program in Vietnam are the ones that do well with the regulations related to child and forced labor. However, other labour-related issues such as occupational health and safety, contracts, human resource development and working time are still areas for improvement to meet the provisions of the law Vietnam's labour (Labour Law, 2019).

5.3.5 Challenges of Environmental Impact Reduction

Currently, the shoe manufacturing industry in Vietnam is mainly using cold glue technology for sports shoes and vulcanization technology for rubber soles/canvas shoes.

These technologies will continue to be used in the near future, so businesses will face difficulties in reducing emissions into the environment, especially emissions.

In addition, the average technological equipment, low automatic ability and asynchronous and limited digital transformation are also challenges for businesses in minimizing environmental impacts.

The scientific research and technology transfer in minimizing the environmental impacts of the leather and footwear industry is still weak and has not met the needs of enterprises. Recently, there have been a few scientific works in Vietnam studying the treatment of scrap leather to make other products, such as artificial leather and plywood flooring. However, the results of these studies have not been applied to industrial production.

6 Some Proposed Strategies to Develop Vietnam's Leather Sector Towards Sustainability

6.1 Market Development Strategies

6.1.1 International Market Development Strategy

To develop international markets, governments and stakeholders should focus on:

(1) developing the capacity to forecast market demand. To facilitate and support leather and footwear enterprises to directly approach potential customers, through activities such as participation in trade fairs and exhibitions; trade promotion, investment and market development; organize training courses on e-commerce development, digital transformation and business connection;
(2) promoting and raising awareness for businesses about the impact of FTAs, helping businesses limit risks and support and encouraging businesses to make the most of the benefits provided by free trade agreements; and
(3) diversifying export markets, and identifying a number of potential markets to orient enterprises to boost exports.

6.1.2 Domestic Market Development Strategy

To develop the domestic market, the government and stakeholders should focus on:

(1) promoting the construction of technical barriers such as the application of technical standards on quality for imported products, anti-dumping lawsuits, environmental tax, etc. to protect domestic production;
(2) developing a plan to exploit the domestic market, and develop forms of e-commerce. Promote the development of the fashion industry in association with the leather and footwear industry. Strengthening international cooperation in the development of the leather and footwear fashion industry; and

(3) building a database system, and share information on raw materials for the leather and footwear industry so that businesses can timely exchange about supply sources.

6.2 Strategy for Developing Raw Materials for Production

The Vietnamese government prioritizes and encourages investment attraction for enterprises with long experience in the field of supporting industries for the leather and footwear industry. In addition, the government should promote the activities as follows.

(1) to promote and support domestic enterprises to enter into joint ventures with foreign enterprises to build investment projects for the production of raw materials and materials for leather and footwear. Priority is given to attracting environmentally friendly eco-leather production projects;
(2) to implement investment promotion programs in markets with a large concentration of enterprises producing leather and footwear materials and accessories such as China, Korea, Taiwan and South America;
(3) to encourage big brands present in Vietnam to participate in the process of developing the supply of raw materials and auxiliary materials, forming the domestic supply chain;
(4) to encourage and have preferential policies for scientific research institutions, research and development enterprises and application of environmentally friendly materials, adhesives, and chemicals such as renewable materials, degradable, water-dispersible adhesive and thermoplastic adhesive; and
(5) to build industrial parks focusing on the production of raw materials for the footwear industry in the South and the North (Lefaso, DTKHCN.05/17, 2018).

6.3 The Innovation Technology Strategy and Digital Transformation

For effective technological innovation and digital transformation in the footwear industry, governments and stakeholders should focus on:

(1) formulating and promulgating mechanisms and policies to encourage the import of source technology, high technology and clean technology in the leather and footwear industry. The government should have preferential policies, encourage businesses to invest in new machinery, equipment and technologies to improve productivity and product quality, minimize negative impacts on the environment, and focus on stages of creating high added value, associated with intelligent and automated production processes;

(2) developing research capacity on design, trial production, product development, quality control, technology transfer, training, etc. for specialized research agencies in leather and footwear;
(3) developing policies to promote the application of technology of the industrial revolution 4.0 (e.g., 3D printing, new materials, automation, robotics and digitization), application of advanced science and technology based on the foundation information technology and artificial intelligence to participate deeply in the global leather and footwear value chain;
(4) promulgating policies to support digital transformation for the entire leather and footwear industry ecosystem, implementing training programs to raise awareness about digital transformation and how to do it, giving priority packages for loans to support the digital transformation of the leather and footwear industry; and
(5) incorporating digital technology applications and digital transformation into human resource training programs in the leather and footwear industry. Domestic technology companies need to develop technology applications suitable to the practical conditions of enterprises in the leather and footwear industry in Vietnam.

6.4 Strategy for Improving Working Conditions and the Quality of Human Resources

6.4.1 Strategies to Improve the Quality of Human Resources

Human resource development is one of the decisive factors of the vision for sustainable development of the leather and footwear industry in Vietnam. Therefore, in the coming time, the government, businesses and stakeholders should focus on:

(1) developing a strategy for human resource development in the leather and footwear industry for the period to 2030 and a vision to 2035 (Vietnam Ministry of Industry and Trade, 2020);
(2) strengthening human resource training, building a team of managers, researchers, technicians, designers and skilled workers to create competitive advantages and apply the achievements of the Industrial Revolution 4.0.
(3) modernizing step by step and improving the efficiency of corporate governance in the industry, and increasing the proportion of technical staff with specialized training in sustainable development, meeting the requirements of customers;
(4) developing and completing training and retraining programs for officials, technicians and technical workers; making in-depth investments and expanding the training system in specialized leather and footwear training schools; and
(5) promoting vocational training in the direction of socialization, as well as encouraging and creating favorable conditions for enterprises to associate vocational training with training institutions throughout the country to meet the needs of enterprises.

6.4.2 Strategies to Improve Working Conditions

Good working conditions are one of the important factors contributing to building a sustainable footwear industry. To achieve this, businesses should focus on:

(1) complying with the law on labour and occupational safety and health (Labour Law, 2019; Law on occupational safety and hygiene, 2015);
(2) enforcing the terms of environmental and labour conditions in the terms of the new generation of trade agreements signed by the Government of Vietnam. Enterprises need to develop standards of decent work and meet the requirements of the Code of Responsible Business, in line with international practices on responsible value chains in the leather and footwear industry;
(3) enhancing the application of automation, digital transformation and lean production to improve working conditions for employees; and
(4) applying green technologies, using less hazardous materials and chemicals to replace highly toxic materials and chemicals.

6.5 Strategy to Reduce the Environmental Impact of the Leather Sector

Environmental issues are a common challenge for the entire fashion industry, and footwear manufacturing is no exception. To reduce their negative impact on the environment, businesses and stakeholders should focus on:

(1) reviewing international standards/requirements to establish and periodically update industry standards and norms on waste, energy consumption, water use, material use and chemicals;
(2) focusing on doing research in order to seek new materials and chemicals, renewable or naturally derived materials to gradually replace unsustainable materials on the basis of a product life cycle assessment, and enterprises should reuse raw materials in the production process;
(3) organizing training courses, disseminating information and improving capacity for management and control of environmental issues at the unit (e.g., chemical control, energy consumption, water consumption) and developing policies to support traceability and transparent value chains in line with international standards; and
(4) developing policies to support and encourage the application of 4.0 technology in the logistics system, domestic distribution and retail network and innovating export marketing methods to apply 4.0 technology.

In particular, businesses that have been more deeply involved in the supply chain, specifically involved in the design and development of products for the export market as well as the domestic market, need to pay attention to the design of ecology products and the sustainable production process. This is the design process in which environmental issues are given special attention, in addition to many other issues that need to be addressed.

7 Conclusion

Sustainable development is the key trend in Vietnam's leather and footwear industry. Leather and footwear businesses develop sustainably when they satisfy economic, human and environmental criteria. To achieve the goal of sustainable development, each business needs to identify the major criteria that need to be addressed depending on the actual conditions of their business. In general, in the current conditions of Vietnam's leather and footwear industry, the major criteria that need attention are: Improving production and business efficiency (reducing waste, improving labour productivity, improving quality) product); improve working conditions and raise incomes for workers; minimize the environmental impact of the product manufacturing process.

The growth potential of Vietnam's leather and footwear industry is still very large. Although there are still many difficulties and challenges, with the results achieved in the pastime, with the expectations of multilateral and bilateral trade agreements in the process of international integration, and especially with the efforts of the business community, Vietnam's leather and footwear industry will have more opportunities for the stronger development in the coming periods.

This chapter provides information on the development situation of Vietnam's leather and footwear industry, forecasts factors affecting its development, assesses opportunities and challenges, as well as proposes some recommendations for sustainable development. maintain Vietnam's leather and footwear industry. We hope that this information will be useful to those interested in the sustainable development of Vietnam's leather and footwear industry in the coming period.

References

Adidas Annual Sustainability Reports. https://www.adidas-group.com/en/sustainability/transparency/reporting/sustainability-reports.

Adidas Annual Sustainability Reports. (2021). Adidas. https://report.adidasgroup.com/2021/en/servicepages/downloads/files/annual-report-adidas-ar21.pdf.

Annual Reports on Vietnam Leather and Footwear Industry 2011–2021, Lefaso. http://www.lefaso.org.vn/.

Assessing the readiness to approach Industry 4.0 of enterprises in Vietnamese industries. Ministry of Industry and Trade-UNDP in 2019 (in Vietnamese).

Bui Van, H., Thuan, N. D., Xuan, P. T. T. (2020a). Research on assessing the impact of the fourth industrial revolution on Vietnam's leather and footwear industry until 2030. *Journal of Science and Technology, 8*, 94–100 (in Vietnamese). https://tapchikhcn.haui.edu.vn/media/30/uffile-upload-no-title30382.pdf.

Bui Van, H., Thuan, N. D., Xuan, P. T. T., & Anh, L. T. V. (2020b). The study applied a shoe factory revolution model in Vietnam the achievements of the Fourth industrial revolution. In *National scientific conference on textiles and footwear, 2nd—NSCTEX2020* (in Vietnamese). https://nsctex.hust.edu.vn/.

Bui Van, H., Xuan, P. T. T., & Hoa, D. P. (2022). Research on building a digital governance model—the basis for setting up an ERP system for shoe manufacturing enterprises. In *3rd National scientific conference on textile and footwear—NSCTEX2022* (in Vietnamese) https://nsctex.hust.edu.vn/.

Deselnicu, I., Deselnicu, D. C., Vasilescu, A. M., & Militaru, G. (2014). EU policy for sustainable comsumption and production—Ecolabel for footwear. In *ICAMS 2014—5th international conference on advanced materials and systems.*

Decision 889/QD-TTg dated June 24, 2020: National action plan on sustainable production and consumption for the period 2021–2030 (in Vietnamese). https://vanban.chinhphu.vn/default.aspx?pageid=27160&docid=200310.

Decision No. 749/QD-TTg approving the "National Digital Transformation Program to 2025, with orientation to 2030". dated June 3, 2020 (in Vietnamese). https://vanban.chinhphu.vn/?pageid=27160&docid=200163.

Decree 08/2022/ND-CP dated November 17, 2020: Detailed regulations on a number of articles of the Law on Environmental Protection (in Vietnamese). https://vanban.chinhphu.vn/?pageid=27160&docid=205092.

Dwivedi, A., Moktadir, M. A., Jabbour, C. J. C., & de Carvalho, D. E. (2021). Integrating the circular economy and industry 4.0 for sustainable development: Implications for responsible footwear production in a big data-driven world. *Technological Forecasting and Social Change, 121335.*

Elkington, J. (1997). *Cannibals with forks: The triple bottom line of 21st century business.* Capstone.

Final report on the topic of Hanoi Department of Science and Technology: "Research on technology of manufacturing flooring materials from leather and footwear scrap for civil works", Project code: 01C-03/ 02-2016-3, 2018 (in Vietnamese).

Global Digital Transformation Conference, FPT, April 25–27, 2019 (in Vietnamese).

Hedaitul Islam, M., Sarker, M. R., Hossain, M. I., Kauser Ali, K. M., & Noor, A. (2020). Towards sustainable supply chain management (SSCM): A case of leather industry. *Journal of Operations and Strategic Plannin.* https://doi.org/10.1177/2516600X20924313.

Hien, N. T. Q. (2020). Evaluation of quality indicators of the working environment in a number of leather and garment factories in Ho Chi Minh City. Master Thesis—Specialization: Natural Resources and Environment Management (in Vietnamese). http://opac.iuh.edu.vn/TraCuuTaiLieuSo2XemChiTiet.aspx?Id=110349.

https://www.adidas-group.com/en/sustainability/managing-sustainability/general-approach/.

Keeble, B. R. (1988). The Brundtland report: "Our common future." *Medicine and War, 4*, 17–25.

Khan, M., Hussain, M., Gunasekaran, A., Ajmal, M. M., & Helo, P. T. (2018). Motivators of social sustainability in healthcare supply chains in the UAE—Stakeholder perspective.

King, P., Tinson, E., & Winston, E. (2010). Sustainability assessment of Nike shoes. For Sustainability Science ENVS 195, Fall 2010.

Labor Code. No. 45/2019/QH14, Vietnam National Assembly promulgated on November 20, 2019 (in Vietnamese). https://vanban.chinhphu.vn/default.aspx?pageid=27160&docid=198540.

Law on occupational safety and hygiene, No. 84/2015/QH13, Vietnam National Assembly promulgated on June 25, 2015 (in Vietnamese). https://vanban.chinhphu.vn/default.aspx?pageid=27160&docid=180606.

Lefaso Annual Report 2020, 2021, 2022. http://www.lefaso.org.vn/.

Marta Hervaa, S., Álvarez, A., & Roca, E. (2011). Sustainable and safe design of footwear integrating ecological footprint and risk Criteria. *Journal of Hazardous Materials, 192*, 1876–1881.

Muthu, S. S. (2020). *Leather and footwear sustainability.* Springer.

Orientation to develop national industrial development policy to 2030, vision to 2045 (Attached to Resolution No. 23-NQ/TW dated March 22, 2018) (in Vietnamese). https://tulieuvankien.dangcongsan.vn/he-thong-van-ban/van-ban-cua-dang/nghi-quyet-so-23-nqtw-ngay-2232018-cua-bo-chinh-tri-ve-dinh-huong-xay-dung-chinh-sach-phat-trien-cong-nghiep-quoc-gia-den-nam-4125.

Project report "Survey and assess the situation of consumption of raw materials and solid waste in Vietnam's Leather and Footwear industry", Code 4123/QD—BTC, Leather and Footwear Research Institute (LSI), 2015 (in Vietnamese).

Project report: "Surveying and assessing the consumption situation of raw materials and solid waste in Vietnam Leather and Footwear industry", SLI, 2015 (in Vietnamese).

Resolution No. 115/NQ-CP dated August 6, 2020, on solutions to promote the development of supporting industries (in Vietnamese). https://vanban.chinhphu.vn/default.aspx?pageid=27160&docid=200721.

Staikos, T., Heath, R., Haworth, B., & Rahimifard, S. (2006). End-of-life management of shoes and the role of biodegradable materials. In *Proceedings of LCE2006*.

Statistical Yearbook of Vietnam 2019, 2020, General Statistics Office, https://www.gso.gov.vn/en/data-and-statistics/2022/08/statistical-yearbook-of-2021/.

Statistical Yearbook of Vietnam. (2021). General Statistics Office, https://www.gso.gov.vn/wp-content/uploads/2022/08/Sach-Nien-detention-TK-2021-1.pdf.

Subic, A., Shabani, B., Hedayati, M., & Crossin, E. (2012). Capability framework for sustainable manufacturing of sports apparel and footwear. *Sustainability, 4*, 2127–2145. https://doi.org/10.3390/su4092127.

Sustainable Production and Consumption. https://doi.org/10.1016/j.spc.2018.01.006.

Sustainable Development Goals (SDGs). (2019). https://about.nike.com/en/newsroom/resources/sustainable-development-goals-sdgs.

Svensson, G., & Wagner, B. (2015a). Implementing and managing economic, social and environmental efforts of business sustainability. *Management of Environmental Quality: An International Journal, 26*, 195–200. https://doi.org/10.1108/MEQ-09-2013-0099.

Svensson, G., & Wagner, B. (2015b). Implementing and managing economic, social and environmental efforts of business sustainability propositions for measurement and structural models. *Management of Environmental Quality: An International Journal*. https://doi.org/10.1108/MEQ-09-2013-0099.

The final report on the topic of the Ministry of Industry and Trade: "Researching solutions and building a roadmap for the development of the leather and footwear supporting industry in Vietnam based on the experience of countries with similar footwear production". Code: DTKHCN.05/17, Lefaso, 2018 (in Vietnamese).

The final report on the topic of the Ministry of Industry and Trade: "Research and evaluate the impact of the Fourth Industrial Revolution on the leather and footwear industry in Vietnam, strategic orientations, policies and solutions for the development of the leather industry. Vietnamese footwear until 2030", Code: DTDL.XH.15/18, Lefaso, 2020 (in Vietnamese).

The World Footwear 2021 Yearbook.

Theories of Sustainable Development. (2015). *Edited by Judith C*. Routledge.

Vietnam Enterprise White Paper 2019. (Ministry of Planning and Investment 2019). https://www.unido.org/sites/default/files/files/2020-03/Final%20Vietnam_White_Paper_2019.pdf.

Ville Matinaro, Y.L., & Tzong-Ru (Jiun-Shen), L., Jurgen, P. (2019). Extracting key factors for sustainable development of enterprises: Case study of SMEs in Taiwan. *Journal of Cleaner Production, 209*, 1152–1169. https://doi.org/10.1016/j.jclepro.2018.10.280.

Wade and Andrea Motawi, *How Shoes are Made*, www.Sneakerfactory.net, Copyright © Wade Motawi 2018.

Wenjie, J. (2019). Sustainable development of supply chain in footwear industry—Take Nike as the case. *Asian Business Research, 4*(3), ISSN 2424-8479, E-ISSN 2424-8983.

Epilogue

On the Road to the Carbon Neutral Leather

Sreeram Kalarical Janardhanan and Luis A. Zugno

1 Introduction

Leather holds an important position in the consumer lifestyle product segment due to its aesthetic and performance properties. Many view leather as an emotional product where the handle (also known as haptics) and durability have defined luxury fashion goods and apparel procurement (Williams et al., 2022). Carbon neutrality achieved through ensuring the net amount of carbon dioxide emissions released into the atmosphere is equal to the amount of CO_2 that is removed from the atmosphere or offset through various measures enables us to limit global warming below 1.5 °C. The two main steps to achieve carbon neutrality are reducing emissions and offsetting emissions. A cradle to gate scenario for leather manufacturing presented in Fig. 1 clearly indicates the challenges that arise from multiple do-undo processes (Saravanabhavan et al., 2006; Thanikaivelan et al., 2003) and varies with the raw material, chemicals employed, mechanical operations adopted, etc.

There are reports that the carbon footprint for a finished aniline leather (leather whose grain structure is clearly visible either without surface coating or with a non-pigmented surface coating of thickness less than 0.01 mm) is around 73 $kgCO_2e/m^2$ (Chen et al., 2014). The reported carbon footprint varies and researchers have classified their studies into six categories, viz., (a) raw hide to tanned hide; (b) raw hide to crust leather; (c) raw hide to finished leather; (d) tanned hide to finished leather; (e) crust leather to finished leather; and (f) tanned hide to crust leather (Laurenti et al., 2017). To mitigate environmental pollution, the industry has adopted several cleaner process technologies (Sreeram et al., 2003) and there have been efforts to carry out life cycle assessment of emerging technologies in the development phase

S. K. Janardhanan (✉)
CSIR-Central Leather Research Institute, Adyar, Chennai 600020, India
e-mail: kjsreeram@clri.res.in

L. A. Zugno
Secretary & Past President, IULTCS, 4052 Basel, Switzerland

S. K. Janardhanan and L. A. Zugno (eds.), *Emerging Trends in Leather Science and Technology*, https://doi.org/10.1007/978-981-99-9754-1_17

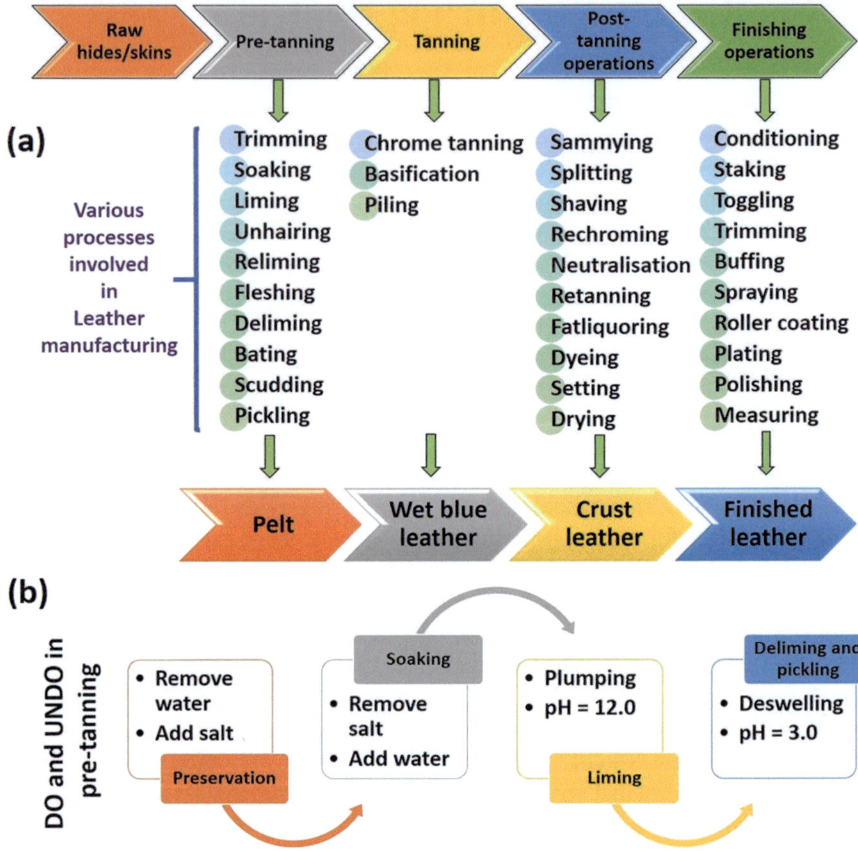

Fig. 1 Cradle to gate scenario in leather manufacturing and challenges associated with multiple do-undo processes

itself (Marrucci et al., 2022). In addition, the generation and disposal of solid waste from leather processing also adds to the CO_2 emission from the leather industry. The details of the solid wastes generated and the respective estimated CO_2 emission are provided in Table 1.

There are arguments that polyurethane (PU) leather (also known as vegan leather; both PU leather and vegan leather are misnomers) has a lower carbon footprint than those manufactured from bovine hides. However, polyurethane and other polymers are synthesized from fossil resources and their incineration could lead to toxic chemicals (Otake et al., 1995). These arguments are based on the fact that the life cycle assessment of plastic products is not yet mature enough (Williams et al., 2022) and the starting point for assessment of leather is not clearly defined.

Methane emission accounts for 55–92% of the carbon footprint of a cattle during its lifetime (Desjardins et al., 2012; Rosa et al., 2018). In terms of mass balance (based on shrunk live weight), the hide contributes to only 4.9%, while the meat

Table 1 Estimated CO_2 emission from leather industry solid wastes (UNIDO, 2017)

Wastes	CO_2 (kg/ton)
Putrefied hides and skins	624
Raw trimmings	624
Hair (Pasting)	373
Fleshing	624
Wet blue trimmings	221
Chrome shavings	221
Buffing dust	221
Dyed trimming	221

Ref: Leather Carbon footprint: Review of the European Standard EN 16887:2017 (UNIDO, 2017)

contributes to 37.8% (which on economic terms is 6.8% and 83.6% respectively) (Desjardins et al., 2012. In spite of this, there is an ongoing demand for considering the cattle life cycle also in the calculation of carbon footprint of leather (Rosa et al., 2018).

2 Way to Carbon Neutrality

2.1 Increasing the Skin Carbon Utility Value

It is now recognized that animal by-products are a significant contributor to circular bioeconomy (both biological resources and waste streams) (Pérez et al., 2022). It has been reported that EU slaughterhouses process 17 million tonnes of animal by-products annually, leading to 3.5 million tonnes of protein and large quantum of wastewater rich in protein (Pérez et al., 2022). Valorisation of animal by-products other than skin has been in vogue. Some possible opportunities that exist (excluding meat and skin) have been detailed in recent literature (Pinto et al., 2022). The value addition to meat by-products has included livestock and poultry feeds, pet foods, energy generation, bioplastics, protein hydrolysates, bioactive peptides and so on (Toldrá et al., 2016).

Effective implementation of these technologies globally would require local government and municipalities to take a prominent role in establishing the infrastructure required. In addition to this, the conversion of animal wastes (feeds, faeces, urine, bedding material, etc.) is also very important. For instance, cattle manure (a product of the animal wastes) is a source for 35.4 wt% carbon (Islam et al., 2018). In recent years technologies such as generation of energy from such wastes have been reported and there is a need for on-site recovery and valorisation of animal wastes through techniques such as pyrolysis (Okolie et al., 2023). High value applications of animal wastes include that of bioactive peptides, hydrolyzed collagen, bio-hydroxyapatite etc. (Tarafdar et al., 2021). An interesting high value application is in the conversion of pork bone to mesoporous bio-apatite, that can be deployed as carriers for Ni-based catalysts for conversion of ethanol to higher alcohols (Xue et al., 2022).

Under circumstances where the cattle lifecycle is included in the carbon footprint of leather, the leather industry would have to actively campaign for such valorisation of animal by products.

2.2 *Value Generation from Skin not Used in Leather Processing*

It is reported that the skin is a valuable resource for several amino acids. For instance, the cow hide has about 8.82 mol% hydroxyproline, 10.80 mol% alanine and 31 mol% glycine. The total protein content on dry weight is around 92.6%, of which 70.9% is collagen. About 0.5% glycosaminoglycans is also present (Naffa et al., 2019). Much of the non-collagenous proteins and glycosaminoglycans are lost during the leather processing and they could serve as valuable resources. To reduce the carbon footprint of leather manufacture, extracting high-end products from skin/leather wastes has been suggested. Elastin, for example, is one such high value products that can be extracted from skin trimmings (Yoseph et al., 2020).

A small percentage of hides/skins sourced from developing and underdeveloped countries are not suitable for leather manufacturing. Due to its biocompatibility, and high biological activity, collagen is widely used in food, medicine, tissue engineering and cosmetics (Liu et al., 2015). Hydrolysis of skin collagen leads to small peptides of 2–20 amino acid length, whose biological activities such as anti-inflammatory properties, and angiotensin-converting enzyme inhibition make it a favourable functional food (Halim et al., 2016). Typical method to prepare collagen and collagen peptides from animal by-products has been detailed by Cao et al. (2022). Adding collagen to meat has been reported to improve antioxidant properties, reduce free water content, inhibit fat oxidation, replace fat in meat products and modulate viscosity of milk.

Collagen also has high utility in food packaging, which includes collagen coating for preservation and casings (films) for packaging. One of the salient advantages of collagen casing is the edible character (Fu et al., 2019).

The varied biological activity of collagen peptides renders them ideal for functional foods (Cao et al., 2022). It is predicted that by 2027, the global market share of collagen in foods and beverages would be around 31.9%, that in healthcare being 49.6%, and cosmetics at around 12.4% (Cao et al., 2022).

Collagen and collagen peptides promote wound healing by alleviating oxidative stress and associated damages from reactive oxygen species. The advantages of collagen dressing on tissue regeneration and collagen deposition for promotion of expression of growth factors has been highlighted by several researchers (Sharma et al., 2022). Similarly, collagen peptides extracted from animal tissues enter the skin fibroblasts as precursor materials to participate in collagen synthesis, leading to protecting skin from aging.

In all the above functions, animal by-products other than hides/skins have been reported as the source. However, with its vast availability there is a potential for skin collagen to serve as a cheaper source for collagen and collagen peptides. In this direction, the use of tannery trimming wastes as bioplastics, and having applications in packaging and wound healing has been reported (Muralidharan et al., 2020). With some process modifications, pickled skin trimmings have been reported to be a suitable raw material for gelatine manufacture (Rahmawati et al., 2019). Collagen extraction and separation techniques have been aggregated and presented in a recent article of Matinong et al. (2022).

2.3 *Tannery as a Part of a Biorefinery*

Energy sector predominantly uses fossil fuel combustion, which is a primary driver of climate change. Energy sector also depletes fossil fuel resources and generates greenhouse gases (estimated to be 80% of the total). To reduce greenhouse gas emissions, four strategies, viz., decreasing energy demand, increasing energy efficiency, end-use electrification and increasing renewable energy share have been suggested (Lund et al., 2022). The goal of ensuring zero greenhouse gas emissions by mid-twenty-first century has led to the concept of carbon neutrality (Huovila et al., 2022). A forward is generation of bioenergy derived from biomass, which consists of three categories, viz., liquid biofuel for transport, bioliquids for energy purposes (heating, cooling, electricity) and gaseous and solid biomass fuels (Daneshmandi et al., 2022). A further development in this area is the biorefinery concept that focuses on converting biomass to bioproducts of increased economic potential and having a potential to achieve carbon neutrality and a circular bioeconomy by integrating available technologies (Duan et al., 2022). Tanning sector with all the green technologies developed to reduce emissions, integrated with animal by-product value addition technologies is an ideal biorefinery as can be extrapolated from the description of circular economy. Figure 2 shows various value-added products that can be developed from the solid wastes obtained from tannery.

Shah et al. describe circular economy as one that maximizes resource efficiency by reducing wastes, conserving the raw material and instituting closed loops and symbiotic relationships between products and materials, leading to social, economic and environmental benefits (Shah et al., 2022). Literature exists for a segment of the industry, viz., tannery wastes being considered as a biorefinery (Aathika, 2018; Appala et al., 2022; Kiran et al., 2014; Mozhiarasi et al., 2021; Nagi et al., 2020). As a circular economy approach, black liquor from paper and pulp industries has been chemically modified to obtain retanning agents, colorants and pigments for use in post-tanning processes (Balasubramanian et al., 2018, 2021; Zakir et al., 2015). Tanneries fit well to the description of biorefinery when along with waste valorisation, other aspects such as producing fuels, oils and other valuable products from a

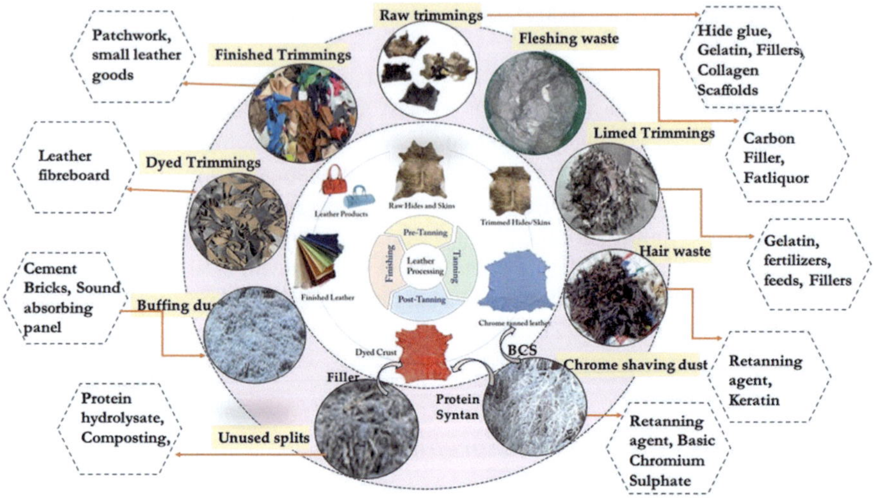

Fig. 2 Value-added products from solid wastes generated from tannery: a showcase of circular economy

single biological feedstock (animal by products) through a series of interdependent processes is also taken into consideration.

The success of a biorefinery would depend on how well the biofuels derived integrate with already existing infrastructure. This then requires four components, viz., platform, products, feedstock and processes (Cherubini et al., 2009). The key to the success of the biorefinery is the platform that is intermediate between raw materials and products. The leather industry needs to develop platforms that are like that reported in the case of cellulose to ethanol with crop and animal system integration (Sendich et al., 2008), lignin/syngas and/or C5/C6 sugars to synthetic liquid biofuels, bioethanol and animal feed.

In an effort to integrate various value-added production systems based on hides/skins, JBS Couros have worked out a production chain where the parts of skin that are not suitable for leather making are removed prior to tannery operations and sent for pharmaceutical and food manufacturing. The company on their website (https://kindleather.com/approved/) claims that they have been able to reduce the global warming potential by 77% and the eutrophication by 81%.

3 Industry 4.0 and Carbon Neutrality

For a typical running shoe of non-leather materials, the carbon footprint has been put at 14 ± 2 kg CO_2-equivalent/pair (Cheah et al., 2013). Other studies have reported the carbon footprint to be in the range of 18–41 kg CO_2-equivalent/pair. It is reported that 25 billion shoes are sold every year and single shoe has 65 discrete parts and

360 processing steps in its assembly line. The complexity of manufacturing calls for automation and integration of assembly chain steps.

A significant shift towards the use of digital platforms, information systems and new technologies into production has been observed in the last two decades. Footwear nesting based on design patterns and manufacturing materials has had a direct influence on product quality and cost (Peng et al., 2013). For leather, image acquisition, processing, feature extraction and classification of defects have been possible through deep learning algorithms (Jawahar et al., 2019, 2023). Intelligent manufacturing as an element of Industry 4.0 aims to provide value-added products and systems using modern technologies. Some of the basic pillars of Industry 4.0 such as autonomous robots, digital twins, system integration, internet of things, additive manufacturing and augmented reality have played a role in improving productivity in the industry. Radio Frequency Identification (RFID) technologies have largely improved the traceability aspects. Scientific and technological challenges encountered by footwear industries that are characterized by product customization and short manufacturing cycles have been reviewed by Jimeno et al. (2021), who conclude that the complexity associated with the manufacturing process makes it expensive for small and medium-sized enterprises (SME) sector to adopt. Co-design and mass customization has been suggested as a way forward (Oliveira et al., 2019). Using a module called shoe cloth, consumers have been provided with the option of choosing colour, texture, embroidery, etc., leading to personalization (Luh et al., 2013). Using design for assembly and design for disassembly strategies cost reduction and recyclability of shoe parts have been suggested to improve sustainability (Ghimouz et al., 2023).

One of the major challenges to carbon neutrality is transport. Transportation and logistics through intelligent vehicle technologies can achieve reduction in carbon emissions (Chen et al., 2023). Literature indicates that 3D-printed shoes can produce less scrap and emissions and transform mass production into mass customization. Even though not implementable for mass scale production, the industry can cut emissions in prototype development through 3D printing.

4 Way Forward

Leather industry needs to move from traditional process–product linkages to one which has a forward and backward integration. A biorefinery platform that looks at a holistic value addition to animal by-products—comprising all forms of value addition to domestic animal products including meat and skin is the way forward. Such a holistic approach will lead to reduction/elimination of process steps including salt-based preservation. An integration would also help better animal husbandry practices that could lead to quality meat and hides/skin and thus reduction in chemical consumption in leather processing. In essence, a biorefinery platform where leather manufacturing is an important component will lead to carbon neutrality and net zero emissions from the sector, which can further be improved through intelligent manufacturing and AI-driven supply-chain logistics.

References

Aathika, A. R. S., Kubendran, D., Yuvarani, M., Thiruselvi, D., Amudha, T., & Karthik, P. (2018). Enhanced biohydrogen production from leather fleshing waste co-digested with tannery treatment plant sludge using anaerobic hydrogenic batch reactor. *Energy Sources Part A-Recovery Utilization and Environmental Effects, 40*(5), 586–593.

Appala, V. N. S. G., Pandhare, N. N., & Bajpai, S. (2022). Biorefining of leather solid waste to harness energy and materials—A review. *Biomass Conversion and Biorefinery.*

Balasubramanian, P., Ramalingam, S., Abu, J. M., & Rao, J. R. (2018). Lignin based colorant: Modified black liquor for leather surface coating application. *Journal of the American Leather Chemists Association, 113*(10), 311–317.

Balasubramanian, P., Ramalingam, S., Sreeram, K. J., & Rao, J. R. (2021). Natural pigment: Preparation of brown pigment from lignin biomass for coloring application. *Dyes and Pigments, 195*, 109704.

Cao, C., Xiao, Z., Ge, C., & Wu, Y. (2022). Animal by-products collagen and derived peptide, as important components of innovative sustainable food systems—A comprehensive review. *Critical Reviews in Food Science and Nutrition., 62*(31), 8703–8727.

Cheah, L., Ciceri, N. D., Olivetti, E., Matsumura, S., Forterre, D., & Roth, R. (2013). Manufacturing-focused emissions reductions in footwear production. *Journal of Cleaner Production, 44*, 18–29.

Chen, K. W., Lin, L. C., Lee, W. S. (2014). Analyzing the carbon footprint of the finished bovine leather: A case study of aniline leather. *Energy Procedia, 61*, 1063–1066.

Chen, Y. Y., Zhang, H., & Wang, F. Y. (2023). Society-centered and DAO-powered sustainability in transportation 5.0: An intelligent vehicles perspective. *IEEE Transactions on Intelligent Vehicles, 8*(4), 2635–2638.

Cherubini, F., Jungmeier, G., Wellisch, M., Willke, T., Skiadas, I., & Van, R. R. (2009). Toward a common classification approach for biorefinery systems. *Biofuels Bioproducts & Biorefining-Biofpr, 3*(5), 534–546.

Daneshmandi, M., Sahebi, H., & Ashayeri, J. (2022). The incorporated environmental policies and regulations into bioenergy supply chain management: A literature review. *Science of the Total Environment., 820*, 153202.

Desjardins, R. L., Worth, D. E., Vergé. X. P. C., Maxime, D., Dyer, J., & Cerkowniak, D. (2012). Carbon footprint of beef cattle. *Sustainability, 4*(12), 3279–3301.

Duan, Y. M., Tarafdar, A., Kumar, V., Ganeshan, P., Rajendran, K., & Giri, B. S. (2022). Sustainable biorefinery approaches towards circular economy for conversion of biowaste to value added materials and future perspectives. *Fuel, 325*, 124846.

Fu, Y., Therkildsen, M., Aluko, R. E., Lametsch, R. (2019). Exploration of collagen recovered from animal by-products as a precursor of bioactive peptides: Successes and challenges. *Critical Reviews in Food Science and Nutrition, 59*(13), 2011–2027.

Ghimouz, C., Kenné, J. P., & Hof, L. A. (2023). On sustainable design and manufacturing for the footwear industry—Towards circular manufacturing. *Materials & Design, 233*, 112224.

Halim, N. R. A., Yusof, H. M., & Sarbon, N. M. (2016). Functional and bioactive properties of fish protein hydolysates and peptides: A comprehensive review. *Trends in Food Science & Technology, 51*, 24–33.

Huovila, A., Siikavirta, H., Rozado, C. A., Rokman, J., Tuominen, P., & Paiho, S. (2022). Carbon-neutral cities: Critical review of theory and practice. *Journal of Cleaner Production, 341*, 130912.

Islam, M. N., & Park, J. H. (2018). A short review on hydrothermal liquefaction of livestock manure and a chance for Korea to advance swine manure to bio-oil technology. *Journal of Material Cycles and Waste Management, 20*(1), 1–9

Jawahar, M., Anbarasi, L. J., & Geetha, S. (2023). Vision based leather defect detection: A survey. *Multimedia Tools and Applications, 82*(1), 989–1015.

Jawahar, M., Vani, K., & Babu, N. K. C. (2019). Machine vision inspection system for detection of leather surface defects. *Journal of the American Leather Chemists Association, 114*(1), 10–19.

Jimeno, M. A., Azariadis, P., Molina, C. R., Kyratzi, S., Moulianitis, V. (2021). Technology enablers for the implementation of industry 4.0 to traditional manufacturing sectors: A review. *Computers in Industry, 125*, 103390.

Kiran, E. U., Trzcinski, A. P., Ng, W. J., Liu, Y. (2014) Enzyme production from food wastes using a biorefinery concept. *Waste and Biomass Valorization, 5*, 903-917.

Laurenti, R., Redwood, M., Puig, R., & Frostell, B. (2017). Measuring the environmental footprint of leather processing technologies. *Journal of Industrial Ecology, 21*(5), 1180–1187.

Liu, D., Nikoo, M., Boran, G., Zhou, P., & Regenstein, J. M. (2015). Collagen and gelatin. *Annual Review of Food Science and Technology, 6*(1), 527–557.

Luh, Y. P., Wang, J. B., Chang, J. W., Chang, S. Y., Chu, C. H. (2013). Augmented reality-based design customization of footwear for children. *Journal of Intelligent Manufacturing, 24*(5), 905-917.

Lund, H., Skov, I. R., Thellufsen, J. Z., Sorknaes, P., Korberg, A. D., Chang, M. G. (2022). The role of sustainable bioenergy in a fully decarbonised society. *Renewable Energy, 196*, 195–203.

Marrucci, L., Corcelli, F., Daddi, T., & Iraldo, F. (2022). Using a life cycle assessment to identify the risk of "circular washing" in the leather industry. *Resources, Conservation and Recycling., 185*, 106466.

Matinong, A. M. E., Chisti, Y., Pickering, K. L., & Haverkamp, R. G. (2022). Collagen extraction from animal skin. *Biology* (Basel), *11*(6), 905.

Mozhiarasi, V., Krishna, B. B., Nagabalaji, V., Srinivasan, S. V., Bhaskar, T., Suthanthararajan, R. (2021). Chapter 10—Leather industry waste based biorefinery. In T. Bhaskar, S. Varjani, A. Pandey, & E.R. Rene (Eds.), *Waste biorefinery* (pp. 267–304): Elsevier.

Muralidharan, V., Arokianathan, M, S., Balaraman, M., & Palanivel, S. (2020). Tannery trimming waste based biodegradable bioplastic: Facile synthesis and characterization of properties. *Polymer Testing, 81*, 106250.

Naffa, R., Maidment, C., Holmes, G., & Norris, G. (2019). Insights into the molecular composition of the skins and hides used in leather manufacture. *Journal of the American Leather Chemists Association, 114*(1), 29–37.

Nagi, M., He, M., Li, D., Gebreluel, T., Cheng, B., & Wang, C. (2020). Utilization of tannery wastewater for biofuel production: New insights on microalgae growth and biomass production. *Scientific Reports, 10*(1), 1530.

Okolie, J. A., Jimoh, T., Akande, O., Okoye, P. U., Ogbaga, C. C., & Adeleke, A. A. (2023). Pathways for the valorization of animal and human waste to biofuels, sustainable materials, and value-added chemicals. *Environments, 10*(3), 46.

Oliveira, N., Cunha, J., & Carvalho, H., (Eds.). (2019). Co-design and mass customization in the portuguese footwear cluster: An exploratory study. In *29th CIRP Design Conference (CIRP Design).*

Otake, Y., Kobayashi, T., Asabe, H., Murakami, N., & Ono, K. (1995). Biodegradation of low-density polyethylene, polystyrene, polyvinyl chloride, and urea formaldehyde resin buried under soil for over 32 years. *Journal of Applied Polymer Science, 56*(13), 1789–1796.

Peng, W. L., Wang, H. Y., Lu, L., (Eds.). (2013). Research on manufacturing materials with footwear nesting system based on AutoCAD. In *2nd International Conference on Intelligent Materials, Applied Mechanics and Design Science (IMAMD2013).*

Pérez, A. H., Lacruz, A. M., & Arán, A. F. (2022). Towards a circular bioeconomy: High added value protein recovery and recycling from animal processing by-products. *Sustainable Chemistry and Pharmacy, 28*, 100667.

Pinto, J., Boavida, D. R., Matos, H. A., & Azevedo, J. (2022). Analysis of the food loss and waste valorisation of animal by-products from the retail sector. *Sustainability, 14*(5), 2830.

Rahmawati, D., Setyadewi, N. M., & Sugihartono, 2019. Extraction and characterization of gelatin from skin trimming pickled waste of tannery. *IOP Conference Series: Earth and Environmental Science, 306*(1), 012022.

Rosa, G., Fontanella, A., Gonzalez, Q. G., Ioannidis, I., Nucci, B., & Brugnoli, F. (2018). Product environmental category rules—Leather. European Commission: Directorate-General for Environment, European Commission. Retrieved from https://green-business.ec.europa.eu/environmental-footprint-methods_en

Saravanabhavan, S., Thanikaivelan, P., Rao, J. R., Nair, B. U., & Ramasami, T. (2006). Reversing the conventional leather processing sequence for cleaner leather production. *Environmental Science & Technology, 40*(3), 1069–1075.

Sendich, E. D., Dale, B. E., & Kim, S. (2008). Comparison of crop and animal simulation options for integration with the biorefinery. *Biomass & Bioenergy, 32*(12), 1162–1174.

Shah, A. V., Singh, A., Mohanty, S. S., Srivastava, V. K., & Varjani, S. (2022). Organic solid waste: Biorefinery approach as a sustainable strategy in circular bioeconomy. *Bioresource Technology, 349*, 126835.

Sharma, S., Rai, V. K., Narang, R. K., & Markandeywar, T. S. (2022). Collagen-based formulations for wound healing: A literature review. *Life Sciences, 290*, 120096.

Sreeram, K. J., & Ramasami, T. (2003). Sustaining tanning process through conservation, recovery and better utilization of chromium. *Resources, Conservation and Recycling, 38*(3), 185–212.

Tarafdar, A., Gaur, V. K., Rawat, N., Wankhade, P. R., Gaur, G. K., & Awasthi, M. K. (2021). Advances in biomaterial production from animal derived waste. *Bioengineered, 12*(1), 8247–8258.

Thanikaivelan, P., Rao, J. R., Nair, B. U., & Ramasami, T. (2003). Stepping into third millennium: Third generation leather processing—A three step tanning technique. *Journal of the American Leather Chemists Association, 98*(5), 173–184.

Toldrá, F., Mora, L., & Reig, M. (2016). New insights into meat by-product utilization. *Meat Science, 120*, 54–59.

UNIDO. (2017). Leather carbon footprint: Review of the European Standard EN 16887:2017. United Nations Industrial Development Organization. Retrieved from https://leatherpanel.org/sites/default/files/publications-attachments/leather_carbon_footprint_p.pdf

Williams, E., Cenian, K., Golsteijn, L., Morris, B., & Scullin, M. L. (2022). Life cycle assessment of MycoWorks' Reishi™: the first low-carbon and biodegradable alternative leather. *Environmental Sciences Europe, 34*(1), 120.

Xue, M., Yang, B., Xia, C., & Zhu, G. (2022). Upgrading ethanol to higher alcohols via biomass-derived Ni/bio-apatite. *ACS Sustainable Chemistry & Engineering., 10*(11), 3466–3476.

Yoseph, Z., Gladstone, C. J., Assefa, D. B., TamilSelvi, A., Sreeram, K. J., & Rao, J. R. (2020). Extraction of elastin from tannery wastes: A cleaner technology for tannery waste management. *Journal of Cleaner Production, 243*, 118471.

Zakir, M. J. A., Ramalingam, S., Balasubramanian, P., Aravindhan, R., Sreeram, K. J., & Rao, J. R. (2015). Innovative material from paper and pulp industry for leather processing. *Journal of Cleaner Production, 104*, 436–444.

MIX
Papier aus verantwortungsvollen Quellen
Paper from responsible sources
FSC® C105338

If you have any concerns about our products,
you can contact us on
ProductSafety@springernature.com

In case Publisher is established outside the EU,
the EU authorized representative is:
Springer Nature Customer Service Center GmbH
Europaplatz 3, 69115 Heidelberg, Germany

Printed by Libri Plureos GmbH
in Hamburg, Germany